Fighting Better

Fighting Better

Constructive Conflicts in America

LOUIS KRIESBERG

Oxford University Press is a department of the University of Oxford. It furthers
the University's objective of excellence in research, scholarship, and education
by publishing worldwide. Oxford is a registered trade mark of Oxford University
Press in the UK and certain other countries.

Published in the United States of America by Oxford University Press
198 Madison Avenue, New York, NY 10016, United States of America.

Library of Congress Cataloging-in-Publication Data
Names: Kriesberg, Louis, author.
Title: Fighting better : constructive conflicts in America / Louis Kriesberg.
Description: First Edition. | New York : Oxford University Press, [2023] |
Includes bibliographical references and index.
Identifiers: LCCN 2022030861 (print) | LCCN 2022030862 (ebook) |
ISBN 9780197674802 (Paperback) | ISBN 9780197674796 (Hardback) | ISBN 9780197674826 (epub)
Subjects: LCSH: Social conflict—United States—History. |
Social classes—United States—History. |
Income distribution—United States—History. |
Democracy—United States—History.
Classification: LCC HN90 .S6 K66 2022 (print) | LCC HN90 .S6 (ebook) |
DDC 303.60973—dc23/eng/20220808
LC record available at https://lccn.loc.gov/2022030861
LC ebook record available at https://lccn.loc.gov/2022030862

DOI: 10.1093/oso/9780197674796.001.0001

1 3 5 7 9 8 6 4 2

Paperback printed by Marquis, Canada
Hardback printed by Bridgeport National Bindery, Inc., United States of America

Contents

Prologue and Acknowledgments

Writing this book was prompted by the recent distressing developments in the United States. The country is fiercely divided. Many people are fearful of further deterioration of the democratic system and are distraught and even in despair. For three primary reasons, I believed I could write a book that would help explain how such unhappy developments happened, and also, importantly, how they might be overcome.

First, I have enjoyed the good fortune of a long life. As a teenager, I was horrified by what I learned of the terrible wars in China and Spain, and then throughout Europe and Asia, soon engaging the United States. I became and remained determined to strive to help avoid such tragedies in the future. I witnessed and to some degree experienced the recovery from the Great Depression and, in later years, the effective struggles for more equality and justice in America. So I came to know that terrible things happened in America, but that triumphs were also accomplished.

Second, I studied and researched international conflicts engaging the United States, finding patterns of conduct that sometimes had terrible consequences but sometimes good consequences.[1] I also studied and researched many kinds of inequality within the United States, again discerning what had good or bad consequences.[2] I did this work utilizing the perspectives of the social sciences, and particularly of sociology. I discovered grand theoretic explanations and many smaller generalizations. I learned the value of gathering and examining all kinds of data. I believed that by comparing variations in consequences, I could discern better policies.

Third, I participated in the creation of the broad field of peace research and conflict resolution.[3] Many people across many countries have learned, by practice and research, what kinds of conduct improve the chances of minimizing destructive violence and maximizing widely beneficial outcomes.[4] It occurred to me that I could apply these findings to a variety of conflicts within the United States during the last seventy-five years.

With this background, I had the audacity to start working on this book, and I have devoted a few years to it. Concerned about the recent troubles in the country, I had decided that a book that examined many conflicts in America since 1945 could help explain how Americans sometimes suffered destructive conflicts, but also had fought constructively to overcome troubles and advance equality, justice, and freedom. I trust that the readers of this book find it interesting and useful in understanding how conflicts can be done better. Perhaps some readers will even apply the findings in practice.

I did not write this book alone, but I do take responsibility for it. Although I officially retired many years ago, I continue to have access to the resources of the Program for the Advancement of Research on Conflict and Collaboration, in the Maxwell School of Citizenship and Public Affairs, Syracuse University. I want to acknowledge particularly the following people who read portions or all of the book drafts, or in other ways helped me in completing it: Michael Barkun, Hans Guenter Brauch, Frederick F. Carriere, Bruce W. Dayton, Todd Dickey, Catherine Gerard, Susan T. Gooden, Joseph Kriesberg, Daniel Kriesberg, Jerry Miner, Tina Nabatchi, Robert A. Rubinstein, Tara D. Slater, and two anonymous readers chosen by Oxford University Press. I thank Angela Chnapko, Senior Editor, Oxford University Press, for her responsive oversight of the complex publishing process. Finally, I want to thank Paula Freedman, my partner, who wonderfully supported the project through all its stages, with fine editing and enlightening conversations.

1
Reflecting Backward and Looking Forward

Americans have often been divided and antagonistic in profound ways throughout their history. Destructive conflicts have been waged many times throughout US history. However, many conflicts were also constructively conducted so that they yielded progress toward greater liberty and justice for all Americans. In this book, I compare conflicts, examining how some conflicts, since 1945, were waged more or less well and other more or less badly.

In the years of Donald J. Trump's presidency, prior increasing political polarization and antagonism further intensified. They were deepened by a violent insurrection in Washington, DC, on January 6, 2021. The political system, at the national level, had become highly partisan and was seriously deadlocked. Profound differences even exist about what is true and what is false, about who is to be believed and who is not. Intense fights occur at the state and community levels. These developments have raised widespread fears about the survival of American democracy. I examine how badly waged conflicts have contributed to getting the United States into this morass and previous troubles, and how constructively waged conflicts in the past have helped overcome troubles and advance democracy. Such an examination can help answer questions about how the United States got into such troubles and how it can get over and past them in the future.

Analyzing Domestic American Conflicts

The turmoil leading to the January 6, 2021—insurrection—was the result of many prior developments in the United States and the surrounding world. These developments include many technological changes in the production of rapidly increasing innovative products and services. They are also related to other trends and forces, such as changes in demography, trade, and communications, and to enduring cultural ways of thinking and believing. The implications of these developments, however, were not inevitable. They were channeled in major ways by broad conflicts or discrete fights about how to deal with them.

This book examines many of those conflicts, focusing on those related to three primary dimensions of all human societies: class, status, and power. These are fundamental analytic dimensions of relations among socially defined groups of

Fighting Better. Louis Kriesberg, Oxford University Press. © Oxford University Press 2023.
DOI: 10.1093/oso/9780197674796.003.0001

people who are differently ranked. They are the arenas in which many conflicts arise and are waged. The conflicts are often about changing the degree of inequality in each dimension. Of course, people's relations incorporate all three dimensions, and people's placement in each dimension affects their ranking in the other dimensions. There are many sociological, social-psychological, and economic processes that drive the course of conflicts, and they will be noted in my analyses of conflicts in this work. The focus of the analyses is to assess how the conflicts did or did not advance equity between contenders in different rankings in class, status, and power. Advancing equity is consistent with realizing basic rights for all Americans, as discussed later in this introduction.

In this book, I examine how conflicts can be waged constructively, by analyzing significant American conflicts that did or did not work out well for the contenders and the country as a whole. The specific objectives set by adversaries and the methods they chose to achieve them are examined. I also note how the contenders' objectives may or may not be consistent with advancing equity in class, status, or power. At times, I offer possible alternative ways of framing the particular objectives and other strategies of contention that might have been deployed to enhance equity. Such analyses should yield guidelines for waging future conflicts constructively. In effect, the book can serve as a coach for people entering, waging, or settling a conflict.[1] Fighting better is a way to overcome contemporary challenges and preserve democracy.

I draw on an empirically derived constructive conflict resolution approach to understand what conditions, goals, and strategies are conducive to waging fights so that they contribute to positive changes. In this introductory chapter, I provide the basic ideas of that approach, which are fully developed in six editions of the book *Constructive Conflicts*.[2] The approach is part of the broad fields of conflict resolution and peace studies as they emerged after World War II and flourished beginning in the 1970s.[3]

The book reviews major domestic conflicts over the last seventy-five years, conflicts with both positive and negative results. Of course, most conflicts have mixed consequences, and assessments may be debatable. Even "successes" have some downsides for diverse people, but they may still be "good enough" to be celebrated for the improvements they constitute. The conflicts I have chosen to focus upon are related to the consequences of the 2016 presidential election in two ways. They help account for the appeal of the assertions Trump made, and even the support for some of the policies pursued by the Trump administration, but also for the resistance to them and their rejection in the 2020 election.

The long-term perspective taken in this book should help to discern better ways to make progress and avoid reactionary setbacks, which is the main purpose of this book. I think such a longer perspective will give readers good reason to expect that many of our present problems will be reduced. I do believe, as

Dr. Martin Luther King Jr. often said, that "[t]he arc of the moral universe is long, but it bends toward justice." I believe this is true, and there is some support of it in this book. In subsequent chapters, episodes of both progressive and regressive developments will be examined.

Long-term trends in class, status, and power inequalities are discernable and underlie specific conflicts. As evidence examined in this book will reveal, class inequality did not greatly change after 1945 until the early 1980s, when it increased at an accelerating rate. On the other hand, status equality significantly increased after 1945, as indicated by the ranking of African Americans, women, and many other people sharing collective identities. The crossing of these two major trend lines generates conflicts and influences how power equality fluctuates and impacts class and status changes.

Although focusing on particular past conflicts in the United States as well as ones that are ongoing, I shall suggest some general insights that help explain why some of those struggles are largely destructive and others are relatively effective and constructive. Some general reasons may be discerned. For example, people successfully making what they regard as progress sometimes expand their goals and overreach. By pushing too far and too fast, they engender resistance that results in considerable destructive consequences.[4] On the other hand, sometimes conditions that should be changed are ignored, due to widespread complacency, and the ultimately resulting conflicts are more damaging than they would have been if needed changes had been recognized and addressed earlier. Constructive changes are also more likely when alternative ways of dealing with the problems needing change are examined before choosing the policy to pursue. This is particularly the case when the examination is based upon solid evidence about the concerns and strengths of the contenders.

Basic American Goals

People in the United States, and people generally around the world, often disagree about what is good constructive progress. There is no total consensus about these matters. I will pay attention in my discussion of constructive conflicts to those matters about which there is considerable consensus, at least in principle. For the purposes of this book, I cite statements that are widely accepted as guides to determine goals that we should try to advance. They are rights Americans claim to have and try to realize. Thus, the American Declaration of Independence proclaimed that all people "are endowed by their Creator with certain unalienable Rights, that among these are Life, Liberty and the pursuit of Happiness." Presumably, it is good to protect and advance such rights. Another guide may be the often-recited pledge of allegiance to the American flag: "I pledge allegiance to the flag of the United States of America, and to the republic for which it stands, one nation under God, indivisible, with liberty and justice for all."[5]

The preamble to the US Constitution states: "We the People of the United States, in Order to form a more perfect Union, establish Justice, insure domestic Tranquility, provide for the common defense, promote the general Welfare, and secure the Blessings of Liberty to ourselves and our Posterity, do ordain and establish this Constitution for the United States of America."

I give special attention to life, liberty, and justice for all Americans. These are broad values or civic rights, and Americans tend to support them in principle. Yet Americans can and do disagree about their meaning, their implications, and their relative priority. They also often disagree about the identities they share in pursuing policies to advance a given value. For example, "liberty and justice for all" may refer to all White American men, to all people in California, to all Americans of one's own generation, to all Republican Party leaders, or to all investors in large US-based corporations. I favor broad inclusiveness, considering efforts to widely advance life, liberty, and justice to be positive and progressive goals, while efforts to restrict or nullify these values as negative and regressive goals.

A Constructive Conflict Approach to Waging Struggles

This book is premised on the idea that conflicts are inherent in social life, but often can be waged with goals and methods that are widely beneficial or, less fortunately, are waged with goals and methods that are largely injurious. This is not a novel idea, but in the last seventy-five years a great many international and domestic conflicts have been studied, and an increasing body of evidence has accumulated that indicates what particular ways of behaving in conflicts tend to minimize their destructiveness. The fields of peace studies and conflict resolution embody that knowledge, and workers in these fields have had experience in applying it. I draw on the work of these fields to understand how many specific fights in several spheres of conflict have been conducted—sometimes well, but sometimes poorly. In doing this, I will also test and enhance that field.

Seven interrelated core ideas in the constructive conflict approach should be stated at the outset. These are recognitions of important, but often neglected, qualities of social conflicts. These qualities make it possible, even likely if recognized, that the conflict will be waged constructively and have positive consequences. Ignoring these qualities will make conflicts more likely to be destructively waged and have negative consequences for the adversaries.

First, what is particularly important for this book is the recognition that conflicts are conducted in varyingly institutionalized manners, which are generally viewed as *legitimate*. Many fights are waged in a highly regulated manner, within the judicial or political systems. However, very often, less-regulated conduct plays an important role in the workings of those systems, particularly of the political system. For example, social movement organizations often engage in legally constrained lobbying or mobilizing of voters for elections. Some conflicts, at some stages in their course, are waged in large measure in nonregulated ways, such as in nonviolent actions that are not legal, but limited.[6]

The United States, as a democratic society, has laws to manage conflicts. However, in some ways the laws may benefit some sides in a fight more than others. America during many decades had laws that upheld slavery and then segregation, and also laws to prevent workers from organizing to bargain with employers about wages and working conditions. Such laws eventually came to be seen as illegitimate, while some laws were changed and became more legitimate to all. American norms provide protection for many nonregulated ways to wage struggles—by holding demonstrations, making speeches, writing books, composing songs, and using social media. There are norms of civil society that usually contribute to waging conflicts constructively. Violations of these norms are often polarizing and destructive.

Second, conflicts are not waged only by relying on negative sanctions or coercion, whether violent or nonviolent; *noncoercive inducements* are also important. Two such kinds of inducements are generally used in conflicts: persuasion and positive sanctions. In persuasion, one side tries to convince the adversary that its members will be better off if they do what is being asked of them, that doing so will better meet their own needs or values. Persuasion is also important in drawing support from people who are not committed to any of the primary antagonistic parties, and is also important in mobilizing supporters and reducing support within adversarial parties. Persuasive arguments may be packaged together into ideologies, which are significant in many large-scale conflicts. Unfortunately, persuasive arguments may be based on fanciful, inaccurate premises, misguided choices, and overly narrow considerations.

Positive sanctions are another kind of noncoercive inducement, where benefits are proffered in exchange for the adversary granting what is being asked of them. Positive sanctions may also be employed by leaders to mobilize and attract more supporters, and to undermine the support of adherents to leaders of opposing parties. They too may be elements in ideologies.

In actuality, some blending of the inducements is common, and the mixture shifts as the conflict goes on.[7] Obviously, in domestic conflicts, positive sanctions and persuasion are very important, as are related inducements when parties and their respective representatives' campaign for votes by promising benefits to some kinds of voters after they are elected to office. Each opposing side in elections also wages persuasive efforts to convince potential and current supporters that their values and interests will be served by its victory.

Coercion, positive sanctions, or persuasion are not inherently constructive or destructive. That depends on how they are fashioned in regard to other features of constructive fighting. Coercion in the form of ruthless violence certainly is destructive, as it denies legitimacy to opponents. Many forms of nonviolent action tend to be constructive, as they acknowledge a possible future of better mutual relations. Even positive sanctions can be destructive, such as when they are covert and used to bribe or reward leaders of the opposing side and divide them from each other and from their constituents. Persuasion can be destructive when it is misleading and not based on truth, or when it serves to dehumanize the opponents, rather than help convince the other side that shared values and interests exist. This is evidenced in the importance of ideologies in many conflicts, not only to rally supporters, but also to gain adherents.

Of course, opposing sides in a conflict may use different blends of inducements. For example, a minority group seeking better legal protections in a country may rely on nonviolent coercion and persuasion, while the country's government may use persuasion and some violent coercion. What is particularly relevant for a conflict to be conducted constructively is that the opposing sides are not greatly unbalanced. In highly asymmetrical relations, the stronger party may tend to behave destructively, relying on violence for self-aggrandizing ends.

Third, parties in conflicts are not unitary actors, meaning that each side in a fight includes multiple perspectives held by different member groups. Each side in a conflict is *heterogeneous*, with many subgroups. The differences between leaders and followers, among leadership factions, between ruling and challenging groups, and among many other divisions can shift in their relative power and their readiness to escalate or to de-escalate a conflict. Recognizing the diversity within an adversarial entity opens up possible bonds and connections across adversarial lines.

Cooperation between segments of opposing camps can generate information about possible conflict transformations, yielding mutually beneficial solutions. Skilled negotiators in adversarial circumstances may search out the relative priorities of various issues in contention by the diverse segments of the opposing camps. This can result in agreements based on trading off the different priorities. This is common in union-management negotiations.

Fourth, members of each side in a conflict tend to *socially construct* their view of the conflict, characterizing who they are and who the other side is, what they want from the other side and what the other side wants of them. These social constructions by antagonistic sides are often matters of contention between the members of the opposing sides. For example, at the height of the Cold War, Americans generally viewed the conflict as between the Free World, led by the

United States, which was threatened by the aggressive totalitarian Communist empire controlled by the Soviet leaders. At the same time, those in the Soviet Union tended to see themselves as leaders of the socialist camp, which was threatened by the imperialist camp led by the rulers of the exploitive capitalist United States. Typically, each side in a destructive fight believes it is good and its cause is just, while the opponent is bad and its cause is unjust, thereby intensifying the conflict. Recognizing that conflicts are socially constructed opens up the possibilities that a conflict can be deconstructed and reconstructed and thereby transformed positively. Destructive conflicts tend to be the result of one or more sides holding dehumanizing images of members of an opposing side.

Fifth, in applying a constructive conflict approach, an adversary will try to understand the opponents' point of view and underlying concerns. That knowledge can contribute to discovering ways of formulating, conducting, and settling a conflict that moderates the fight and provides some *mutual gains*. This could contribute to reducing polarization. Formulating goals that have some attraction for the adversary is helpful in this regard. Discovery of possible mutual benefits is critical for constructive conflict transformations; fights are not necessarily zero-sum conflicts, in which one side's gains must be at the expense of the other side. Too often, people in a conflict assume that any gain their opponents may make will be a loss for them. Even when negotiating, some people neglect possible trade-offs and shared gains. Put another way, expanding the size of the pie can make it possible for all slices to be larger. There is a large literature about how to negotiate so that mutual benefits are maximized.[8]

Looking for mutual gains does not negate opposing sides having contrary goals. That is the essence of a conflict. Indeed, it is constructive for each side to be clear about what it and the adversary seeks. Too often, one or more side expresses its feelings about some condition; it expresses its anger, grievance, or resentment without being clear what changes are desired by some specific people. That posture hampers discovering what problem underlies a particular conflict.

The very search for common benefits makes for constructive fighting and even conflict de-escalation. Doing so encourages taking the perspective of members of the adversary, learning what their concerns are and how they are likely to respond to one's own actions. Knowing the relative priorities of the adversary's multiple goals facilitates making deals involving trade-offs, which is crucial for avoiding unintended destructive escalations. This is also related to effective persuasive efforts.

It should also be obvious that using language that dehumanizes the adversary hinders fighting a conflict constructively. Demonizing the enemy is conducive to polarization and destructive escalation. Since conflicts are socially constructed, they can be framed in different ways, including in ways that incorporate possible mutual gains.[9]

Conflicts that are waged within a society, unlike conflicts between societies, can be expected to have potential benefits for the contending parties. They are unlikely to be wholly zero-sum conflicts. The pledge of allegiance in America celebrates that, with its great diversity, it is one indivisible country.

Sixth, conflicts are not isolated and bounded. They are *interconnected* over time and social space. Conflicts are often nested in larger conflicts, such as when proxy wars during the Cold War were sustained by the United States and the Soviet Union. Furthermore, each side in a fight often has one or more other adversaries on other fronts. In that case, the changing salience of one conflict will change the salience of the others. Cross-cutting conflicts and associated identities tend to counter the intensity of any one of them, while multiple conflicts that coincide with each other tend to increase their intensity and intractability. Multiple conflicts and identities that coincide reinforce the intensity of all the identities and conflicts. The concept of intersectionality refers to mutually reinforcing the systems of oppression, most notably of race, class, gender, and sexuality.[10] Again, recognizing these realities can provide paths to constructively transforming a destructive relationship.

Conflicts are linked through time. Major conflicts are rarely resolved and disappear. They usually are transformed, for better or worse. Often, one side in a conflict makes some gain at the expense of an opposing side. That gain may overreach and spur a backlash, a renewed struggle, and even a worsened circumstance for the prior winner. To fight better, it is advisable to consider how to make changes that are broadly acceptable.

Finally, seventh, conflicts are *dynamic*. They are constantly in flux, with the relations between adversaries shifting constructively or destructively. Relations among subgroups within each side are also changing, and the surrounding context of the conflict is generally in flux. A broad way of viewing the dynamic nature of conflicts is to view them as moving through many stages. This is not a simple linear process, with all constituent elements moving at the same speed, distance, and direction. Broad stages, however, may be distinguished, using terms such as underlying conditions, emergence, escalation, stalemate or turning point, de-escalation, transformation, settlement, reconciliation, or revival. I will discuss these stages briefly here and return to many of them while examining the major conflicts that are discussed in this book. Different constructive options can be generated at different stages and transitions.

Conflict Stages

Broad stages in a conflict may be distinguished, such as latent, emerging, escalating, de-escalating, and settling. *Latent* refers to circumstances that have the potentiality to become manifest in a conflict that is perceived by antagonists. The recognition of that potentiality is in the mind of conflict analysts or of persons inside one of the potential sides in conflict. The latent conditions may increase inequalities between parties in a relationship, which is likely to be viewed as a grievance by members of the declining side or viewed as a rightful opportunity to get even more by the members of the rising side.

A conflict emerges when significant members of a potential or overt adversary group meet three conditions. First, they see themselves as having a shared identity based on a variety of possible characteristics that distinguish themselves from another group with a different identity. Second, they must share a belief that they are suffering an injustice inflicted by members of the other side. Finally, they must believe that they can act to induce the opposing side to act differently and so end or lessen their grievance; more extremely, they may believe they can destroy and replace their enemy. Herein lies the importance of politicians, religious and ethnic leaders, and intellectuals in constructing stories and theories that convince believers about their identity, grievance, and capacity to gain redress. The resulting social constructions may appear credible to people who find the interpretations to be congenial, despite evidence that belies them. A conflict is more likely to be waged constructively insofar as it is based upon well-grounded analyses by the adversaries.

A conflict begins to escalate as members of one side make their demands clear to the other side and begin to try to win the other side's acquiescence. If the other side ignores, rebuffs, or answers with counterdemands, the conflict is likely to escalate further as greater inducements are brought into play. Coercion is often employed and increased when other inducements fail. Often, each side increasingly sees its opponent as hateful and collaborative bonds are cut. Escalation can be done constructively, for example, by communicating the interest in ultimately reaching a mutually beneficial outcome of the conflict. Avoiding destructive excesses of escalation also helps in keeping a fight more constructive. For example, if one side believes it is winning, it may expand its goals, resulting in a disastrous overreach.

Sometimes one side is able to decisively defeat an adversary, but in most large-scale conflicts, one or more sides begin to recognize that an outright "victory" is not within reach. Often, then, de-escalating moves are ventured or intermediaries explore possible mediating interventions. Mediation in many spheres is institutionalized or may be introduced in an ad hoc manner. De-escalating actions, to be effective in constructively transforming a destructive conflict, require careful analysis. The actions might be perceived by the opponent as a sign of weakness, and the opponent may then increase its demands, or the actions may be perceived as a trick to gain time and be brusquely dismissed. Tentative, exploratory exchanges through informal intermediaries sometimes help bridge these pitfalls; this may be undertaken by religious figures, academicians, or politicians.

Many conflicts are settled, at least for a time, by direct negotiations and written agreements or treaties. There is a vast literature about negotiations, including research about reaching mutually beneficial agreements and about the great variety of possible mediation contributions.[11]

Major conflicts are rarely entirely ended. After all, conflicts occur between parties who have a relationship that continues when their conflict subsides. That is why workers in the field of conflict studies refer increasingly to *conflict transformation* rather than conflict resolution.[12] The term usually refers to a conflict that has been waged with severe harms and changes so that it continues in a moderate way, by means that the adversaries regard as legitimate and acceptable. In some cases, this breaks down and injurious fighting is then renewed. However, there are cases in which processes occur that transform the relationship between former adversaries into one of friendship, fulsome collaboration, and even unity.

The Major Conflicts Discussed in This Book

In the following chapters, I will examine particular conflicts that have occurred since the end of World War II, in a few different domestic conflict realms. International affairs are not focused upon, since I have written extensively about international conflicts and America's various engagements in them. This was the primary subject of my *Realizing Peace: A Constructive Conflict Approach*, and of numerous other writings.[13] In this book, I discuss how constructively various parties in American domestic conflicts conducted themselves. The conflict arenas stressed in this book are in three primary dimensions of rankings: class, status, and power. Of course, particular fights usually are conducted to some degree in more than one arena, but usually adversaries view their fight to be primarily in one or another dimension of ranking.

The three dimensions of ranking were notably distinguished by Max Weber, who believed that Marx's emphasis only on class was overly simplistic.[14] Many sociologists and other social scientists have tended to stress the primacy of class or one of the other dimensions. I believe, however, that focusing on all three dimensions best provides a comprehensive, three-dimensional view of society. Attention to status is getting more attention as issues regarding respect and recognition have become more salient and culture wars are fought.

In this work, I consider the many ways rankings in each dimension reverberate with rankings in the other dimensions. Theories have been developed to explain the determinants and the consequences of different degrees of inequity, but my focus here is on the conflicts various groups wage about changing their degree of equality.[15] The analyses in this work will reveal how conflicts in one dimension impact on changes in other dimensions. I rely more on analysts who examine particular fights rather than those who examine the general sources of inequality.[16]

Chapters 2–4 focus on conflicts about inequalities in class relations—that is, differences in income and wealth—in three different time periods. In chapter 2, I discuss conflicts in the years between 1945 and 1969 that were related to labor-management struggles and the fight against poverty. Chapter 3 discusses the period from 1970 to 1992, examining conflicts and rising class inequality. Chapter 4 considers the period from 1993 until 2021, focusing on contentions about inequality.

Chapters 5–7 relate to social status or prestige, particularly ascribed rather than achieved status. Chapters 5 and 6 discuss the series of conflicts about gaining more equality of status between African Americans and Whites. Chapter 5 examines the period from 1945 through the 1960s. Chapter 6 discusses the period from 1970 until 2021. Chapter 7 considers the series of fights relating to collective identities, including about women gaining more equality with men, about LGBTQ identities, and about various White identities, all from 1945 to 2021.

Chapters 8 and 9 discuss conflicts since 1945 for more political power to shape governmental policies, and examine how the way the conflicts were waged undermined constructive norms and practices and produced a highly antagonistic system. I will also discuss power differences in nongovernmental settings. The concluding chapter 10 proposes ways to overcome the destructive forms of conflicts regarding class, status, and power that were evident in the 2020 presidential election and afterward. It includes possible ways to enhance American life, liberty, and equality for the future. I am not writing the history of developments in each realm. Many lengthy periods of time are given scant attention. My interest is focused on contentious events, many of which were highly consequential, and some of which had relatively few lasting consequences. Some consequences were progressive, and others were regressive. That variety can help discern how particular fights may be waged constructively with enduring effect. To that end, the strategies used by opposing contenders will be analyzed and often critiqued. I acknowledge that my assessments are made with hindsight and reflection. The partisans in each conflict often lacked important information, faced immediate pressures to act, and needed to respond to multiple pressures. My second-guessing should not be seen as denigrating people struggling to advance their cause as best they can. Yet progress in fighting better depends upon reflecting on how past efforts made things worse, or at least inadequate, as well as often better.

My choice of particular conflicts to discuss from the innumerable conflicts in America's past was guided by four considerations. First, some conflicts were chosen because I was engaged in them, if only as a witness. Second, some conflicts had great visibility and enduring implications. Third, some conflicts were readily recognizable as having been waged either constructively or destructively. Finally, some conflicts were chosen due to their relevance to the problems of America in the 2020s. Of course, many conflicts were chosen because they possessed more than one of these characteristics.

It is obvious to most Americans that the country is beset by many serious conflicts, many of which are being conducted badly. The political system should provide institutionalized ways to manage national, state, and local conflicts. However, it is widely recognized that the system is highly polarized, and many leaders act as if they were tribal leaders, fighting primarily to defeat an enemy tribe. There is a low level of trust across most dividing lines in society. Recourse to various forms of violence in conducting official and nonofficial contentions has become highly concerning. Work that needs to be done cooperatively but is not undertaken results in profound shared losses. This has been true in dealing with COVID-19, global warming, mental illness, failing infrastructure, home-lessness, and on and on.

This book offers some explanations for what has gone wrong and some-times gone well. In reflecting on such explanations, it provides ideas about how conflicts can be done better. It points to particular practices that yield good conflicts, ones that should be fought, and fought constructively. Those ideas should stimulate reflection in shaping specific actions in circumstances that we face, which are unique and new, as they always will be.

2
Class-Related Conflicts, 1945–1969

Major divisions in social relations are related to unequal class or economic ranking. This occurs in different degree and various forms among diverse entities, including persons, political groups, corporations, and countries. The interactions across class divisions may occur in market transactions and also in struggles between occupational strata. Societies with highly unequal incomes, wealth, and control over conditions of economically productive work are often subjected to severe fights about the allocation of such matters.

In this work I discuss latent as well as manifest conflicts. Latent or dormant conflicts exist in the relations between people with different levels of income or wealth, or different employment positions. In this work, such major differences are viewed as latent or underlying class conflicts, which sometimes erupt into overt fights about taxes, welfare benefits, or union organizing. Admittedly, some people may view economic inequalities as inevitable aspects of free markets or of differences in intelligence and diligence and not inherently conflicting. In this work, I focus on two kinds of class conflicts. One is about people with different amounts of money (income or wealth), exemplified by fights relating to taxes or reducing poverty. The other is between employees and their employers or between people at different hierarchical positions, exemplified by battles over union rights.

Three chapters in this book are devoted to analyzing US struggles about major class matters. In this chapter I discuss the period from the end of World War II until 1969, a period that begins with significant progress toward greater equality. The next chapter covers a period when economic differences became more unequal, with destructive consequences and emerging resistance. The third chapter deals with the arrival of hyper-inequality and possible redress.

Overall, this book is focused on major contentions among US groups. Keeping the constructive conflict approach in mind, however, I emphasize that the struggles over dividing desirable goods are not always zero-sum conflicts. That is, the gain for one side need not be entirely at the expense of the other side. The fights may have mutual losses or mutual gains for the opposing sides. Indeed, in any large conflict, it is likely that both sides will suffer some losses and enjoy some benefits, although not to an equal degree.

Fighting Better. Louis Kriesberg, Oxford University Press. © Oxford University Press 2023.
DOI: 10.1093/oso/9780197674796.003.0002

Worker-Management Contentions

I begin examining labor-management conflicts by considering labor disputes manifested by workers striking against their employers. Work stoppages in the United States were relatively frequent in the late 1940s and early 1950s; for example, in 1950 there were 424 work stoppages. Presumably, the high frequency was related to the restraints on striking during World War II and the reduced capability of workers to exert pressure against management. Many latent problems were left unaddressed during the war. Also, after the war, the demand for consumption goods was high and labor availability was tight, giving workers leverage to strike effectively.

National work stoppages are one indicator at one level of labor-management conflict and one manifestation of the conflict. Labor-management relations, however, exist at the macro level, overlapping with political relations at the national, state, and city levels. They also exist at the level of industries and of individual plants. They occur, too, at the micro interpersonal level. Conflicts at different levels, times, and social spaces are interconnected. As discussed in the first chapter, since conflicts are interconnected, escalation in one set of conflicts can exacerbate and escalate other conflicts or, contrary-wise, decrease their salience, as enemy number one becomes less significant when a new enemy rises into the highest prominence. Very often, as well, diverse conflicts coincide and reinforce each other, as when ethnic and class divisions are the same. Employee-management relations have often been conventionally viewed and experienced as antagonistic.

Certainly, many corporate heads shared a fierce antagonism to unions, the federal government, President Roosevelt, the New Deal, and the Democratic Party. For many of them, these entities were viewed as overlapping antagonists. When Harry Truman became president upon Franklin Roosevelt's death in 1945, they regarded him with similar hostility.

In 1947, the Republican-controlled Congress overrode President Truman's veto and enacted the Labor Management Relations Act (Taft-Hartley Act). This act reduced the protections and capacities of workers and their unions that had been established in 1935 by the National Labor Relations Act (Wagner Act). It permitted union shops only where state law allowed it and a majority of workers voted for it. Taft-Hartley required unions to give sixty-day advance notice before a strike and authorized eighty-day federal injunctions against a strike, when a strike threatened to endanger national health or safety. It also narrowed the definition of unfair employer labor practices and restricted union political contributions. To raise the specter of a Communist threat from trade unions, it required all union officials to deny under oath that they had any Communist affiliations.

At the same time, official action was taken to better conduct collective bargaining and improve its effectiveness. The Federal Mediation and Conciliation Service was created as an independent agency of the US government. The agency was given the mission of preventing or minimizing the destructive impact of labor-management disputes by providing mediation, conciliation, and voluntary arbitration services. These are important elements in conducting many kinds of conflicts constructively.

Indeed, the growth of collective bargaining and improvements in its practices was an important source for the field of conflict resolution that began to emerge in the late 1950s.

Earlier research and writing had stressed the possibilities of cooperation and mutual benefits between different levels of a corporate hierarchy. For example, Mary Parker Follett developed the idea of participatory democracy and of society as integrative.[1] She pioneered in recognizing horizontal interactions within hierarchical organizations, the importance of informal processes within organizations, and the idea that authority could be based on expertise. Follett contributed greatly to the win-win concept—that opposing sides could reach benefits that were mutually beneficial. Her approach embraced conflict as a way to enhance diversity and an opportunity to forge integrated solutions rather than unsatisfactory losses and compromises.

Another impetus to the field that came to be called "human relations in industry" was the Hawthorne experiments, which were conducted from 1927 to 1932 at the Western Electric Hawthorne Works in a suburb of Chicago. The study was intended to assess the impact of different working conditions on employee productivity. The analysis of the findings by Roethlisberger and Dickson stressed that changes in lighting, scheduling, or other such conditions had little effect compared to the impact of the attention given to the workers by the experimentation.[2] The human relations in industry approach emphasizes that the emotions of the workers affect their productivity. This includes feeling good or bad about family life and social relations with superiors and colleagues. This interpretation of the Hawthorne studies, however, was rebutted by critics who re-examined the experiments and noted the importance of monetary incentives for productivity.[3] Treating subordinates respectfully and listening to their concerns could indeed improve relations, create shared responsibilities, improve productivity, and be mutually beneficial. But it can be falsely done and essentially manipulative, and therefore ineffective or worse.[4]

These discussions and observations about collective bargaining made important contributions to the field of conflict resolution. Negotiations as a way to settle disputes was a salient feature of the emerging field. A central conflict resolution idea in conducting negotiations is that many concerns may underlie positions taken in a conflict and efforts to settle them. Learning the adversary's underlying concerns has many possible benefits. I mention only three at this point.

First, if one side's representatives think about their adversary's concerns, they may think of ways to satisfy some of those concerns that are actually mutually beneficial. For example, union representatives may be holding out for a large increase in wages, compensating for workers feeling stuck in boring or otherwise unpleasant jobs. Offering job training leading to better jobs or arranging for revolving job transfers might improve overall compensation in ways both sides benefit.

Second, as each side recognizes multiple concerns on the other side, more trade-offs become feasible. That is, each side is likely to have somewhat different priorities, and therefore each side can concede on lower-priority matters in trade for items of higher importance. Third, cooperation can increase productivity, and greater output. A larger pie can yield a larger slice for everyone.

The realities of collective bargaining make clear another important matter about negotiating settlements of conflicts: the power balance between adversaries. Workers' struggles to organize unions and conduct strikes at the end of the 1800s and in the early 1900s, in the United States, were often violently suppressed. Many employers were opposed to workers organizing unions to advance their interests, but some came to favor collective agreements as stabilizers of industrial relations.[5] Where and when market conditions and political climate were supportive of unions, collective bargaining functioned, but when and where they did not, labor rights suffered, and possible mutual gains were lost.

Ideas and practices about how to mobilize members of the weaker side to empower themselves were improvised, tested, and passed on. Heroes were celebrated in stories and songs, and battles lost and won were discussed. Saul Alinsky wrote two major books about how to organize people in a community to formulate short- and long-term goals and various strategies to achieve them.[6] His writing and his actions have remained influential guides to create pressure in order to make a deal with an adversary. He fostered enduring organizations that enhance citizen participation and democratic governance, notably the Industrial Areas Foundation.[7] With greater power symmetry, mutual gains are more likely to be attained.

Labor Unions

The story of the vicissitudes of the United Automobile Workers (UAW) provides a good case to understand the changing relations between corporate managers and US workers.[8] Prior to the UAW's formation, unions had been largely organized by crafts or trades, such as electricians or plumbers, and they were organized together in the American Federation of Labor (AFL). In 1935, a caucus of industrial unions emerged and soon left the AFL, forming the Congress of Industrial Organizations (CIO), which included the UAW. In 1936, the UAW successfully instituted sit-down strikes, winning recognition with General Motors (GM) in February 1937. In the next month, a sit-down strike won recognition for the UAW by Chrysler.

The Ford Motor Company leadership, however, was fiercely resistant to recognizing the UAW. Henry Ford was against unions for several reasons. He reasoned that unions would restrict productivity so as to necessitate more employment. Ford also believed that union leaders would foment crises to maintain their own power to reach an agreement. Finally, Ford was distressed by the long closure of the River Rouge Plant and considered breaking up the company. Clara Ford, his wife, insisted she would leave him if he destroyed the family business. In June 1941, Ford recognized the UAW and signed a contract with highly favorable terms for the union.[9] He claimed that, in doing so, the union workers would cooperate with him to compete better against rival companies.

During this period, Walter Reuther led many of the UAW's organizing, striking, and negotiating activities and was the president of the UAW from 1946 until his death in 1970. He survived two assassination attempts, which were not solved, and he died in a charter plane crash at the age of sixty-two.[10] He led in reaching innovative contracts with the major automobile companies that established significant benefits for the workers in their plants.[11] He also was a major public figure, striving to advance progressive social policies relating to the civil rights movement, women's rights, universal health care, affordable housing, environmental concerns, and opposition to the Vietnam War. The UAW and other unions had important roles in the Democratic Party. Such collaboration helped win the UAW support in return.

Reuther's activities cost him the enmity of President Richard Nixon, Senator Barry Goldwater, and other conservatives who considered him to be dangerous. The Republican Party overall was hostile to labor unions and acted to curtail their capacities for a few reasons. Many Republicans could reasonably believe that such anti-union policies would weaken the Democratic Party's electoral strength. Furthermore, many corporate managers and investors, believing their interests would be better served if workers were less able to fight for their interests, would contribute a good deal of money to the Republican Party. Finally, as will be discussed in later chapters, there is some conservative ideological consistency in supporting policies that minimize government expenditures for social welfare benefits for working-class people, reducing taxes, and generally weakening government regulation and oversight of private business. This seems illustrative of zero-sum, short-term calculations. Instead, Republican leaders might have competed to win labor votes.

Clearly, occupations vary greatly in their proportions of union memberships, but even farmworkers were able to form a union in 1962, the National Farm Workers Association (NFWA), with the leadership of César Chávez and Dolores Huerta.[12] In 1965 they joined a strike begun by Filipino American grape pickers in Delano, California. Following a dramatic march to the state capitol in Sacramento, the union launched a national public boycott of table grapes. After five years, the union, now called the United Farm Workers (UFW), gained recognition and negotiated improved wages and working conditions. Winning widespread public support worked.

In the 1950s and 1960s, the gains of the industrial working class contributed to the high rate of economic growth in the United States and increased income equality during this period.[13] Trade union membership was relatively high compared to later decades, as examined in the next chapter. Union membership as a percentage of employed workers peaked in 1954 at 28.3 percent.[14] Controlling for relevant factors, union workers' wages are 10 to 30 percent higher than the wages of nonunion workers. The role of cultural and political factors in affecting union membership is indicated by the large variation in the percentage of employed workers in unions among the states, with southern states having very low rates.

Unions and the collective bargaining associated with them tend to provide corporate and societal benefits. Thus, companies with union contracts have higher retention rates for workers and the companies have lower recruitment and training costs. Information about collective bargaining and the agreements reached in the 1950 and 1960s between blue collar workers and large companies were publicly known. White-collar workers knew that their pay grades were affected by those agreements.[15] Corporate leaders sought good government relations and good public regard, and many seemed to think the country's economic well-being coincided with their companies' well-being. Giving themselves very high salaries would have been unseemly in this interdependent context. There was some common good sensibility.

The War on Poverty

During the 1950s and 1960s, the US economy was growing rapidly, raising the belief that an age of affluence was arriving, and even concerns that the country was facing the problem of what to do with the leisure time that growing productivity would produce.[16] Despite this bounty, over 20 percent of the population lived in poverty. This dire circumstance was examined vividly in Michael Harrington's 1962 book, *The Other America*, which quickly drew great public attention, including President John F. Kennedy's notice.[17] Kennedy's chair of the Council of Economic Advisers, Walter Heller, convinced him that a large, broad-based tax cut would produce economic growth.[18] He and his advisers also thought that a focus on poverty would balance the middle-class tax cut and help the people who were left behind even with economic expansion. Heller formed a task force to determine what programs could lift people out of poverty.

Also, in the early 1960s, African Americans and some Whites had begun to organize to expand and improve welfare benefits. The National Welfare Rights Organization (NWRO) was organized under the leadership of George A. Wiley to take direct actions to enroll eligible people in welfare programs.[19] Earlier, when Wiley was a colleague at Syracuse University, I was one of several other colleagues who supported CORE (the Congress of Racial Equality) and met with Wiley to discuss the civil rights campaign. He often initiated discussions by posing the question: "What do we want, if they say, Yes?" I realized that was a great question. Setting the goal should be the first step, with strategies then chosen to reach the goal.

When Lyndon B. Johnson became president, following Kennedy's assassination on November 22, 1963, Heller brought the emerging antipoverty plans to his attention. That policy matter was very attractive to Johnson for a few reasons. He grew up in a poor area in Texas and taught in schools with children of migrant workers; he admired the achievements of the New Deal in recovering from the Great Depression, and he was proud of his own work as Texas director of the National Youth Administration during the New Deal. Moreover, he saw reducing poverty as part of responding positively to the demands of the civil rights social movement organizations that were escalating. One manifestation of that was the massive, resounding August 1963 "March on Washington for Jobs and Freedom," where Dr. Martin Luther King. Jr. made his "I Have a Dream" speech.

Heller's staff was tasked with specifying particular programs to reduce poverty by a wide variety of strategies. The idea of "opportunity theory" provided an encompassing perspective to lend coherence to the War on Poverty. Opportunity theory was developed by two sociologists at Columbia University, Richard A. Cloward and Lloyd E. Ohlin, who were studying juvenile delinquency and ways to counter it.[20] The idea stressed that juveniles who acted delinquently wanted much the same things as nondelinquent youths, but they saw the ways to get what they wanted blocked. If opportunities to get what they wanted were opened up, they would use them. This was the guide for experiments funded by the Ford Foundation and the New York City program Mobilization for Youth.[21] That idea was compatible with the idea that mutual gains were attainable for parties with different interests as well. Enabling poor people to flourish would grow the economy for all.

Johnson presented the initial set of antipoverty programs in the Economic Opportunity Act of 1964 (EOA) in March 1964. The programs particularly targeted urban centers and the black ghettos in them. It passed with the Democratic votes in the Senate in July, and the Democrats in the House passed it in August. Many southern Democrats voted no because of its racial integration provisions. Republicans attributed their opposition to the bill to states' rights concerns, its costs, and already existing programs that might be improved. The bill was part of a broad strategy to prevent people from becoming and remaining poor. The EOA included the Job Corps, Volunteers in Service to America (VISTA), and the Community Action Program. In addition, the Food Stamp Act was begun in August 1964, Head Start began in the summer of 1965, and the Social Security Act of 1965 established both Medicaid and Medicare. The Medicare bill was passed overwhelmingly, with a significant number of Republicans also voting for it. Notable advances to reduce poverty and class inequality were made.

All these programs were presented as ways to provide opportunities for advancement for people whose opportunities were obstructed. It is noteworthy that the programs did not undertake structural changes that would ensure jobs would be available. No great undertakings were made that would ensure employment or minimal income to everyone. The programs were mostly designed to prepare able-bodied persons to get jobs. This is the way Johnson viewed the program, and it also appealed to American values that would help win widespread public support. In this fight to reduce poverty, broad appeals to build wide coalitions were likely to be more effective than framing the fight as being against enemies who would portray the programs as giving money away to people who wouldn't work. Nevertheless, there was some opposition to the programs of the War on Poverty. The conflict was largely pursued within the established national political institutions, but direct actions in the streets and at welfare-related offices kept up pressure for expanding benefits. The fight was waged on two closely related fronts, which was effective for the would-be change-makers.

A hallmark of the War on Poverty was to strive for the maximum feasible participation of the poor in shaping and administering the programs.[22] Although ridiculed by some critics, this feature had enduring benefits in the struggle to reduce poverty. The Office of Economic Opportunity invited cities to submit proposals to establish community action centers. This tended to bypass the state and local governments by providing funding more directly to beneficiaries. The established, mostly White, officials were less able to control the spending of new funding.[23]

My experience at Syracuse University offers a local perspective of the workings of community action centers. In 1962, I had joined a team of sociologists and social workers who had established the Youth Development Center (YDC) in 1960, with support from the Ford Foundation.[24] I worked on a large study of social mobility and public housing, comparing the effects of living in public housing projects located either in low-income or middle-income neighborhoods.[25] Much other YDC research was focused on juvenile delinquency and poverty-related matters, so we were ready to apply for a Community Action Center grant, which we received. Two other agencies in Syracuse were already working to improve the quality of neighborhood life: The Mayor's Commission on Youth and Crusade for Opportunity.

One of the major activities of the Syracuse University community action program was to train tenants in public housing and other people living in poor neighborhoods about ways to organize and take actions to improve their conditions. Warren Haggstrom, from the School of Social Work, led that effort. He brought in Fred Ross, from the Industrial Areas Foundation (founded by Saul Alinsky) to help provide training in organizing. Residents in each public housing project organized and selected representatives, a practice that continues throughout the country. The skills acquired and practiced by many people in poor neighborhoods have helped them in taking on leadership roles. When federal funding for the Syracuse University community action program ceased, PEACE, Inc. was created to oversee public housing and other social welfare programs. It, like other such entities across the country, has representatives who are members of the communities receiving benefits on the board of directors.

I, along with others associated with YDC, and many other people engaged in antipoverty work, emphasized structural conditions that tended to keep poor people poor or ones that helped poor people to move out of poverty. Changing the actual circumstances meant more than reducing the barriers to seizing opportunities, useful as that may be.[26] After all, during World War II, the vast number of jobs that were opened by the absence of workers who were in the armed forces were quickly filled by people who were often considered unqualified earlier (including women). The demand for workers, if great enough, will bring workers to meet the demand. Circumstances that are congenial to preparing people for employment or entrepreneurial undertakings are more effective than punitive conditions that would coerce people to behave better.

The contributions of the programs to reducing poverty and reducing the adverse consequences of poverty have been researched. First, the changes in the rate of poverty should be noted. The rate was declining in the period immediately before the EOA was begun, but its establishment increased the decline. In 1959, the overall poverty rate was 22.4 percent, and it fell to 11.1 percent in 1973.[27] The aged (over 65 years) experienced the greatest decline in poverty, falling from 35.2 percent in 1959 to half that by 1973. The rate of children (under age 18) in poverty in 1959 was 27.3 percent, and it fell to 14.4 percent in 1973.

The direct action of many local groups agitating for welfare benefits to be made available to those who were eligible were effective in expanding the welfare rolls. Piven and Cloward argued at the time that "expansive relief policies are designed to mute civil disorder."[28] For example, applications for Aid to Families with Dependent Children (AFDC) rose greatly between 1960 and 1968. Nationally, they increased by 85 percent and the proportion of accepted applications rose from 55 percent in 1960 to 70 percent in 1968.[29]

Nixon's inauguration in January 1969 would result in cutting back provisions of the War on Poverty, which Republican Party leaders claimed had failed. However, in August 1969, President Nixon announced the Family Assistance Plan (FAP), promoted by Daniel P. Moynihan, his Assistant for Urban Affairs. All families with children would be eligible for a minimum stipend, and the absence of a "man in the house" would no longer be a precondition for welfare. FAP won some public support, but it failed to be approved by the Senate.[30] Many Democrats believed the stipend was too low and some welfare benefits would be lost. Many Republicans were ideologically opposed to the concept. Mutual mistrust hampered formulating a mutually acceptable FAP. Jill Quadagno provides other considerations to explain the failure to pass an FAP.[31] In a detailed analysis, she found that a class analysis did not explain the outcome, but status considerations were also important: some Whites (notably southern Democratic senators) believed that it empowered African Americans, and some men thought it would reduce men's dominance over women. In any case, a broad coalition of supporters did not arise and mobilize enough people to establish a stipend for all low-income families.

Nixon went on to impose more restrictions on poor people's claims to obtain welfare benefits.[32] Nevertheless, some basic programs have survived, with modifications. Clearly, there has been wide enough public awareness of the programs' benefits to have sustained them. For example, the value of Head Start programs has been well documented.[33]

The Equal Opportunity Act of 1964 and related subsequent legislation did more than reduce the proportion of people in poverty. The legislation expanded the safety net for all who were in poverty or might fall into poverty as a result of accidents, medical crises, old age, or low earnings. In addition, other government-supported programs for veterans aided young men from low- and middle-income families to attend institutions of higher learning and to get assistance in purchasing a home. The GI Bill and other veteran benefit programs contributed to reducing income and wealth inequality in later years.

Unfortunately, there was considerable racial discrimination in the administration of these programs. I delay discussion of the discrimination until chapter 4, when White and non-White relations are examined.

Turmoil and Regression

The conflicts related to labor unions and those related to reducing poverty could not be isolated from the many other conflicts that became salient in the 1960s and 1970s. That the successes of the War on Poverty were stifled and the potentialities of building the Great Society envisioned by Johnson were unrealized is attributable to other simultaneous conflicts. Primarily, the tragically escalating US military engagement in Vietnam became increasingly divisive within the country and the Democratic Party. It drained federal resources and attention. A peace movement grew, mobilizing opposition to the US war in Vietnam and fueling student insistence on greater autonomy and freedom in colleges across the country. In addition, other conflicts related to social status erupted. They include the struggle for civil rights for African Americans in the southern states with their Jim Crow laws, which overlapped with the struggle for greater economic equality, but the campaign itself suffered as it moved northward. The women's movement also emerged and began to transform the United States. These are discussed in chapters 5–7.

As is discussed in detail in chapter 5, expectations of greater progress among African Americans in urban centers across the country were unmet. High rates of poverty and unemployment persisted. One kind of response, triggered by bad incidents, often erupted with counterproductive, destructive riots. Many major riots in black neighborhoods occurred in the 1960s. They occurred in Birmingham, Alabama, in 1963; in Los Angeles, California, in 1965; and in Chicago, Illinois, in 1966. Then, in 1967, they broke out in Tampa, Florida; Cincinnati, Ohio; and Atlanta, Georgia. More than 110 cities had riots on April 4, 1968, the night Martin Luther King Jr. was assassinated.

Dr. Martin Luther King Jr. recognized the disastrous implications of the many interconnected conflicts in the United States. In February 1967, King joined four senators who opposed the US war in Vietnam in a daylong symposium in Beverly Hills, California.[34]

King analyzed the casualties of principles and values of that war, including the Great Society and the reduction of poverty. Then, in April 1967, King delivered his notable sermon at the Riverside Church in New York, where he spoke somberly of the immorality of the US war in Vietnam. In December 1967, King announced plans for the Poor People's Campaign and to hold a poor people's march on Washington, uniting Whites and Blacks. However, King was assassinated in April 1968 before the march. Nevertheless, the march was held, and the campaign was renewed in 2018, as discussed in chapter 4.

The US engagement in the war in Vietnam aroused great dissent. The protests and associated divisions escalated and certainly drained attention away from efforts to reduce class inequality. The institutionalized political system did not function well to manage the country's divisions and disorders. The 1968 election process added its own problems. Johnson could not find a way to an end of the US military engagement in Vietnam, which he deemed acceptable. One problem in doing so was the fear that he and other Democratic leaders would be charged with "losing" Vietnam, as right-wing Republicans charged President Harry Truman had "lost China" to Chinese Communist forces. One strand in the right-wing social movement in the US has been the belief in Communist conspiracies.[35] This danger was magnified in the early 1950s, when Senator Joseph McCarthy aroused fears of widespread Communist subversion, which he falsely alleged played a role in the country's loss of China.

Finally, in March 1968, Johnson announced he would not seek another term as president. This followed Eugene McCarthy's strong challenge to Johnson in the New Hampshire primary, running as an anti–Vietnam War candidate. Former attorney general Robert F. Kennedy then entered the campaign for the Democratic Party's presidential nomination. Kennedy appealed not only to the anti-Vietnam voters, but also to progressives on economic and race matters.

Vice President Hubert Humphrey was the third major candidate in the Democratic primaries. Kennedy was gaining support when he was tragically assassinated in June 1968. Ultimately, Humphrey won the presidential nomination at the 1968 Democratic National Convention in Chicago. Massive crowds outside the convention hall protested and the Chicago police "rioted," furiously attacking the protestors. The Democratic Party was in bitter disarray.[36]

Nixon entered the 1968 Republican primaries as the front-runner, and he defeated his major competitor, the conservative Ronald Reagan, on the first ballot at the Republican Party Convention. Nixon ran as a "law and order" candidate representing the "silent majority." In the election campaign he was generally in the lead, but as Election Day approached his lead fell. Humphrey's standing rose when the North Vietnamese leaders accepted Johnson's proposal for peace talks in Paris in return for a bombing halt. Publicly, Nixon supported the bombing halt and the negotiations, but secretly he urged South Vietnam's government not to participate in the talks until he was president and would get them better terms.[37] South Vietnam complied just days before the US voted, and Nixon was elected president.

Once elected, Nixon continued the war in Vietnam, with some escalations. On April 30, 1970, he announced he was sending troops into Cambodia to cut off supply routes to North Vietnamese forces fighting in South Vietnam. Students at colleges and universities across the country launched large protests, including at Kent State University, in Ohio. On May 2, 1970, Governor James A. Rhodes (R-OH) vehemently denounced the students: "They're worse than the brown shirts and the communist element and also the night riders and the vigilantes. . . . We're going to eradicate the problem, we're not going to treat the symptoms."[38] On May

4, 1970, National Guard troops opened fire on the protesting students at Kent State; four students were killed and nine were wounded.[39] Students across the country began occupying campus buildings and shutting university operations.

The shock of National Guardsmen shooting down unarmed students might have generated an outcry that mutual regard for civil life needed to be restored. Instead, some political leaders chose to intensify the conflict, approving of the violence. Governor Ronald Reagan said, "If it takes a blood bath, then let's get it over with." Nixon ordered intensive surveillance by the FBI and other intelligence agencies to investigate the assumed connections between antiwar demonstrators and Communism.

How the university shut-down protests were conducted and handled varied greatly from campus to campus.[40] Teaching at Syracuse University then, I can attest to a relatively constructive happening at Syracuse University when students closed down the university and occupied several buildings night and day. The protesting students sought to arouse pressure to force an end to the US war in Vietnam. They also sought a greater role in university governance and to close the university and end the semester, with students receiving course credits. Some faculty spoke at the rallies in support of the national protest movement. Many faculty members wore arm bands, indicating their readiness to help mediate or ameliorate disputes. Large open meetings of faculty were held to discuss whether and how the semester could be ended, and students graded. The university administration decided to end the semester. Interestingly, many students also demanded they be taught about nonviolent protest, and workshops did begin that summer. Throughout the student occupation, the Syracuse city police chief, Tom Sardino, took a careful and respectful approach. He kept uniformed police off the campus and talked and listened to all sides; he spent a couple of nights with the students occupying the administration building.

In the country as a whole, however, the division intensified. The Democratic Party turned more to the left, nominating George McGovern to run against Nixon in the 1972 presidential election. Nixon won overwhelmingly, and the country seemed to have moved to the right. As progress for peace, justice, equality, and liberation were seemingly blocked, marginal groups related to such causes became more radical in their goals and extreme in means of struggle. Many people were offended by these developments, opposition to the war was regarded as unpatriotic, and the conduct of "hippies" was viewed as outrageous by many. Fights about cultural, identity, and status differences distracted attention from class inequality.

Some liberals were disenchanted with the New Left, the counterculture, and the Vietnam War protests, and some even with the domestic policies of the Johnson era. They had become neoconservatives (neocons), and were increasingly hawkish in foreign policy. The US government and local officials imposed more and more suppressive efforts. In response, some of the protesters escalated the conflicts, threatening and in some cases resorting to violence.

Nixon finally did signal the end of US military engagement in Vietnam. In April 1973, an agreement was signed on terms that might well have been feasible much earlier. US troops withdrew and, not long afterward, South Vietnam was taken over by the North Vietnamese military forces. Nixon, however, was no longer president. He was engulfed in the Watergate scandal, was about to be impeached, and resigned. The scandal was another blow to American confidence in government, but Nixon's resignation might have been celebrated as a triumph of the political system, and it did result in legislation that increased limits on government surveillance.

Gerald Ford succeeded Nixon and spoke about unifying the country and overcoming the deep divisions. Jimmy Carter won the next election and also tried to overcome the earlier divisions. They made some progress, yet also contributed to the legacy of violence associated with protests, intense ideological animosities, and different views of reality that lingered and contributed to the polarization in the United States, eventuating in the January 6, 2021, insurrection.

Conclusions

The 1950s and early 1960s are generally seen as years of economic growth and increased well-being, at least for White Americans. Pent-up consumption demand from the war years was complemented by higher wages. The federal government further spurred the economy by constructing large-scale infrastructure projects such as the interstate highway system begun during Dwight D. Eisenhower's presidency. In addition, US government expansion of a safety net for people in need, partly prompted by protests, initially reduced poverty and economic inequality. More discussion about the important roles played by social movement organizations in conflicts related to inequalities in class, status, and power are provided in later chapters.

Substantial progress in countering severe class inequality was made in the early years after World War II. Many elements of a social welfare safety net were made and remain. During the later period, in the context of many other issues, some gains were cut back by opponents of such New Deal programs. The later 1960s and early 1970s, however, are generally viewed as wild and chaotic, with ongoing, progressive, and liberating consequences. Undoubtedly, people in the United States were looser and freer after the 1960s. But conservative views and policies also intensified as some people sought to restore greater order. The social conflicts related to these developments were often pursued constructively and the outcomes had mutual benefits. Conflicts do not always have to be zero-sum, with the gains for winners made at the expense of the losers. Too often, however, conflicts were not constructive and resulted in backlashes. The ground was shifting to open the path toward increased inequality. In the next chapter we see how that path was taken and with what consequences. The next chapter begins with Ronald Reagan's presidency and increasing class differences.

3
Rising Class Inequality, 1970–1992

In the late 1970s, a profound shift toward ever greater income and wealth inequality began that became a source for major social problems and conflicts in the United States. Initially, this shift received little attention, except from some researchers and intellectuals, but it is now widely recognized as a grave issue in the United States, exacerbating many other immense challenges.

This chapter first describes the evidence of the growing inequality in the United States and discusses many of the consequences of great income and wealth disparities. Second, it examines competing explanations for the very high and increasing inequality. Many explanations stress large, immutable forces, but other explanations attribute the rising inequality to policy choices resulting from contentions or other considerations not intentionally raising inequality. Third, I examine some manifest conflicts related to the rising degree of class inequality. The manifest conflicts are usually framed in terms of issues such as wages, immigration, tax policy, and safety-net provisions. These issues may be bundled together in ideological packages that are adopted and advanced by various groups. The ideological struggle is examined, as it is waged outside as well as inside established political institutions. Since the struggle is largely conducted within the context of US political institutions, I focus on Ronald Reagan's presidential terms and a survey of nongovernmental actors. Chapter 4 will examine the struggles relating to hyper-inequality from Bill Clinton's presidential administrations to Joe Biden's first years as president.

Rising Class Inequality

The United States presently has exceptionally greater economic inequality than most other countries with highly developed economies. It was not always so.[1] Income and wealth information is collected in several ways in the United States.[2] The annual Census Bureau's Current Population Survey (CPS) and the Internal Revenue Service's (IRS) Statistics of Income (SOI) are the most widely used sources of data. In addition, the Congressional Budget Office (CBO) combines CPS and SOI data to estimate comprehensive household income data before and after taxes. There are several complexities in gathering and interpreting data on income inequality. These include sampling limits,

Fighting Better. Louis Kriesberg, Oxford University Press. © Oxford University Press 2023.
DOI: 10.1093/oso/9780197674796.003.0003

including all sources of income, and benefits in kind, such as food benefits and subsidized housing.

There are also different ways of calculating inequality. One method is to compare the amount of income in terms of the proportion of income received by each quintile of the population; for example, the proportion of the total income in the country that the top 20 percent of the population received can be compared to the other quintiles. Another measure is the Gini coefficient, which calculates the degree of inequality across the whole society. If one person had all the society's income, the coefficient would be 1, and if the income were equal for all persons, it would be 0. Actual Gini coefficients vary between 0.3 and 0.5.

For the purposes of this book, it is not necessary to analyze the advantages and disadvantages of various ways of measuring economic inequality. The results are roughly similar. Whatever measures I might adopt, two findings are clear. Economic inequality has greatly increased in the United States, beginning in the late 1970s, and it is now strikingly greater than the inequality in other economically advanced countries. Some degree of inequality is inevitable, and the level during the 1950s and 1960s was associated with strong economic growth. The new heights of inequality, however, are associated with many social and political problems.

I will mention here several measures of the degree of income and wealth disparity in the United States. Using CBO data, in 2014 the top 1 percent of households had 17 percent of the country's income, while the bottom 20 percent received only 4 percent.[3] Taking into account taxes and federal transfers, however, the inequality is lessened. The top 1 percent received 13 percent and the bottom 20 percent of households received 7 percent. Taxes and federal transfers are somewhat redistributive in the United States, but they are so to a much lower degree than they are in other economically developed countries. An analysis of data from the thirty-four countries in the Organization for Economic Co-operation and Development (OECD), mostly economically developed countries, demonstrates that.[4] Without considering taxes and transfers, the United States ranks tenth in inequality, measured in Gini scores, but when taxes and transfers are considered, it increases in relative inequality to rank second most unequal among OECD countries.

Income inequality had been relatively static since 1945 until it began to rise at the end of the 1970s. For example, considering wage and salary income, the share of all such income received by the top 1 percent of earners was 5.1 percent in 1970 and it rose sharply to 12.4 percent by 2007.[5] The increase of the next 4 percent of top earners was steady but much more modest. Increasing income disparity is attributable to the much greater increases in income of the very highest earners compared to the small or null increases of the people with lower levels of income.

Wealth is immensely more unequally distributed than income, since many people have no net assets and some are in debt, while a few persons own huge amounts of assets.[6] Using CBO data, in 2013, the top 10 percent of the wealth distribution held 76 percent of all family wealth while the lowest half of the distribution held only 1 percent.[7] Wealth inequality had been even greater before the Great Depression, after which it sharply declined until the 1940s, and then it very gradually declined until wealth inequality, like income inequality, began to increase markedly in the late 1970s.[8] For example, from 1989 to 2013, the wealth of families in the wealthiest 90th percentile had increased 54 percent more than the increase in wealth of the poorest 90th percentile.[9]

Wealth inequality in the United States was less than the inequality among European countries for most of its history.[10] Inequality in the US and in the European countries markedly declined from the height it had reached before the crash of 1929. After 1970, wealth inequality began to rise, but the rise was consistently greater in the United States than in Europe. US wealth inequality is exceptionally high among economically developed countries.

Problems Arising from High Class Inequality

Very high levels of economic inequality, it turns out, are associated with many health, social, and political problems. Thus, among economically developed countries, life expectancy tends to be higher in countries with greater income equality.[11] People living in the United States, despite high per capita income, tend to have lower life expectancies than do people living in countries with less income inequality. Indeed, higher income inequality is associated with higher rates of infant mortality, poor self-reported health, and many kinds of illnesses. Higher income inequality is also associated with various social problems among economically developed countries. What is even more striking is that low life expectancy and the incidence of many medical and social problems are associated with higher income inequality among the US states. For example, residents in relatively egalitarian Utah, New Hampshire, and Wisconsin have relatively long life expectancies, while residents in relatively more unequal states, such as Louisiana, Mississippi, and Alabama, have shorter life expectancies. Of course, the greater income inequality may tend to be associated with people with low income who also tend to be of low status in "race" and ethnicity. As discussed in chapter 5, people with low statuses tend to face discrimination, which itself can be a source of circumstances that reduce life expectancy.

It may be that high inequality produces deleterious effects. Or it may be that various social policies or local cultural or other conditions influence both inequality and problems related to health and social life. Possibly, both interpretations have merit. In many places in this book, I will examine the various relations between economic inequality and other social and political phenomena. At this point, I discuss how high inequality, in itself, can have ill effects. To understand this, it is important to realize that relative standing inherently matters.[12] In a poor society where hardly anyone has an automobile, or a telephone, having one is not necessary to get along. But as more stuff becomes used in a society, lacking that stuff is more and more burdensome. It is harder and harder to find a way to have what is increasingly needed or desired. Particularly for people growing up or living in relatively very poor circumstances, a personal, familial, or societal blow can have terrible consequences. That is stressful. They may become burdens and threats to their society. It serves the interest of more well-to-do citizens to ameliorate such circumstances. Fights to reduce extreme poverty and inequality can readily be framed and conducted constructively.

Interestingly, a review of 269 comparable studies of anxiety levels in the United States between 1952 and 1993 reveals a trend of increasing anxiety among men and women.[13] A similar increasing trend was found for mental depression. Significantly, in recent decades, there have been numerous studies of narcissism, an insecure self-esteem, marked by a tendency to be insensitive to others and being preoccupied with oneself. The studies reveal that there has been a trend toward increased numbers of people with such narcissism. A main source of anxiety and narcissism is high levels of social stress. Many aspects of recent social changes are stressful. Furthermore, stress contributes to ill health and life-threatening conduct. This may be indicated by the declines in life expectancy in the United States. Life expectancy in the United States had gradually increased until very recently. Thus, increases occurred steadily, with slight fluctuations from 1980 to 2014, and remained about the same until 2014, after which it began a steady decline into 2021.[14]

High economic inequality is associated with many other unfortunate consequences, as discussed further in subsequent chapters. Later chapters discuss the burdens that extreme inequality imposes on good race relations, gender relations, and equitable political governance.

Explaining Rising Inequality

Explanations for the rising economic inequality in in the United States either emphasize major technological or market changes that are largely uncontrolled or they emphasize public and private policies that may be contentious. I will discuss both kinds of explanations. Persons who are not highly concerned about rising economic inequality tend to emphasize technological and global market developments, which they tend to treat as impersonal and immutable. Persons concerned about the rising inequality are more likely to emphasize specific practices and policies that are chosen by particular persons and organizations relating to taxes, social welfare benefits, labor relations, trade and tariffs, immigration, and foreign investments. Different willful policies associated with any of these developments can tend to increase or decrease economic inequality.

To begin, I discuss changes in the technology employed to produce goods and provide services. Undoubtedly, increasing automation has affected how people engage in their work, and some tasks that were done by humans are increasingly done by machines, with implications for wage inequality.[15] Thus, many semi-skilled jobs in the automobile manufacturing industry were increasingly automated and the workers performing those tasks lost their jobs. Many of them left the industry and accepted lower-skilled and lower-paid jobs. Some workers may have moved into jobs that required more skills. For those workers living in cities that had been centers of automobile manufacturing, plant closings meant the loss of alternative jobs as well, which were in other businesses associated with the auto plants and their workers.

The set of technological developments in manufacturing and mining jobs might be expected to increase productivity, and therefore the workers who remained in the industries should draw higher wages. Indeed, between 1948 and 1973, worker compensation increased along with increasing productivity.[16] However, between 1973 and 2016, worker compensation stagnated, increasing very slightly despite steadily increasing productivity. Thus, productivity between 1973 and 2013 rose 74.4 percent, while hourly compensation for workers rose only 9.2 percent.[17] Salaries of upper management personnel, however, rose substantially.

Before discussing the many specific policies and developments that help account for wage stagnation, on the one hand, and spectacular increases in income for relatively few people, on the other, I will discuss another set of developments that are often said to depress wages and increase inequality, and are included in the term *globalization*. Thus, increased world trade, investment, and communication placed the higher-paid US workers in competition with low-wage workers in many developing countries for goods and services sold in the United States. That might depress wages in some US industries, but it does not seem to have depressed salaries for higher management. Indeed, the proportion of people in relatively well-paid employment has risen. As observed earlier, the increasing income inequality in the United States is attributable to the much greater rate of income increases in the top 1 percent and the top 10 percent of earners, rather than the decreasing income at the bottom.

Globalization encompasses an intensification and expansion of international trade and the expansion of transnational corporations. The expansion of foreign imports relative to exports has been widely viewed as a cause of lower wages and higher unemployment, at least in some industries.[18] For many reasons, imports became greater in dollars than exports. The automobile industry, for example, had long been dominated by General Motors, Chrysler, and Ford, and their leaders gave little attention to possible foreign competition. Those three corporations had 89.6 per cent of the US car market in 1966.[19] Small foreign cars, however, began to make increasing inroads in the United States in the 1960s. This was greatly spurred further by the 1973 Arab oil embargo and resulting higher gasoline prices. The US companies began to expand sales of smaller cars, but they made costly mistakes in manufacturing and handling their new models,[20] and imported foreign-made automobiles continued to increase in sales.

The number of automobile workers declined and the United Automobile Workers (UAW) lost members. Foreign auto manufacturers began to invest in constructing plants in the United States, and they chose to locate them in southern states, where they hired workers who were not in unions. Some US auto companies followed and built their new plants in non-union states in the south. These developments weakened the UAW in bargaining for worker wages and in political influence in fending off anti-union legislation.

The expansion of US-based corporations into immense transnational organizations with plants operating in many countries also has implications for increasing inequality. Many parts of a product assembled in the United States may be produced in low-wage countries, which enhances the corporation's earnings. The pricing of transferring the parts of a product from one country to another, within the corporation, can allocate profits in one country or the other, depending on tax rates in different countries. The taxing and financing of investments in one country or another necessarily influences decisions about the location of plants. These matters contribute to the vast expansion of the relatively well-paid financial sector of employment.

Leaders in transnational corporations are not likely to share the concerns that the leaders of a corporation headquartered and operating in one or a few US cities have. Those locally based leaders are likely to know more and have more concerns about the life in their localities, including cultural attractions, schools, environmental conditions, and even their own employees, than the more mobile transnational leaders. As noted in the previous chapter, corporate leaders, when they were more locally based, acted to some degree to maintain good standing in the eyes of the people of their communities—for example, by contributing to local charities or to funding civic improvements.

The contributions of innovations in production technologies and expanded trade do not necessarily enhance inequality in the form of extremely high gains by a small percentage of people, or of declining relative incomes of middle- and working-class people in the United States. How they are managed matters. There has been a tendency in the United States to barely acknowledge that technological changes and shifts in the global context produce winners and losers, and in any case to not do very much to ease the hardships of the losers. This might be seen as a cultural matter. US individualism does not favor collective responsibility for such misfortunes. People in the United States, however, also embrace communal values. At least in times of crises, Americans pitch in to help the less fortunate. This is true after large floods, fires, or major economic crises, such as the Great Depression of the 1930s.

What must be considered are not general cultural patterns, but rather specific policies that increase or decrease economic disparities, when major events and processes are raising them. As indicated in the previous chapter, many policies contribute to lessening inequality. On the other hand, other policies tend to enhance it, which some winners may pursue to take advantage of the shifts for their self-aggrandizement. Clearly, many developments and policies, in addition to advances in technology and globalization, help to account for the exceptionally high inequality that has emerged in the United States since the end of the 1970s.[21]

Policies associated with technological change and globalization often enhance inequality in the United States. Three sets of additional policies directly impact economic inequality. One set of policies pertain to labor unions, the second to tax policies, and the third to governmental and nongovernmental policies to provide benefits that reduce inequality. These policies are often quite contentious and manifested in overt conflicts. Many episodes over the previous decades are discussed next.

Conflicts Related to Rising Class Inequality, 1970s–1980s

Having discussed technological developments that may influence income disparity, I turn to examine contentions regarding the rising inequality since the end of the 1970s. In the preceding chapter, most attention was given to the conflicts regarding reducing inequality, from the mid-1940s to the mid-1970s. Since then, governmental programs have less often striven to reverse the rising inequality, and nongovernmental social movements have been less successful. There have been major governmental and nongovernmental efforts to reduce inequities between Whites and non-Whites and between men and women. Those efforts clearly have implications for the level of economic inequality, and they are discussed in the next two chapters.

This chapter examines fights relatively directly related to economic equality issues such as wages, taxes, and social benefits. These conflicts are generally conducted within the context of governmental institutions, but nongovernmental actions influence what happens there and also have direct impacts. The fights are conducted within larger contexts of varying ideologies and identities, which are related to the large-scale developments discussed earlier: globalization and technological changes, including in communication as well as in production. The fights are also interconnected, linked over time as well as over social space. Everything cannot be discussed simultaneously, and I choose a rough chronological sequence, loosely marked by presidential administrations, as the best ordering. This survey starts at the time when income inequality had begun its steep rise, and it considers the role of public policies in that climb.

Ronald Reagan's Administration (1981–1989) and George H. W. Bush's Administration (1989–1993)

Ronald Regan was nominated, campaigned, elected, and governed as a conservative Republican.[22] The Republican Party had been moving to the right on many issues, including ones relating to economic inequality. Reagan and the conservative Republicans did not claim to be striving for more economic inequality, yet the domestic policies they formulated and enacted clearly tended to increase inequality. A premise of conservative Republicans was that federal government operations should be cut back, which meant reducing taxes and social welfare provisions and generally avoiding income redistribution.

Reagan's administration proposed policies that it asserted would overcome the previous conditions of stagflation, slow economic growth, and inflation in President Jimmy Carter's administration.[23] Reagan adopted a convenient "supply side" strategy, claiming that large tax cuts would free up money for investments and consumption and thereby spur economic growth.[24] The new tax bills made deep cuts in taxes on corporations and wealthy income earners. The large tax cuts were accompanied by increases in military expenditures, albeit with substantial decreases in other federal spending. The result was an increased budget deficit, which countered the incentive for business leaders to borrow money in order to expand their investments.

The policies most directly related to inequality are found in budget and tax bills and in specific social welfare legislation. In President Reagan's 1982 budget, the largest reductions in expenditures were the deep cuts in the money provided to the states. During the 1960s, the federal government had increased funding for mandated programs to deal with many social problems, by subsidizing housing, nutrition, healthcare, and job training. Such expenditures had helped reduce poverty and counter growing inequality. Providing money to states also made possible transferring some funds from states that were economically thriving to states suffering long-lasting economic problems.

Welfare legislation has three major components in the United States: means-tested, social insurance, and improving human capital.[25] Providing cash or benefits in the form of food, healthcare, or housing subsidies, on a means-tested basis, has long been a form of welfare for people with low income. Reagan argued that such government welfare programs often contributed to "dependency" and a disinterest in working. He often referred to "welfare queens" to suggest widespread misuse of welfare benefits. Reagan's budget director, David Stockman, worked skillfully to use congressional budget reconciliation procedures and to win over enough Democratic House members to achieve cutbacks in all three kinds of programs in the Omnibus Budget Reconciliation Act (OBRA) of 1981. Not only were expenditures reduced, but programmatic restrictions were increased.

Further welfare retrenchments were made in the Family Support Act (FSA) of 1988, after lengthy negotiations between the White House, Senate, and House of Representatives. The primary focus was on enabling welfare recipients to become independent of welfare. Programs were mandated for all states to provide Aid to Families with Dependent Children (AFDC) recipients with job opportunities and skills training. The FSA increased the amount an AFDC recipient could keep of his or her earnings, without losing benefits. Although there were improvements in many procedures relating to AFDC, the participation rates declined. States varied greatly in funding and implementing the new mandated programs. At the same time, federal expenditures did not expand, but rather declined a small amount.

Early in his presidency, Reagan sought Social Security cuts, but there was popular resistance, although not on the scale discussed in the previous chapter. Moreover, the Democratic Party continued to hold the majority in the House of Representatives, with Thomas (Tip) O'Neill as Speaker, and Reagan's initial cutting efforts failed. Facing problems in meeting Social Security benefit obligations, the administration established a bipartisan National Commission on Social Security Reform. The legislative result in 1983 was an increase in revenue and greater restrictions on benefits. Payroll tax rates were raised equally for employees and employers and at a combined rate for self-employed persons. In addition, some classes of organizations previously exempt would be required to be covered. Benefit reductions included provisions that were made to lower automatic monetary increases, and long-term problems would be resolved by increasing the retirement age to 67 in 2022. The total tax revenues that were to result from these changes equaled 2.7 trillion dollars, which would be set aside in the Social Security Trust Fund and provide Social Security surpluses for thirty years. But the money was diverted to pay for tax cuts and other expenditures.[26]

Overall, during the years of the Reagan administration, the number of people in poverty in the United States increased by more than 3 million from 1980 to 1988.[27] If the goal of the welfare program of the Reagan administration was to reduce the proportion of the population in poverty and improve their well-being and independence, it failed. If the goal was to reduce federal and state expenditures to that end, it succeeded. A legacy of the reform experiment was to change the political conception of the problem of poverty from what it was in the years of the New Deal and the War on Poverty. Earlier, it was widely understood that poverty resulted in good measure from lack of opportunity. The circumstances in which poor people lived hampered overcoming their poverty and ill-served the society as a whole. The new doctrine tended to blame the poor and seek to drive them to change their behavior.

In addition, changes in tax policy, like changes in welfare programs, contributed to increasing income inequality. Passage of the Tax Reform Act of 1986 was Reagan's highest priority in his second term, and it had support from a mix of Republicans and Democrats in Congress, including Tip O'Neil, the Speaker of the House, and a Democrat. Like Reagan's 1981 tax bill, it contained tax cuts, but its provisions were much broader. It reduced the number of tax brackets, eliminated many deductions, and removed many low-income taxpayers from the tax base. What had the greatest impact on increasing income inequality in the country was greatly reducing the marginal tax for the top-income households from 50 percent to only 28 percent.[28] The evidence indicates that the great increase in the amount of money received by the top 1 percent of households was not driven by growing wage inequality.

This circumstance certainly indicates that a single policy change could change a major driver of the extremely high proportion of the country's income that is received by a tiny percentage of people. That is, a reduction in extreme inequality would be achieved by raising the cap on the tax rate for households receiving extremely high incomes. This possibility was becoming highly salient and contentious. Of course, recipients of high incomes themselves, or their agents, engaged in considerable lobbying of members of Congress and other officials to cut their taxes.[29] There was no comparable lobbying group lobbying or fighting for higher taxes for high-income recipients.

Reagan and Republican Party leaders did not frame their policies in terms of seeking greater class inequality. Many people enduring the consequences of rising inequality did not frame the events as grounds for a fight. The conflict largely remained latent until years later.

One other realm of public policy relating to conflicts and inequality relates to labor unions. In the previous chapter, I noted the tendency for corporate and Republican Party leaders to oppose workers organizing in unions. Ronald Reagan, as governor of California, acted against the farmworkers' organizing efforts, and he vetoed a bill that would grant farm workers the right to engage in collective bargaining. Most famously and influentially, on August 5, 1981, he fired 11,345 air traffic controllers who had ignored the federal law banning government workers to strike. This action emboldened private employers to act similarly and replace strikers rather than negotiate with them. This seriously weakened the power of unions. Furthermore, Reagan's action was subsequently used to justify refusing to even negotiate with public employee unions, which was not Reagan's actual policy.[30]

Very significantly, Reagan appointed three management representatives to the five-member National Labor Relations Board (NLRB), established in 1935, to oversee elections of union representation and labor-management bargaining. He appointed Donald Dotson as NLRB chair. Dotson had asserted that "unionized labor relations have been the major contributors to the decline and failure of once-healthy industries" and have caused a "destruction of individual freedom."[31] NLRB decisions reflected that view.

The Reagan administration's domestic policies to reduce taxes on high-income individuals and corporations, to cut welfare benefits, and to weaken trade unions were not opposed by mass demonstrations or social movement organizations contesting them. Perhaps, as a legacy of the anti–Vietnam era social movement activities, peace movement networks, and experience provided a base for action to confront Reagan's initial militant anti-Soviet foreign policy. In any case, Reagan's military buildup and foreign interventions aroused great popular resistance and mass demonstrations, and foreign conflicts had relatively high salience.[32] Nevertheless, on domestic matters, there was some resistance to Reagan within the official political system, since the Democrats controlled the House. This is indicated by the relatively large number of vetoes Reagan executed.[33]

Furthermore, there was growing ideological controversy as partisans on each side argued to win adherents. In the period examined in this book, two broad streams of thought have been salient; simplistically put, they have been the contention between the left or liberal side and the right or conservative side. Many important disagreements occur within each side, and many quarrels cut across it, but it refers to a significant division that is commonly used by people on both sides. Economic theory provides a good starting place for discussing many political policy conflicts. Keynesian economics, developed by John Maynard Keynes (1883–1946), was the standard economic model for understanding and responding to the Great Depression and managing the postwar economic expansion, until the early 1970s. It emphasized the active role the government needs to play to manage the market, including increasing government expenditures when the economy is weakening. This view was challenged, beginning in the 1950s, by advocates of free market policies, believing that the government role should be largely limited to making monetary adjustments. Milton Friedman, at the University of Chicago's Department of Economics, was a highly influential analyst and advocate of this approach. His brother-in-law, Aaron Director, also was an economist with similar convictions. He was a professor in the University of Chicago Law School and contributed importantly to the highly conservative approach of many of the Law School's graduates, who went on to distinguished, influential careers.[34]

Of course, academic economists locate their reasoning within widely shared evidence and theorizing. Once they are discussed in the public sphere, however, they are simplified and sometimes reduced to slogans. Some formulations can become popular with political leaders, even when the theoretical setting and empirical evidence is slim and not supported by many academic economists. This was the case, for example, with Reagan's adoption of "supply-side economics." It was asserted that cutting taxes would so spur economic expansion that total tax revenue would increase. The political virtue of this assertion is that it justified cuts of working-class benefits, and not for ideological reasons and payoffs to rich political supporters.

Following the oil shock and resulting stagflation of the 1970s, Ronald Regan found a highly conservative path to win the nomination and the election by promising to revive the economy by applying supply-side economics. Competing unsuccessfully for the nomination, George H. W. Bush referred to Reagan's views as "voodoo economics." Reagan chose Bush as his vice president, whereupon he converted to "Reaganomics." As president, Reagan did implement a large tax cut, and the economy did improve. But tax revenues did not increase, and the result was a large increase in the national debt, since government expenditures were not reduced, given increases in the military budget.[35] The actual policy employed turns out to have been Keynesian, increasing expenditures to spur economic growth.[36] The way this was structured in conjunction with social welfare changes, however, resulted in rising inequality and a rising proportion of people living in poverty. Nevertheless, Republicans ignored those changes and the large increase in deficit spending as they praised supply-side economics.

After Reagan's second term as president, Vice President George H. W. Bush won the Republican nomination for president. In his campaign, he famously pledged, "no new taxes," but he was concerned about the huge deficit of $2.8 trillion, triple what it was in 1980.[37] Bush wanted to reduce the deficit by reducing the federal budget, but the Democratic-controlled Congress would not support that, and Bush eventually increased taxes. The financial problem had been enhanced by the Savings and Loan collapse as a result of the industry's riskier investments, following state and federal deregulation. A federal bailout cost more than $100 billion. International events drew great attention from the George H. W. Bush administration, particularly the ending of the Cold War. No important changes in policy checked the trend toward increasing economic inequality.

The Escalating Ideological Contest in the 1980s

Conflicts about economic equality and labor rights are also waged in the public sphere by nongovernmental actors. The partisans in the conflicts generally use various noncoercive means as well as negative and positive sanctions. I note some important actors in such struggles in which partisans strive to sustain those on their side or to win over those who are on opposing sides or who are generally indifferent.

Research and Policy Institutions

Academic institutions are certainly major sources of thinking about economic inequality, class conflict, labor relations, social welfare, and public policy relating to those topics. Academic scholars approach these matters from the perspective of various disciplines, including economics, sociology, political science, psychology, and anthropology. They do research and publish accounts of their findings and they teach students. To varying degrees, some also attempt to inform and influence the general public and policymakers. Think tanks, more directly, are important bases for examining policies to deal with economic, social, political, security, and other challenges. They have staff members who write reports on specific issues, they sponsor events to gain attention to their work, they provide spokespersons for the news media, and some provide havens for persons who are in-between political or other prominent positions. They vary greatly in size, influence, and partisanship.

Many such think tanks have been and remain essentially nonpartisan, centrist, and independent in their work. This has long been a hallmark of think tanks associated with universities, but also of independent institutes and foundations.[38] For example, one of the oldest and largest think tanks is the Brookings Institution, founded in 1916, with operating expenses of roughly $98,000,000 in 2017. The work it produces is highly regarded and its staff members' reports, counsel, and media appearances are broadly utilized. Other early think tanks include the Russell Sage Foundation (1907), the Carnegie Endowment for International Peace (1910), and the Council on Foreign Relations (1921). They tend to be independent and mainstream.

In the 1970s, several very wealthy individuals led in establishing think tanks to advance public and official thinking in support of unfettered corporate conduct, and to some degree have acted as advocacy organizations. Three such institutions have been highly influential. The oldest of the three is the American Enterprise Institute (AEI); founded in 1938, it became very active in the 1970s. The Heritage Foundation, the largest of the three, was founded in 1973, differentiating itself from the AEI by its Christian conservatism.[39] Both were very influential in and with the Reagan administrations and later with the administrations of George

H. W. Bush and George W. Bush. The smallest of the three is the Cato Institute, which was founded in 1974; initially named the Charles Koch Foundation, it was renamed in 1976. It is libertarian in its political philosophy. All three have many staff members who develop conservative policies regarding a wide range of issues and comment through many media regarding current issues.

Starting in the 1980s, clearly progressive, liberal think tanks were established to counter the conservative think tanks. They include the Center on Budget and Policy Priorities (1981), the Economic Policy Institute (1986), the Open Society Foundation (1993), the Center for American Progress (2003), and the Third Way (2005). They support analyses and specific policies that can be used in debating and defeating the conservative ideas. To some degree they also compete with each other as well as collaborate in seeking to influence the public at large.

Advocacy Groups

Among the many voluntary associations that characterize the United States, advocacy groups have always been important. They operate at the city, state, and national level and vary in size and matters being advocated. They engage in lobbying, media campaigns, pressuring opponents, and other tactics. I will discuss three such groups that work to change policies that pertain to increasing economic inequality: Americans for Prosperity, the American Legislative Exchange Council, and the State Policy Network. They cooperate and are highly influential in promoting right-wing legislation and fighting against liberal legislation, they pursue court cases to advance their policies, and they generally move the Republican Party rightward. There is nothing comparable among advocacy groups that work for reducing inequality, but I discuss two that make some progressive contributions: the Ford Foundation and the Roosevelt Institute.

Americans for Prosperity (AFP) was formed in 2004 and funded by the brothers David H. and Charles Koch as an extremely conservative public advocacy group.[40] Some matters about which AFP advocates seem to relate to short-term concerns about the Koch brothers' investments in carbon-producing companies, such as opposing a cap and trade bill and global warming regulations. AFP also worked against President Barack Obama's Affordable Care Act and his efforts to expand Medicare and economic stimulus during the recession. Most pertinent here, AFP acts in ways that tend to increase inequality. It advocates against trade unions, supporting right-to-work laws as well as limitations on collective bargaining rights by public-sector trade unions. It has also advocated against legislation to increase minimum wages.

The American Legislative Exchange Council (ALEC) was established in 1973, as the Conservative Caucus of State Legislators, and took the current name in 1975. By 2011, it had 2,000 legislators as members and introduced about 1,000 bills annually on diverse conservative issues in state legislatures, obtaining enactment on about 20 percent of them.[41] Many of the bills relate to limiting the role and organizing strength of public sector unions as well as to reducing government regulations. Major corporations are supporters, such as AT&T, ExxonMobil, Coco-Cola, Pfizer, and Koch Industries.

The State Policy Network (SPN) was founded in 1992, focusing on reducing state programs related to health and welfare, and on expanding charter schools. It also operates in opposition to collective bargaining and other matters pertaining to public sector unions. It is a member of ALEC and has hundreds of affiliated local organizations, including sixty-five think tanks. It receives funding from foundations and many corporate donors, including the Koch Brothers, Phillip Morris, Kraft Foods, and GlaxoSmithKline.[42] It also makes grants to member groups for specific undertakings.

These three advocacy groups often collaborate, as they did in advancing anti-union rules across the country.[43] Beginning in 2003, two large financial entities, the Bradley Foundation and the Donors Trust and Donors Capital Funds, began serving as primary sources of funds for the campaigns. In states with Republican governors, ALEC, SPN, and AFP have provided guidance and funding to achieve legislation or executive orders to restrict collective bargaining by public-sector workers. This was often cloaked as a necessary way to save public money during an alleged financial shortfall. In states with Democratic governors, legal challenges to laws protecting union rights were made. In conjunction with these efforts, they helped pursue an important US Supreme Court case: *Janus v. American Federation of State, County and Municipal Employees (AFSCME)*.

In June 2018, by a 5–4 vote, the Supreme Court decided that government workers who choose not to join unions cannot be required to contribute to a union to help pay for the union's benefits, known as agency fees.[44] The decision was based on First Amendment grounds, compelling nonunion workers to subsidize speech with which they may disagree. This overruled the Court's 1977 decision on this matter that distinguished between compelled payments for political matters and payments to help meet the costs of the union's collective bargaining efforts, which it deemed prevented freeloading. Clearly, the Supreme Court had changed in its composition between 1977 and 2018, when it decided to weaken unions. Indeed, following this decision, AFSCME agency fee payers declined from 112,233 in 2017 to only 2,215 in 2018, causing a $4.2 million drop in revenues.[45]

I do not regard the struggles as conducted by these advocacy groups to be very constructively waged. The failures to be more constructive relate to the narrow beneficiaries of the groups' efforts, at the expense of large numbers of fellow citizens, rather than engaging in discovering mutually beneficial outcomes. When their efforts are successful, they contribute to increased inequality in the United Sates, not to liberty and justice for all. Furthermore, their actions are pursued in ways that often are obscured and unrecognized, thereby undercutting US democracy. Moreover, the arguments used in these groups' ideological struggle are often derived from dubious information rather than empirically based policies and principles.

I now discuss two advocacy organizations that tend to be liberal, with a particular focus on inequality: the Ford Foundation and the Roosevelt Institute. The Ford Foundation was established in 1936 as a small Michigan charitable foundation, led by Ford family members. It soon became the largest foundation in the country, remaining so for many years. It funds programs around the world and is run by an independent board and staff. It supports work in many arenas, but it has generally included major grants related to advancing human welfare and social justice.[46] For example, as noted in chapter 2, during the late 1950s, the Ford Foundation provided grants to learn how to open opportunities that would reduce juvenile delinquency and poverty.

In 2015, Darren Walker, Ford's president since 2013, announced that the foundation's crucial task was to disrupt the drivers of inequality. Program officers discussed what drivers there were and how they might be overcome.[47] They stressed five drivers: entrenched cultural narratives that undermine fairness, tolerance, and inclusion; failure to invest in and protect vital public goods; unfair rules of the economy; unequal access to government; and persistent prejudice and discrimination. They make grants to respond to these drivers in seven interconnected ways, including civil engagement and government. Their grants tend to support activities that sustain and build the infrastructure that responds to and disrupts the drivers of inequality. For example, the foundation regularly provides grants to NPR and PBS, and also to organizations that analyze drivers of inequality and policies to overcome them, such as the Roosevelt Institute. The Ford Foundation does not fund individual issue-based policy campaigns, direct services, or get-out-the-vote efforts. Significantly, other US foundations have been prompted by Ford Foundation's focus to also work against extreme inequality.

Some criticisms have been made, however, about overly long-range, infrastructure-building approaches, which avoid contentious confrontations, compared to the great successes of the direct concrete actions of the conservative foundations.[48] A crucial criticism is the failure to recognize the importance of the political economy. Specifically, there are failures to focus on workers' rights, fiscal policies, and Wall Street. The right-wing conservative advocacy groups have worked persistently for immediate policy goals: weaken workers' collective rights, decrease taxes for the wealthy, and weaken regulations that would protect workers and consumers.

It is as if foundations, even if seeking to advance social justice, are cautious about engaging in conflicts; contention and partisanship are often avoided.

I think one of the virtues of recognizing that conflicts can be waged constructively is that contention may be more often ventured when that is recognized.
The recognition that conflicts can be conducted to mutual benefit was greatly
enhanced by programs of the William and Flora Hewlett Foundation.[49] In 1984,
the foundation launched a strategy to build the field of conflict resolution by providing long-term grants to applying universities to form centers for support of
conflict analysis and resolution theory, practice, and infrastructure. In the same
year, it initiated the publication of the *Negotiation Journal*. By the end of 1994,
eighteen centers on conflict analysis and resolution had begun to be funded.
Practitioner organizations pertaining to conflicts in the environment, community, and many other sectors were also awarded grants. The infrastructure for the
field was strengthened, primarily by supporting professional organizations and
meetings. Graduate programs in the field have grown greatly since 1989, spurred
by the rising demand for training in negotiation and mediation.[50]

The Roosevelt Institute, a relatively small but influential liberal advocacy
group, was constituted in 1987 by the merger of the Eleanor Roosevelt Institute
and the Franklin D. Roosevelt Institute. Donors include the Ford, MacArthur,
and William and Flora Hewlett Foundations. A main focus for the institute is
to solve major economic challenges and develop ways that the government can
help solve them. The chief economist at the institute is the distinguished professor of economics at Columbia University Joseph Stiglitz.[51] In 2015 he was
joined by Nell Abernathy, Adam Hersh, Mike Konczal, and Susan R. Holmberg
in a Roosevelt Institute report titled *Rewriting the Rules of the American
Economy*, which was published by W. W. Norton and attracted considerable attention in progressive circles. The book examined how the great inequality in
the US economy resulted from choices, and it provided better choices that could
be made. The substance of the argument and its influences are discussed in the
next chapter.[52]

Media Channels

Newspapers and then radios were the primary conveyors of information to the
public about political issues before the end of World War II. Since then, television and then social media have complemented, and to some degree have become the primary sources of, political information. Initially, a few news channels
and radio networks were dominant, and they sought to provide a broadly shared
perspective or different sides of controversial matters. This approach was fostered by the Fairness Doctrine implemented by the Federal Communications
Commission (FCC) between 1949 and 1987. It required that radio and television
stations provide free airtime for responses to controversial statements that were
broadcast. President Reagan opposed the doctrine, and his appointments to the
FCC resulted in the ending of that doctrine in 1987.[53] When Congress passed a

bill to restore it, he vetoed the bill. Subsequent congressional efforts to restore the doctrine have failed.

The end of the Fairness Doctrine "liberated" Rush Limbaugh. On August 1, 1988, Limbaugh opened his national radio show on WABC. He was then on the air for three hours, every weekday, drawing a huge following until his death in 2021. A 2008 Zogby poll found that he was the most trusted news personality in the country, being so selected by 12.5 percent of the respondents.[54] His popularity opened the way for many other extreme conservative radio talk shows. Unfortunately, he also opened a path to normalizing as "news" what often was radical right-wing opinion, which included distorted, false information and conspiracy theories.[55]

About eight years later an even more profound change in mass media news coverage appeared, which contributed to the polarization of politics in the United States In October 1996, the media mogul Rupert Murdoch launched Fox News to 17 million cable subscribers, with Roger Ailes as CEO. Murdoch already had succeeded with a 24-hour news channel in Europe, and Fox News quickly proved popular in the United States.[56] At the time of the 2000 presidential election, it was available in 56 million homes across the country. It had big stars, including Bill O'Reilly and Sean Hannity, for many years.

The programming was decidedly right-wing and Republican.[57] In addition to being partisan, it demonstrably and influentially often conveyed erroneous information and unfounded conspiracy theories.[58] It is noteworthy that Rush Limbaugh and the Fox News stars were not only highly partisan as advocates for right-wing Republican persons and policies, but they would often report inaccurate information and denigrate persons with whom they disagreed.

Social Media

By the end of the 1990s, Internet-based social networks had begun to emerge and rapidly expand, particularly as they were utilized on mobile devices. Facebook and many other platforms were soon used by millions and millions of individuals. They became an important source for news, and every social network on a platform carried the news people in each network wanted to know.[59] One attractive aspect of this was that people could share news about their children and their travels with their friends. It could also include news of disasters and local social problems, as well as efforts to mitigate them. However, with no oversight, the news may be erroneous, unwittingly or sometimes willfully. False news and other dirty tricks had a new channel through which they might be applied.

These many actors and the societal changes they helped bring about constitute the environment within which traditional US political institutions functioned at the end of the twentieth century and the early part of the twenty-first. Gradually, the two major political parties had become highly polarized, as had the country as a whole. As discussed in the next chapter, the stage was set for hyper class inequality, and regressive consequences.

Conclusions

The rise in class inequality beginning in the late 1970s was remarkably sharp but was also somewhat masked, so that conflicts regarding the rise were muted. I documented the sharp increase in class inequality and also the societal ill effects of inequality at the outset of the chapter. Explanations for the increase may be disputed. Acknowledging the role of worldwide technological and market developments, I emphasize the roles of US corporate practices and national governmental policies in accelerating class inequality.

Conflicts related to class inequality were channeled within the US political power system. Changing class- and status-ranking systems also influenced the political power system. However, insofar as the government is democratic, it was guided by the public. In this chapter, the important role of organizations and mass media in influencing the public in debates about ideological issues were noted. Clearly, persuasion was a primary mode in waging conflicts during this period.

Another matter should be recognized. The contentions of the 1960s and 1970s and their consequences demanded more attention than they received, as they reduced the salience of rising class inequalities. These matters related to foreign wars and policies and also to domestic contentions relating to Black-White relations and the women's movement, which are discussed in chapters 5, 6, and 7. In chapter 4, I continue to focus on class hyper-inequality and related conflicts after 1992, which set the stage for the extreme societal division and splintering that erupted in 2021.

4

Hyper Class Inequality, 1993–2022

Extreme class inequality can generate even greater inequality. People who are extremely wealthy can use their wealth to get even more money. The people who are very poor have difficulty getting by; living precariously, they find it very difficult to improve their economic position. This chapter begins with an examination of the degree of mobility between classes. Then I discuss US government policies related to class inequality during the presidencies of Bill Clinton, George W. Bush, and Barack Obama. I go on to examine the role of important nongovernmental actions and actors in conflicts affecting economic inequalities. I next consider the extreme level of inequality that had arisen, even before Donald J. Trump's election, and proceed to discuss Trump's presidency, when economic inequality reached new heights, antagonism between political parties intensified, civil discourse deteriorated, and destructive conflicts became customary. I conclude with observations about major changes and aspirations following the election of Joe Biden to the presidency.

The extremely high degree of income and wealth inequality that has been reached in the United States was discussed in the previous chapter. This situation might seem acceptable to many people if the opportunities to become rich were widely shared and people could rise in their class level through their intelligence and skill. There is considerable research relevant to these beliefs about mobility within generations and intergenerationally. Within the same generation, there is some upward mobility in absolute incomes. But *relative* income is different. Movement from one income quintile to another is not large, particularly at the top and bottom quintiles, within which about half of the families remain in the top or bottom quintile into which they were born.[1]

Movement between those in the top and bottom quintiles is rare, with 3 percent rising and 6 percent falling between those two quintiles.[2] There was increasing downward mobility in terms of absolute income between 1990–1991 and 2003–2004; the share of households who had their income decline by $20,000 or more increased from 13.0 percent to 16.6 percent.[3] For the middle class, there was an increase in financial distress.

There is abundant evidence that intergenerational mobility in the United States is lower than it is in other economically developed countries.[4] For example, this is documented in a report of the Center for American Progress, specifying the elasticity or correlation of incomes between the parental and children's

Fighting Better. Louis Kriesberg, Oxford University Press. © Oxford University Press 2023.
DOI: 10.1093/oso/9780197674796.003.0004

generations in nine countries.[5] The ordering of the nine countries in the correlation or inelasticity between father and son earnings is as follows: United Kingdom, United States, France, Germany, Sweden, Canada, Finland, Norway, and, with the greatest mobility, Denmark.

The degree of mobility varies among different levels of parental incomes. A 2012 Pew Charitable Trust study found that the amount of money a person acquires in excess of what their parents acquired varies by the quintile levels, with the smallest amount being in the lowest quintile and the largest amount in the highest quintile.[6] This means that while most persons may have a higher absolute income than their parents, they do not have more relative income.

The pattern of intergenerational wealth mobility is similar to within-generation movement. It is very sticky for the bottom and top quintiles: 41 percent of each parental generation had children in the same quintiles. Greater upward and downward mobility occurs for persons in the middle quintiles. However, the absolute amounts of wealth possessed by different levels of the children's generation is quite different depending on the parental generation. Wealth has decreased at the bottom and middle and has greatly increased at the top. For example, the wealth held by the lowest quintile in the parents' generation decreased by 63 percent in their children's generation while it increased by 27 percent for the parents who were in the top quintile of wealth.

If a person comes from a low-income family, it is particularly dispiriting to live in the reality that the chances of rising above that income level are low, and they are vulnerable to great setbacks in the event of serious health events or national economic downturns. If they could believe political or social contention might help improve their circumstances, they might struggle to do so, which raises resistance from those who feel their relative well-being will be lessened. Such conflicts are often framed as zero-sum fights in which one side gains at the expense of the other. But hyper-inequality can be rightly framed as widely harmful so that it would be reduced by increasing equality through constructive fights and gaining common benefits. If low-income earners made greater contributions to the societal GDP, all society members could benefit.

Several variables help explain the transmission of class ranking across generations in the United States.[7] Two highly significant variables are education and race. I discuss education briefly here, and examine both in greater detail in the next two chapters.[8] Clearly, the higher the income of the parents, the better are the educational services they can provide for their children. This is true for all stages of education, from preschool to graduate school. It is also true for parents making use of public schools. The quality of educational resources among public schools varies greatly among the states and school districts, due to the dependence of the schools on local funding, particularly local property taxes and sales taxes. Overall, the sources of funding are 47 percent from the state, 45 percent from localities, and 8 percent from the federal government.[9] The variation in funding has important implications for the quality of the educational experience and the consequent learning of the students. Schools with smaller budgets have teachers with lower salaries and length of experience. The schools with low budgets have higher turnovers of teachers and larger class sizes. New and old research makes it evident that spending more money does improve student outcomes, particularly where there are well-established Head Start programs.[10]

A fundamental way to improve the educational resources in public schools for all children is to increase the funds for educational resources in public schools in poor neighborhoods in cities and poor rural localities. High-income families tend to live in neighborhoods with excellent, well-resourced schools. Fights about equalizing the quality of schooling for all people are waged in the courts, in political elections, and in the streets, as will be seen in chapters 5 and 6.

Class Conflicts during the Administrations of Bill Clinton, George W. Bush, and Barack Obama

As previously noted, major conflicts in the United States are generally conducted by institutionalized political processes: elections, legislation, lobbying, judicial decisions, and presidential orders. At times, however, the contentions are conducted in less institutionalized manners, by street protests, strikes, and occupations. In either case, they vary in their degree of destructiveness and constructiveness for different parties.

Bill Clinton's Administration, 1993–2001

Running as a centrist Democrat, Bill Clinton defeated President George H. W. Bush in 1992. He faced a large budget deficit left by the Reagan and Bush presidencies and an economy still recovering from the recession of the early 1990s. Many of his advisors supported tax increases and spending cuts to reduce the deficit, arguing that this would encourage Federal Reserve chairman Alan Greenspan to lower interest rates; this, together with increased investor confidence, would produce economic expansion.[11] Secretary of Labor Robert Reich argued that stagnant earnings were a bigger problem. Clinton believed cutting the deficit would offer other political benefits and took that path in fashioning his budget.

Republicans, nevertheless, fiercely campaigned against Clinton in the 1994 midterm congressional elections and captured both Houses of Congress. The Republicans in the House of Representatives were led by its Speaker, Newt

Gingrich, who sought a conservative "revolution," aiming to cut taxes, reform welfare, and cut government spending. Clinton was open to some Republican ideas and in many ways opted for a "Third Way," triangulating between the traditional Republican and Democratic views. Nevertheless, the Republicans often resisted and fought against Clinton's policies, seeking even more drastic moves. For example, the Republicans tried to cut Medicare spending and cut taxes for the wealthy, which led to government shutdowns before they failed.

Budget fights between the Clinton administration and the Republicans in Congress, which had implications for poverty and income inequality, recurred throughout Clinton's presidency. One budget provision that is generally well-regarded by Republicans and Democrats is the Earned Income Tax Credit (EITC), a refundable tax credit for low- to moderate-income working individuals and families. The amount of the refund depends on income and number of children. It was introduced in 1975 under President Gerald Ford and was made permanent by the Revenue Act of 1978 under President Jimmy Carter. The EITC was expanded as a part of the comprehensive Tax Reform Act of 1986 and signed by Reagan, who praised it. Clinton succeeded in expanding the program, while the increasingly right-wing Republicans, with Newt Gingrich's leadership, fought to decrease its benefits.[12] The Republicans also continued trying to cut Medicare benefits. They happily agreed to ending AFDC (Aid to Families with Dependent Children), replacing it with TANF (Temporary Aid to Needy Families), which would provide block grants to states.[13] Recipients could not receive more than five years of benefits in their lifetime, and some states set fewer years for limits to lifetime benefits. Recipients must be working or in training or doing community service. Before the law was passed in 1995, more than 13 million people received cash assistance, but by 2016 the number of recipients had fallen to 3 million.

In his January 27, 1996, State of the Union Address, Clinton proclaimed that the era of big government was over. Relatedly, Clinton reduced governmental regulations in several arenas. Notably, this included the financial services industry, where he agreed to repeal the Glass-Steagall Act provision that banks must choose to be either commercial or investment banks. Commercial banks were subject to government oversight and protections such as deposit insurance. Freeing investment banks to make loans without such safeguards permitted weak loans, which contributed to the financial crisis of 2008.

Clinton did, nevertheless, strive to provide universal healthcare, and he charged Hillary Clinton to lead a task force that would develop a plan to achieve that goal. The task force rejected a single-payer system; rather, the plan would extend employer-based health insurance, with persons not covered by employers to be insured by a government plan. The government would set a minimum level of benefits that each insurance plan would provide. This proposed universal healthcare plan was furiously opposed by the Republicans. Clinton insisted upon universal coverage, but the Republicans would not even offer amendments to the proposed legislation.[14] In addition, health insurance companies supported a large advertising campaign against the proposed plan. Clinton eventually gave up on healthcare reform, but he went on to support a bipartisan plan, the Children's Health Insurance Program (CHIP), which would provide healthcare coverage for children of the working poor.

Overall, the economy is generally viewed as doing very well during Clinton's presidency. There was low unemployment, steady growth of GDP, and steady decline of the federal debt as a percentage of GDP.[15] But the rich got richer while wages and working conditions continued to stagnate, as discussed in the previous chapter. Economic inequality continued to grow. Attention to the growing plight of many blue-collar workers was slipping away in the Democratic Party, and there was a tendency instead to celebrate meritocracy.[16] Some talented people, even with humble backgrounds, might be successful in elite schools and be rewarded with very successful careers in government or business. They were deemed to be very smart and their expert views about how to deal with social problems became very influential. This has had some unfortunate consequences, as is discussed later in this chapter.

The way conflicts were conducted during Clinton's presidency deserves consideration. It might be argued that Clinton's approach of accepting some Republican policies and ideas reveals the weakness of the constructive conflict approach. Such conciliatory actions by Clinton were not reciprocated by the Republican Party leadership; indeed, he was subjected to severe hostility.

I think that a broader and more calculated constructive conflict approach might have been more effective. No doubt, Clinton was a superb and often effective political tactician as president. But his progressive goals might have been better advanced if they were fought for in the context of confronting the Republican priorities that protected the interests of narrow interest groups, rather than to the interests of the country as a whole. That would have entailed arousing and working with nongovernmental actors and the people who were adversely impacted by the changing circumstances and growing inequality.

In framing the conflict with Republicans, Clinton's acceptance of the priority of shrinking the government was not productive. The Democrats might have framed the contentions in terms of the goals that would be met by government action that would benefit all the people in the United States. The common benefits might be stressed: if poor people had the resources to be economically productive, all the people in the country would benefit. The GDP would increase, and tax revenues would rise.

Leaders of the Republican Party also might have conducted their conflict with the Democrats much more constructively. They might have acknowledged and welcomed the Democratic policies that were similar to Republican preferences, avoid nasty name-calling, and not persist in personal investigations of the Clintons that were not well grounded. Instead, they chose fierce opposition and name-calling. Unfortunately, there was little pressure from proponents on either side favoring more constructive strategies.

Latent economic conflicts did not become more salient in the 1990s, despite the realities that wages of blue- and white-collar workers had become stagnant, while the income of the very highest earners had soared. There were growing disparities in income, and many impoverished people remained so. On the contrary, the earlier fights had not lessened inequality or greatly reduced poverty after its initial reductions. The disorders of the 1960s exhausted some people. Others felt defeated and retreated to struggles at the local community level, where they thought they could have more impact. Still others turned more to personal fulfillment and self-improvement.

For latent conditions to become overt conflicts, some people with complaints or grievances are not enough. The people with complaints must also believe that certain other people are significantly responsible for the conditions that cause their grievances. They may be substantially correct in making that attribution, but they also may be mistaken, attributing the fault to scapegoats. In either case, they must also believe that they themselves can take actions to induce those others to change and improve the deleterious conditions that they suffer. Hope is necessary. For many people suffering at this time, hope that they could change the people who were responsible for their complaints was lacking. The times were dispiriting, including the enduring social discord, the assassinations of Martin Luther King and Robert Kennedy in 1968, and the declining trust in the government. Some people reacted by expressing their feelings of anger and despair by withdrawing from collective action, and even increasing anxiety and recourse to drugs.

Several obstructions may have blocked the path to constructive conflicts that might have turned back the rising class inequality. Many people may have believed that vast immutable factors caused the growing class inequality, such as international trade, foreign investments by US corporations, or technological changes in production. Some people may have blamed scapegoats, such as immigrants or even adopted delusional conspiracies. Many people may have felt too weak and hopeless to bring about the conditions that would ease their grievances.

Yet there were possibilities that conceivably could have been brought together so that more constructive conflicts would have been effectively waged. One development that had great potentialities was the end of the Cold War in 1989. The dissolution of the Soviet Union and the democratic transformation of the Eastern European countries occurred with remarkable little violence. People in the United States could reasonably feel pleased and hopeful about their capacities.

Furthermore, the end of the Cold War potentially could entail reduced military expenditures and the appearance of a peace dividend.[17]

It could have been a time for broadly sharing newly gained resources. Various coalitions might have been forged to improve needy public schools, help provide child care, or repair aging infrastructure, particularly in localities and regions that had fallen behind as the US economy had undergone great changes. Peace movement activists could have helped mobilize people to draw on the peace dividend to help build the infrastructure for a secure peace and advance liberty and justice for all the people in the United States. Collaborations might have been possible with organizations of African Americans and of women seeking equality in earnings and entry into occupations from which they had been excluded.

None of this happened. Part of the reason is that the peace dividend turned out to be smaller than it reasonably could have been. President George H. W. Bush, a proponent of it during the 1992 presidential election campaign, proposed a savings of only $66 billion from previously planned defense expenditures over the next five years. Bill Clinton and other Democratic candidates proposed savings between $180 billion and $200 billion, about a third of the then current budget. The Pentagon struggled to justify a budget that was to be cut by a third.[18] Military expenditures, however, had a momentum recognized by Congress and its lobbyists, regardless of great reductions in threats.

George W. Bush's Administration, 2001–2009

The designation of George W. Bush as president rather than Al Gore cannot be viewed as a mandate for the Republican Party. Al Gore, the Democratic candidate, won the popular vote. Although the Republicans retained a narrow majority in the House, they lost five seats in the Senate, resulting in fifty seats for each party. At the end of election night, Florida was the crucial state in the election, since whichever candidate won it, would have the electoral votes to be president. Bush held a tiny lead in the vote, which would require a recount. The Florida Supreme Court ordered a partial manual recount, which might give Florida to Gore. However, the US Supreme Court ordered an end to this process, leaving Bush with a victory in the state and therefore in the national election.

After such a victory, one might think that the victor would demonstrate some regard for the concerns of the majority of the population, which had not voted for him. Bush, however, pursued much of what had become standard Republican policies. He had campaigned to cut taxes, and that was his initial highest priority—the best use of the surplus built during Clinton's presidency, according to Bush.[19] However, economic growth had slowed by the time he took office, and the reason for tax cuts was as a stimulus to the economy. The form of the proposed cuts, however, were not Keynesian. That would have cut taxes to lower- and middle-income citizens and so increase consumer demand, which would spur expanding supply. Republican tax cuts to corporations and to wealthy citizens would incentivize them to invest more and thereby grow the economy. Of course, the money might well be spent in other ways. Bush wanted more but settled for a $1.35 trillion tax cut over twelve years. The 2001 tax bill reduced the highest tax rate from 39 percent to 35 percent and also lowered the estate tax. In a 2003 bill, taxes on capital gains and also on dividends were lowered. These provisions obviously would tend to increase, not decrease, income inequality. Bush did expand Medicare to include the cost of prescription drugs, a progressive policy. Bush, however, also tried to partially privatize Social Security, but all Democrats and several moderate Republicans opposed his idea, and he abandoned the effort.

The major economic matter during Bush's presidency was the financial crisis of 2008, which followed several years of financial deregulation.[20] Banks lent more and more for subprime house mortgages. This contributed to more people seeking home ownership and therefore to increases in house prices, creating a housing bubble. Many banks also invested in derivatives and default swaps, betting on the value of mortgage packages that included subprime loans. When the bubble burst, and house prices dropped, homes were foreclosed, many banks were in deep trouble, and the Great Recession began. The Economic Stimulus Act of 2008 was enacted, and actions were taken to avert the failure of Bear Stearns and of Fannie Mae and Freddie Mac. Lehman Brothers did fail, and stock prices dropped. The Troubled Asset Relief Program (TARP) was created, with $700 billion to buy toxic assets and provide loans to General Motors and Chrysler.[21] All this helped end the financial crisis, but the recession went on into Barack Obama's presidency. The burdens of the recession were particularly heavy for people who were economically vulnerable. Suffice to say here, the recession pushed economic inequality even higher.

The major event in Bush's presidency was the terror attack of November 11, 2001. Many of the responses taken by the Bush administration aggravated the tragedy with self-inflicted harms. In another book, I examined explanations for the failure to avoid the misguided decision to invade Iraq.[22] Obviously, the invasion of Iraq and subsequent wars and disasters have harmed US international standing and security. Here, I will only note some of the ways the very long wars in Afghanistan and particularly in Iraq contributed to the increase in US class inequality and increased its harms.

As was the case with the Vietnam War, the wars in the Middle East drew government money away from possible domestic applications that would reduce poverty and extreme class inequality. Furthermore, the wars intensified divisions within the United States, which in turn hampered cooperation between the two major political parties. They also increased public distrust of the government and politicians, increasing people's belief in the incompetence and/or dishonesty of public officials. Such sentiments discouraged turning to the government to help solve pressing social problems. I close these reflections about the relationship between US wars and US class inequality with one final observation. Casualty rates for persons in military service in wars since World War II have been higher among persons of lower socioeconomic standing than among those of higher standing.[23] Interestingly, class was not associated with military service. For young men from low-income families, even more hazardous military service seemed better than employment alternatives.

Barack Obama's Administration, 2009–2017

Barack Obama and the Democratic Party won the November 2008 elections overwhelmingly. The financial crisis was severe, the United States was mired in wars in Iraq and Afghanistan, and the Republican Party leadership was not unified. The Democrats won control of both Houses of Congress, including the filibuster-proof 60-seat supermajority in the Senate. John McCain and Sarah Palin, as the Republican candidates for president and vice president, respectively, proved to be weak candidates pitted against the young outsider Obama, running with Joe Biden for vice president. Obama offered the people in the United States the hope of desired change and unity.

Actions to recover from the Great Recession had the highest, immediate priority for the Obama administration when it took office in January 2009. In February, Obama signed an economic stimulus bill valued at $787 million; it included various tax breaks and incentives and spending for education, healthcare, infrastructure, and direct assistance to individuals.[24] He soon sought an additional stimulus bill, but the Republicans, in unyielding opposition to Obama, prevented its passage.

Within his first 100 days, Obama also signed two bills that would have side effects that would help reduce inequality. One was the Lilly Ledbetter Fair Pay Act, which extended the period for filing pay discrimination lawsuits. The other was the reauthorization of the Children's Health Insurance Program (CHIP) and its expansion of coverage from 7 million children to 11 million.

Obama's administration took many actions to revive the economy and to make finance and other changes that would help manage the losses or to mitigate any recurrence. Many homeowners faced foreclosures on their homes, as house values and employment fell. The Home Affordable Modification Program (HAMP) and the Home Affordable Refinance Program (HARP) were established to limit foreclosures.[25] These programs helped some homeowners, but many lower-income households suffered severe losses.[26] In addition, Obama provided a bailout to the General Motors and Chrysler auto companies, which had been done first at the end of Bush's presidency. By the end of Obama's presidency, all the loans had been repaid.

Many changes in taxes were made during the Obama administration, which reduced the great income disparity in the United States a little. Overall, tax changes increased the share of income received by the bottom 99 percent of families than was the case in any previous administration since 1960.[27] The earned-income tax credit was expanded, the child tax credit for working families was increased, and a new tax credit was established for students and families paying for college. In addition, Congress did agree to restore Clinton-administration tax rates for high-income taxpayers, to reinstate the estate tax, and to apply Medicare taxes to the investment income of high-income recipients.

When Obama and his administration considered which major domestic program to undertake, they decided upon healthcare reform and increased medical care coverage. Expanding the number of people receiving medical care protections would help reduce inequality, when in-kind benefits are taken into account. In July, the Speaker of the House, Nancy Pelosi, and a group of Democrats put forward their plan, the Affordable Health Care for America Act. The death of Senator Ted Kennedy, a major advocate of healthcare reform, on August 25, 2006, slowed the final passage of the bill.[28] It was not passed until March 21, 2010, and signed by Obama on March 23, 2010. Republican members of Congress fiercely opposed the Affordable Care Act (ACA), often called Obamacare, and they would not participate in fashioning it.

Overall, not only did Obama's administration set recovery from the Great Recession in motion, but it also contributed to reducing the hyper income inequality. When Obama took office, the unemployment rate was high: 7.8 percent; and it rose to 10 percent in October 2009, but then fell to 4.8 percent by the time he left office. That was a drop of 3.0 points.[29] Moreover, household income rose 5.3 percent (measured by median income, inflation adjusted), and poverty declined slightly, by 0.5 percent.

The forces and structures making for the hyper-inequality, however, were barely influenced by federal governmental policies. Political polarization and Republican Party opposition to Obama limited what other governmental policies might have accomplished. Many other actors and factors, however, are powerful, and they have been indicated, but a fuller understanding requires a broader examination, which I continue in this and later chapters of this book.

Nongovernmental Struggles, 1993–2017

Major nongovernmental events and developments were undertaken from the right and the left that might have modified the level of economic inequality. They occurred at the city, state, and national levels in varying degrees of linkage to institutionalized political processes. I discuss the large-scale protest in 1999 in Seattle, the Tea Party movement, the Occupy Wall Street manifestations, and the constructive transformation of labor relations in the auto industry.

Protest in Seattle, 1999

At the end of the twentieth century, a global social movement emerged opposing globalization from above; that is, the dominance of multinational corporations and of international institutions, which controlled world markets to serve their own interests. The International Monetary Fund (IMF) and the World Trade Organization (WTO) were major actors whose policies needed to be changed. The policies they deemed objectionable related to patterns of trading and investments that served corporate interests rather than people's interests.[30] Early demonstrations occurred in Berlin, in 1988, and in Paris and Madrid, in 1989.

On January 1, 1994, the day that the North American Free Trade Agreement (NAFTA) was signed, a well-organized uprising began in Chiapas, Mexico, opposing corporate intrusion in Chiapas. That uprising was a forerunner and model for the "Battle in Seattle," a direct-action effort to block a WTO meeting in Seattle. A broad coalition of organizations with diverse concerns participated in that large-scale Seattle protest.[31] Many groups focused on opposition to WTO free trade policies, which were seen as harming workers' livelihood; other organizations were concerned with environmental matters, and some groups had pro-labor and anti-capitalist agendas.

Between 50,000 and 60,000 persons joined the protest, which occurred between November 28 and December 1, 1999. This followed months of mobilization, preparation, and training. Protesters came from many diverse organizations, including the AFL-CIO, Sierra Club, Friends of the Earth, International Workers of the World, and National Lawyers Guild. Different people used different tactics. Many protestors, mostly workers, marched down an avenue, some groups tried to block entry to the WTO meeting, some distributed flyers presenting reasons for the protest, and a few AFL-CIO leaders met with President Clinton. Nonviolent street obstruction was dealt with fiercely by police, who were well prepared for violence. They used tear gas, clubs, rubber bullets, and arrests to suppress the protests. On November 30, the mayor called in the National Guard and imposed a curfew from 7:00 p.m. to 7:00 a.m. The extreme conflict escalation was later repudiated: the chief of police resigned, the mayor was defeated in the next mayoral election, and the city paid damages to people who proved to have been arrested without cause.

This was not a constructively conducted conflict episode. The media focused on the violent confrontation, providing little attention to the complex issues in contention. Discussion of the issues in advance of the WTO meeting and planned protest might have helped diminish destructive escalation. Representatives of the WTO might have indicated their willingness to listen to the dissatisfactions expressed by the groups planning protests. They might even have communicated their interest in learning more about their critics' grievances. Some of their goals may have been better achieved by such openness. On the other side, the protesters might have proposed specific changes they sought. As Saul Alinsky pointed out, a community organizing effort will be stronger and gain supporters by winning something for which its members fought.[32] Admittedly, it is difficult for broad coalitions, seeking to make decisions by democratic consensus, to choose precise demands. Nevertheless, some efforts in this regard can be usefully attempted by fashioning specific short-term desired steps. Related actions might have been organized in cities and states adversely affected by the policies of international corporations. Finally, other parties might try to foster communications between opposing sides and explore possible points of potential agreement. Other parties also might study the issues in contention and provide possible mutually acceptable elements in resolving some disagreements in the conflict. Possible plans to help the displaced workers in the Rust Belt cities might have provided goals for workers to demand corporations and governments meet. Their misfortunes were not of their making, and financial assistance might assist in new training and relocations or investments for new industries.

Too often, leaders on each side in a conflict think too much about how to express and demonstrate how strongly their side feels about their demands. They give too little attention to the likely responses of their adversaries. This may help leaders mobilize their supporters, but it is often counterproductive in trying to influence the opposing side or outside observers.

The Tea Party Movement

The Affordable Care Act (ACA), or Obamacare, became a primary issue of continuing contention due to a right-wing social movement. The movement began in early 2009, and gave birth to the Tea Party, which proved to be highly influential. The first reference is attributed to Rick Santelli, a CNBC reporter, who delivered a tirade against the Obama administration's consideration of a plan to relieve mortgage foreclosures, on February 19, 2009. He called for a Chicago

Tea Party to protest measures to "subsidize the losers' mortgages."[33] The mean-spirited dismissal of people experiencing troubles that they had not created was not unusual. The term "Tea Party" was seized upon as a rallying call for a growing number of protests. Fox News gave the protests great attention. A social movement was born.

Three circumstances converged that enabled the Tea Party to adopt a strategy and use tactics that resulted in its rapid growth and enduring destructiveness.[34] First, numerous citizens felt some discontent about taxes, governmental "interference," and dissatisfaction with politicians. Some incitement of these sentiments was the hallmark of right-wing media figures and even many Republican Party leaders, as previously discussed. These sentiments tapped into some common US views regarding individualism and liberty, and also served the interests of corporate leaders who desired governmental assistance at times but sought low taxes and minimal regulation. Moreover, many persons who felt they had worked hard felt bitterly that liberals scornfully pushed them aside to benefit people on welfare, Blacks, and immigrants. For example, this is documented in a lengthy in-depth study of the conservatives in Louisiana.[35]

Second, already existing adversary organizations pitched in to help local Tea Party groups, providing funds, links, and information about how to mobilize and gain influence. These connections included Freedom Works, Americans for Prosperity, and, importantly, the Republican Party. The very wealthy Koch brothers, David and Charles, devoted vast funds to mobilize opposition to Obama, support right-wing persons, and foster their favored policies in the Republican Party.[36] The Tea Party received considerable assistance from them.[37]

Third, the strategies adopted were relatively effective at the local and state level, as well as nationally. The very term adopted, Tea Party, resonated with the US struggle for liberty and fighting nonviolently against taxes. Going to congressional town halls and acting disruptively to demonstrate opposition gave expression to feelings of anger and disrespect of politicians and regular order. Furthermore, not much commitment of time or labor was entailed by such actions. The town hall protests drew much media attention, magnifying their scale and significance. Indeed, the disruptive protests at some town hall meetings held by members of Congress were staged to draw cable news attention. Thus, stars on Fox News would announce when in the future they would be at a particular protest, ensuring attendees could see a star and maybe be on television. Fox was not the only network paying attention. In the spring of 2009, the April 15 rallies received even more attention from CNN than from Fox, but Fox covered Tea Party protests often, before and after that day. In addition, the threat to challenge elected members of Congress in primaries sometimes won compliance.

Obamacare made a good target for right-wing ire. It signified increased federal government services, expanding New Deal government benefits rather than cutting them. It was deemed to increase governmental control of individuals' lives. Of course, for most persons, who are not so ideologically guided, the Obamacare benefits would prove attractive. The Republican Party leaders were not able to fashion a better alternative and repeal and replace Obamacare, even as they did act to undermine it, in some ways and in some states.

The Tea Party movement is viewed by many observers as having been very effective in achieving its objectives. It had access to large resources, influential links, and the simple goal of blocking Obamacare. But it failed to do so; indeed, it aroused support for the ACA and even new support for a more comprehensive single-payer approach. A less radical stance and readiness to consider coopera- tive actions to produce a mutually acceptable healthcare reform might have been less destructive and more successful.

Nevertheless, the appearance of power produced actual power. The Tea Party did shift the Republican Party to the right, and right-wing ideologues un- derstandably claimed the results of Republican victories in the 2010 midterm elections validated their approach. Subsequently, their right-wing ideology was blended with right-wing populism, which was carried to extremes by Donald J. Trump.[38] Ideologies can be deployed in efforts at persuasion, an element in the core ideas of the constructive conflict approach. During Obama's first admin- istration, the Republican Party's determination to deny Obama a second term solidified a radical right-wing approach rather than broadening its appeals and policies for a wider, less destructive long-term future.

Occupy Wall Street, 2011

The Tahrir Square uprising in Cairo, early in the Arab Spring, inspired the pro- test tactic of occupying public space and the Occupy Wall Street (OWS) move- ment.[39] Kalle Lasn and Micah White, editors of *Adbusters*, a Canadian magazine, had previously tried and failed to initiate a large public protest in Canada. During the summer of 2011, they and other activists took steps, using Twitter and Facebook, to mobilize support for an "occupy Wall Street" protest in lower Manhattan on September 17, 2011, and an occupation of Wall Street for a few months. On that day, a large rally and a march was held, and a temporary city was set up in Zuccotti Park, a city park in the middle of the Wall Street area that was open overnight.

In the following weeks, marches and demonstrations continued in New York City, as labor and other groups held demonstrations supporting the Occupy Wall Street action. Many clashes with police occurred, and police made hundreds of arrests. Many activists filmed these events so they could be shown on YouTube.

Very quickly, similar occupations occurred in cities across the United States and in many other countries.

I visited the Occupy Wall Street encampment in Zuccotti Park on November 3, 2011. It was well organized, including tightly packed small tents, displays of books and information about various organizations, and boards listing training and educational workshops, scheduled speakers, and relevant events. While I was there, speakers included Cornel West and Chris Hedges. The occupiers with whom I spoke expressed their pride and pleasure at their actions, some concern about how it might end, and the conviction that they were provoking conversations about the overly high income and wealth inequality in the United States.

The occupation ended abruptly and violently when police burst into the park encampment during the night of November 15, 2011. Many protesters were injured, and many were arrested. Such crackdowns were also made in Oakland, California, and Tampa Bay, Florida. At the time of the fourth anniversary of the onset of the Occupy Wall Street protest, the *International Business Times* reported that the city had paid out $1.5 million to settle eighty lawsuits (with more still pending at the time).[40] That cost does not include overtime for police officers who were required to do patrol work around Zuccotti Park during the occupation.

In the course of the Occupy operation, no specific demands were advanced, and no particular policies were put forward.[41] The broadest constituency was claimed: the 99 percent as opposed to the 1 percent of rich persons in the United States. The extraordinary amount of income and wealth of the richest 1 percent was decried, and perhaps proclaiming any specific policy or goal would have divided the protestors. This strategy, however, would preclude some tangible victory. As it happened, all the protest occupations of public spaces were suppressed or dissolved in the face of colder and colder weather and police closures. Nevertheless, small groups had long continued in small local protest and learning activities. The OWS activists could claim to have raised the public's consciousness of inequality in the United States, as did their counterpart in many other countries. The official Occupy Wall Street website continues: http://occup ywallst.org. Micah White went on to become a well-known activist and the author of *The End of Protest: A New Playbook for Revolution.*[42]

Occupy Wall Street did not have the impact in US politics that the Tea Party movement had. Of course, the OWS protesters lacked the support of well-established organizations, wealthy donors, or political party connections that the Tea Party movement had. OWS protests also lacked traditional mass media coverage. What attention they received was focused on police actions. Social media were growing in use, but not so much among older citizens. To some degree, however, as with the Tea Party activists, citizen discontent with a circumstance that OWS activists wanted to correct was present. Income inequality had been growing, since workers' wages were stagnant and corporate officers' incomes were rapidly rising. The banking crisis, the recession, and the bailout of banks to avert their failure generated great popular discontent.

The choice of occupying public space as a way to build a social movement and to reverse increasing income and wealth inequality was dramatic, it gained attention, and it was replicated in 950 other cities in the United States and abroad. However, it also posed some problems. The commitment of camping out in some city park or square was too extreme for most people to undertake. Being open-ended, it did not provide guidance on next steps for the protesters. Setting forth a variety of plausible policies toward greater equality, and paths to implement those policies, might have yielded more immediate and lasting progress.

Significantly, a CBS/*New York Times* poll, conducted in October 2011, found widespread public disapproval of the extreme economic inequality in the country.[43] Asked whether money and wealth were distributed fairly in the United States, 66 percent answered no and only 26 percent said yes. The poll also found that 43 percent of the public agreed with the views of the Occupy Wall Street movement, 27 percent did not agree, and 30 percent did not know. However, these views did not mean that the public regarded the movement positively; the poll reported that 25 percent of the respondents said they had a favorable impression, 20 percent had an unfavorable impression, and 53 percent said they were undecided or had not heard enough to decide.

Political disapproval of the high US income and wealth inequality also could be heard at the time.[44] President Barack Obama, in a speech on October 16, 2011, honoring Martin Luther King Jr., said, "Dr. King would want us to challenge the excesses of Wall Street without demonizing those who work there." On an earlier occasion, Obama said that the protests "express the frustration" of ordinary people with the financial sector. More surprisingly, also on October 16, 2011, Eric Cantor, then the Republican majority leader in the House of Representatives, on Fox News, said Republicans agreed that there was too much income disparity in the country. He went on to say "that there is a growing frustration out there across the country and it is warranted. Too many people are out of work."

In retrospect, many broad policies to mitigate the financial injuries suffered by many persons after the 2008 recession might have been fought for and perhaps implemented. That would include assistance to homeowners to avert many mortgage foreclosures, which would have limited increasing and solidifying wealth and income inequality. Numerous national and local actions were taken that mitigated the crisis to some degree. For example, the US Department of Justice sued the Bank of America for fraud in relation to foreclosures; the case was settled in 2014.[45] The Bank of America set aside money for redevelopment and legal assistance to take on systemic issues to help homeowners avoid foreclosures in stricken areas in Philadelphia, implemented by local community organizations.

Constructive Conflict Transformation within the Automobile Industry

The initial resistance of the leading US automobile manufacturers, especially Ford Motors, to recognize the UAW as the auto workers' representative for collective bargaining was discussed in chapter 2. I also noted there that the labor-management conflict began to be transformed and waged constructively by the opposing sides. I now examine the collaborative nature of that relationship in the face of two great crises in the US automobile industry: first in 1979–1982 and then in 2006–2009. The benefits of that collaboration for the associated national economic circumstances deserve attention.

In both crises, auto sales fell precipitously.[46] In January 1979, Ayatollah Khomeini took power in Iran, after the Shah of Iran was driven from his throne. The Ayatollah cut Iran's oil production and the price of gasoline soared, driving the US economy into a recession. Sales of the more fuel-efficient Japanese autos increased greatly, while Ford's sales dropped 47 percent between 1978 and 1982, and hourly employment dropped similarly (about 100,000 jobs). UAW contract negotiations with US companies were opened six months early. The UAW agreed to the automakers' demand to terminate the Annual Improvement Factor (AIF), which had provided a 3 percent yearly wage increase to correspond with year-by-year 3 percent increase in productivity.

Another UAW-automaker response to the recession was to establish joint programs to retrain displaced workers. Ford, General Motors, and Chrysler created joint training funds, supported at the rate of 5 cents for each hour worked by union members, later increased to 10 cents or more an hour, plus premiums for overtime hours. Union and management oversaw expenditures for establishing retraining programs at community colleges and other providers. Significantly, in two years more than 90 percent of the unemployed autoworkers had new jobs. A commitment to help laid-off workers can mitigate the severe repercussions of large-scale cuts in employment for the workers involved, their families, their communities, and the country as a whole. Additionally, Mutual-Growth Forums were initiated in the UAW-Ford contract, and to some degree in other contracts. They engaged labor-management in dialogues about the challenges in the auto industry. They improved the quality of the automobiles they produced and improved the flexibility in shifting models of cars to be produced.

Those constructive actions proved highly useful in responding to the later collapse in US auto sales, in 2006–2009, and the accompanying great reduction in the hourly workforce. For example, Ford reduced its hourly workforce from over 90,000 to about 40,000 during 2006–2007. Collective bargaining resulted in reducing hourly wage workers by 60 percent for new company entries; but once 20 percent of the workforce was at the entry wage, those first hired would move up to the full wage. All three major auto companies went further than most companies in mitigating the impact of layoffs on employees. Ford went furthest, providing more than a dozen options so that no employee left involuntarily.[47] Options included early retirement programs providing $100,000 and six months of medical coverage. Another nonretirement option offered four years of college tuition (up to $15,000 a year), at half salary and full benefits for four years.

Significantly, such collaborative corporate programs demonstrate that technological change or industrial setbacks due to major shifts in the market need not necessarily reduce displaced workers, their families, and communities into hapless victims. Corporations, governments (state and national), and philanthropies can provide support for decent transitions. What happens when military bases are closed in the United States is instructive.[48] Even when localities are highly dependent on the base for employment, disastrous losses can be averted by planning, leadership, engagement, and creativity, and can even lead to better economic circumstances.

The cooperative arrangements between UAW and Ford and the other major automakers helped the industry resurge. It improved the productivity of the automakers. It improved profits and ensured that those gains were shared. For example, "Ford enjoyed profits of $6.2 billion in 2011, $7.2 billion in 2012, $8.3 billion in 2013, and $6.9 in 2014. As a result, the company's workers received profit sharing checks of $6,200 for 2011, $8,300 for 2012, $8,800 for 2013, and $6,900 for 2014."[49] Disappointingly, the story of collaboration between the UAW and the corporate managers of Ford and other automakers is not widely celebrated or even known.

Donald J. Trump's Presidency and Joseph Biden's 2020 Victory

Many conflicts related to class disparities occurred during Donald Trump's presidency, some hidden and others in response to provocative government actions. Of course, many government actions reflected long-standing Republican Party ideas, but to an unusual degree Trump himself or his close aides decided the actions. This section of the chapter focuses on this highly consequential president in two regards. It begins with a discussion of Trump, and then provides a survey of the actions taken by him and his administration that especially pertain to economic inequality in the United States. I then review some efforts to resist and counter actions that increased inequalities and efforts to increase equalities.

This section concludes with a discussion of the results of the struggle between Trumpism and opposing forces.

To an unusual degree, Trump imposed his views on the Republican Party after winning the party's nomination for president. Therefore, I start with a discussion of Trump as a person and his presidential orientations in general and then discuss his administration's policy actions relating to class inequalities. Trump has many highly unusual personality characteristics.[50] He is extremely narcissistic, which tends to make him concerned with almost everything in terms of his personal interests and glory. He presents himself as highly brilliant, which is coupled with discounting the knowledge of experts and ignoring information that might counter whatever he feels is true. This contributes to his dismissal of scientific reasoning and findings. He seeks and is susceptible to adulation from others. He has fashioned a highly coercive and zero-sum way of thinking and behaving, which is readily seen as bullying and authoritarian. He enjoys appearing to be tough and being a crude, macho man.

These qualities have been attractive to some people, even to the extent of cultish admiration. Other people are put off by his behavior but let that pass for the sake of advancing desired policies. Others thought they could use him for their own interests. Finally, some people simply feared differing from him. Of course, many other people abhorred him and his conduct, thinking he was a con man, a sociopath, or simply evil. For the purposes of this book, I focus on the actions he took and their consequences.

Trump's way of acting as a politician builds on his personality, but is also influenced by his prejudices and calculations about what will get him votes. He enjoys and sees utility in exercising power to intimidate others, to demonstrate taking vengeance, and even to reward those who honor him. These patterns of thinking and acting are evident in his political rallies and actions taken as president. Trump's obsession with Barack Obama is illustrative. Trump's racist prejudice played a role in his charging what was obviously false: that Obama was not born in the United States. Then, at the 2011 White House Correspondents' Association dinner, President Obama ridiculed Trump at length for his political ambitions.[51] Trump listened with clenched teeth and was stony-faced as Obama spoke. As president, Trump went to great lengths trying to undue everything that Obama had done in his presidency.

Turning now to Trump's words and deeds regarding US class inequality, I note that they were often at variance with each other. In his 2016 campaign for the presidency, Trump channeled the grievances of many in the working class who suffered from the injuries resulting from the country's extremely high income and wealth inequalities. He used some of the rhetoric of left-wing populism, but the policies he pursued were a blend of right-wing nationalism, right-wing conservatism, and ill-informed prejudices.[52] The actual policies he implemented expressed his prejudices and the impulses of his personal character. Sometimes they mostly served an element in his coalition of supporters or the interests of his enterprises and those of his rich admirers, and not the interests of those workers, miners, and farmers he favored in speeches at his rallies. He also tried to present policies as if they were consistent with his campaign slogans. He constantly claimed that he was accomplishing much, that his achievements were successful outcomes of fights he had won. His approach in word and deed was quite contrary to the constructive conflict approach.[53]

For examples, I will start with foreign trade. Trump, in his nationalistic mode, believed that US workers and farmers were hurt by foreign trade. He brushed aside the great interest in expanding foreign trade of many large corporations and the Republican Party. He seemed fixated on US "trade deficits" and got into trade fights and raised tariffs to negotiate new deals to fix the problem. But his bullying style was self-defeating. Thus, Obama had believed that China's trade practices raised some concerns. He led in building a coalition of other traders with China to strengthen the country's bargaining position. Obama established the Trans-Pacific Partnership (TPP) in February 2016, bringing together Australia, Brunei, Canada, Chile, Japan, Malaysia, Mexico, New Zealand, Peru, Singapore, Vietnam, and the United States.

Upon his election, however, Trump immediately withdrew from the TPP agreement, even before it had been ratified. Trump, to end what Obama had accomplished, weakened his own bargaining power. He went on to conduct bilateral negotiations with China, which led to costly tariff fights and an agreement providing few gains. US soybean growers were harmed seriously, and Trump had to give them money in compensation. Trump engaged in other bilateral trade wars with traditional allied countries, with no more success.[54] Moreover, the tariffs were paid by domestic importers to the government as revenue.[55]

Trump's major legislation was the 2017 tax bill, named the Tax Cuts and Jobs Act (TCJA). Trump and the Republican Party leadership promised it would have many benefits, based on greatly reduced taxes for the wealthy, consistent with their favored supply-side economic theory. Large tax reductions for wealthy persons, for corporations, and for large estates would greatly stimulate the economy and raise investments and wages, or so they claimed.[56] Of course, cutting revenues to the federal government was desired by conservative Republicans as a way to shrink the government and its services and regulations.

There was, however, widespread opposition to the extreme nature of the proposed act. Democratic Party members totally opposed the bill because it cut taxes so greatly for the wealthy and for corporations, but not for the middle class. They sought to defeat the bill, and then work together for an acceptable bill.[57] Even thirteen House Republicans opposed it because the bill eliminated the tax deduction for state and local taxes.

Over four hundred millionaires and billionaires signed a letter urging congressional rejection of the act because of its disproportionate tax cuts for the wealthy and the huge national debt that would result. Warren Buffett and Bill Gates were publicly critical of the proposed bill.[58] Nobel Prize in Economics winners Joseph Stiglitz, Paul Krugman, Richard Thaler, and Angus Deaton were very critical of the legislation.[59] There also were street protests. The bill was viewed negatively by the general public, before and after it was passed.[60] Despite all the widespread opposition, however, the TCJA passed, winning with only Republican votes: in the House: 227–203 and in the Senate: 51–48.

The TCJA did not deliver the promised benefits. A Congressional Research Service report found that the 2017 Tax Act had little measurable effect on the overall US economy in 2018. The tax cuts did not come close to paying for themselves by turbocharging the economy, as President Trump had repeatedly promised. Trump boasted of blue-collar wages rising, but the claims ignored increased living costs and the decline in benefits. Taking those into account in fact reveals declines in wages.[61]

Alan Blinder and Mark Watson calculated the annual growth rate of nonfarm jobs and the annual growth rate of GDP during Democratic and Republican administrations, starting with Franklin Delano Roosevelt.[62] They found that, generally, Democrats had higher growth rates by far than did Republican presidents. Trump had the lowest rates of all his predecessors, being negative in the growth of nonfarm jobs. The only plausible explanation they offer is that Republicans tended to cling to their preferred theories, like the power of tax cuts and deregulation, while the Democrats have been more pragmatic and heeded lessons about what policies have actually worked in the past. Attention to evidence does improve the likelihood of achieving what is wanted.

Trump, like his Republican predecessors, has been hostile to trade unions. One manifestation of that is the way the National Labor Relations Board (NLRB) functioned during Trump's presidency.[63] The NLRB was established by Congress to implement the National Labor Relations Act, giving workers the right to join together and take actions to improve their pay, benefits, and working conditions. This became particularly urgent during the COVID-19 period with its wave of strikes and job actions by healthcare workers and workers at Amazon and other companies. Workers took to the streets, started petition drives, and made bargaining demands in an effort to get safety equipment and institute other measures to avoid workplace exposure to COVID-19. The board was composed of three members and the general counsel, all Republicans without experience representing workers or unions, while both Democratic seats were vacant. Consequently, the NLRB stripped workers of their protections under the law. It restricted their ability to organize at their workplaces, repealed rules holding employers accountable for their actions, and undermined workers' bargaining rights. Citing the COVID-19 crisis, the board unilaterally halted all elections by workers seeking to form unions.

Assessing the Trump Administration's Effects on Class Inequality

Overall, unwittingly, Trump's speeches increased the public's awareness of the nation's great class inequality and some of its ill effects. The unequal bad effects of COVID-19 made the inequalities even greater and more evident. The reality of Trump's failure to mitigate the occurrence of increasing inequality and his denial of the unfolding tragedies helped shift the public and the Democratic Party to the left. Trump and the Republican Party's claims of economic success seems to be based on stock market prices, indicating good profits for corporations and investors. That would indicate increasing class inequality, however, which would entail more people suffering reduced *relative* class standing,

Actions of Resistance and Opposition

Trump's approach to waging conflicts was clearly contrary to the constructive approach. Although he and many of those associated with him claim great results for the US from acting as they did, these claims can be questioned on two grounds. One is that some of the results they seemed to seek would enrich corporate investors and reduce funds for future government spending. But those goals and attainments served only a small proportion of the people and were contrary to advancing life, liberty, and justice for all. Second, they failed to enduringly achieve the populist goals they claimed to be seeking. A somewhat more constructive approach might have achieved more constructive results.

On the first ground, it should be evident that in seeking to increase inequality, they overreached, as they aroused resistance and mobilized opposition. The results, to some degree, were counterproductive for the extreme goals they set. Even a somewhat more constructive approach in passing the TCJA might have had broader and more enduring consequences. The Republican Party efforts to destroy the Affordable Care Act (ACA) failed and harmed the party as well as ACA beneficiaries. A somewhat more collaborative approach might have been more productive for everyone. Trump's harsh anti-immigration policies, particularly at the US-Mexico border, were touted as helping employment of US citizens. The at times brutal implementation, however, aroused great public opposition—outcries that produced some administration modifications. Negotiated agreements with Mexico were more effective. The policies did result in large declines in immigration. Those reductions, however, hurt the national economy in both the short and long term.[64]

Several state and city governments opposed many of the Trump administration's policies and conduct. Regarding matters such as deporting undocumented persons, there was no cooperation with federal agencies. In many realms of government policy, some government employees quietly abstained from taking actions they believed to be improper.[65] In some cases, people chose to resign rather than engage in conduct they thought wrong. Finally, and importantly, numerous nongovernmental actions were also taken. These included traditional actions of protest and nonviolent operations. They also included teacher strikes and public pressure for raising the minimum wage. Such actions were related to the general increase in popular sentiment for reducing class inequality. One indicator of that was the popularity that Bernie Sanders gained in his run for the Democratic Party's nomination for president in 2016 and again in 2020. His primary theme was to greatly increase income and wealth equality. Trump also had expressed the resentments of working people who felt they were being left behind. However, he did not present policies that might overcome the reasons for the resentments by reducing class inequality. Instead, he blamed immigrants and bad international trade deals.

The hyper-inequality of income and wealth in the United States has many unfortunate consequences for the country. It distorts the political system to serve the interests and concerns of the wealthiest people, neglecting the needs of most others. It warps the economy by advancing production and profits for the interests of the wealthy at the expense of production and incomes for most other people. More and more people are relatively worse off than the ever-richer wealthy people. Social problems, including physical and mental illness and crimes, are associated with income inequality among the US states and among the economically advanced countries in the world.

In the 2018 midterm elections, the Republican Party suffered numerous defeats.[66] The Democratic Party gained a net total of forty-one seats above the seats they won in the 2016 elections. That gain was the Democrats' largest gain of House seats since the post-Watergate 1974 elections. Trump may have had appeal for his base. However, the Republican Party lost in an election that had the highest turnout for a midterm election since 1914. The defeat did not, however, result in any change in direction of Trump or the Republican Party. If anything, Trump assumed greater authoritarian prerogatives, bolstered by falsehoods.

Despite the political polarization and the different political orientations in response to COVID-19, major bills pertaining to the pandemic were soon passed by the Democrat-controlled House and the Republican-controlled Senate and signed by President Trump. The bipartisan qualities in the legislation that was passed to meet the economic consequences of the pandemic are notable.

Initially, on March 4, 2020, $8.3 billion was provided for public health agencies and vaccine research.[67] Next, on March 18, 2020, $192 billion was provided for economic support for people in need resulting from COVID-19. On March 27, 2020, the Coronavirus Aid, Relief, and Economic Security (CARES) Act, providing around $2 trillion, was signed by President Trump. Further expansions of unemployment benefits and aid to healthcare providers were made and the Consolidated Appropriations Act, 2021 was enacted on December 27, 2020. The Biden administration undertook to pass legislation with even broader social welfare benefits, and that will be discussed in chapter 10, on the future possible gains in class, status, and power inequalities.

Given the high degree of political antagonism and polarization in the United States, and their intensification regarding public health guidance in responding to COVID-19, this extent of bipartisan legislation is surprising. Discernable constructive conflict elements help to explain what happened. The gravity and urgency of taking action when so much of the economy shut down was undeniable for both Republican and Democratic leaders. Trump's lack of ideology made it easy for him to spend freely, uninhibited by the Republican doctrine to avoid increasing government debts. Republicans joining in large-scale government spending would ratify Democratic deficit spending for other programs and for economic stimulus purposes. The CARES legislation would offer benefits to businesses and individuals. Each side could claim they were serving their constituencies as the 2020 elections loomed. The legislation provided some elements of a win-win payoff. As negotiations for more stimulus and relief bills were pursued and passed, the concerns of each side were recognized and to some degree addressed, consistent to some extent with a constructive conflict approach.

In any case, the United States experienced a high rate of COVID-19 fatalities among the high-income G7 countries.[68] That tragedy had many causes, including inadequate preparation and limits of the medical care system. It was also due to the great class and ethnic inequalities revealed by the distribution of COVID-19 cases and deaths. Low-income people were more likely to become ill and die because they were more likely to be exposed to the virus due to their kind of employment, crowded living quarters, vulnerability due to underlying health problems, and inadequate medical care services. This revelation might have been an excellent time to take actions to redress the deadly circumstances of low-income people. Some of them in low-paying jobs and/or having inadequate protection against the pandemic did mount protests and even strikes. This proved effective to a limited degree, and federal stimulus programs were of some assistance for newly unemployed workers. Substantial improvements, however, were not made during the Trump administration.

In the 2020 elections, Donald Trump and Mike Pence were decisively defeated by Joe Biden and Kamala Harris. Moreover, the Democrats won enough Senate elections to have fifty seats in the Senate against the fifty Republican senators. Since the vice president can vote to break a tie vote, the Democrats controlled the Senate. Trump, however, declared that he had won the election and resorted to numerous judicial challenges to overturn the outcome. There was no evidence of any fraud or error that would change the election outcome, but he continued to assert that the election for president was stolen from him. Consequently, on January 6, 2021, an insurrection at the Capitol building attempted to block the certification of the electoral votes, which failed. This matter is discussed in detail in chapter 9. Biden and Harris were inaugurated as president and vice president on January 20, 2021.

President Biden nominated remarkably diverse and experienced persons to fill the numerous cabinet and agency positions constituting the federal government's leadership. He moved immediately to place public health experts in charge of guiding how to overcome the COVID-19 pandemic, and he set national policy to administer the new vaccines against it. The large stimulus bill was designed to make a full economic recovery swiftly. Importantly, consideration was given not only to a reduction of the terrible damages suffered by low- and middle-income persons due to the havoc caused by COVID-19, but also to an improvement of the poor circumstances revealed by it.[69] The $1.9 trillion US Rescue Plan, signed on March 11, 2021, did include many such provisions, including child-care tax credits and a child tax credit expansion. Public opinion surveys reported that passage of the American Rescue Plan was popular. This was helped because Biden calmly explained the reasons for going big and including a wide range of beneficiaries, including poor people, tribal people, and women. The Republican Party's total opposition seemed partisan and lacking in imagination.

Conclusions

This chapter has discussed the extent and the implications of the rising class inequalities since the developments examined in the prior two chapters. Great class disparities continued to increase. When asked whether the present economic system is generally fair to most people or unfairly favors powerful interests, 70 percent of the people in the United States say it is unfair.[70] That varies little by income level, but it differs considerably by political orientation. Among Republicans and Republican leaners, 50 percent say it is unfair; while among Democrats and Democratic leaners, 86 percent say it is unfair. A full examination of this information must await the analyses of status and power inequalities in the following chapters. It is noteworthy here that class-related conflicts seem muted, despite the widespread feeling that the US economic system is unfair.

Many large-scale developments in technology and globalization have contributed to increasing income and wealth inequality in the United States and to the ill effects of those increases. Importantly and tragically, however, little was done to mitigate the inequalities and their ill effects. Indeed, many social policies have exacerbated the magnitude of increasing economic disparities. The deterioration in the way economic conflicts have been waged in the United States has been great, and the destructive consequences have been widespread.

Recent political developments signaled by the erroneous policies of the Trump presidency have aggravated the damages, which is discussed in chapter 7. In a struggle for greater equality, many advocates tend to see their cause as a fight for justice, and they feel self-righteous about their effort. Too often they may seek to advance their cause by demonstrating their righteousness, which is not appealing to their opponents. The ravages of the Trumpian governance upon US norms, values, and popular well-being has heightened widespread political concerns but also political engagement. Their disruptiveness is spurring fresh thinking, as is examined in later chapters.

5

Reducing African American Inequalities, 1945–1969

As discussed in the first chapter, this book focuses on domestic conflicts related to three primary dimensions of social ranking: class, status, and power.[1] Chapters 2–4 were focused on class, measured particularly by income and wealth. The two other broad dimensions of ranking are examined next; this chapter begins the discussion of status inequalities, focusing on one of two major subdimensions of status: ethnicity and gender. Status ranking depends upon societal evaluations of individuals and categories of people in terms of the honor, respect, or esteem accorded them. In this and the next chapter I focus on African Americans, particularly in relationship with Whites. In conventional American speech, they are usually referred to as "races," which is problematic, as must be examined.

Conceptual Issues

Race is a very important kind of status dimension in America, but the concept of race is illusionary in three primary ways. First, popularly, races are viewed as purely a biological matter, but race identifications are socially assigned and chosen, and therefore often disputed. They are not inherent but are socially constructed. Second, the definitions of races have some fluidity, and how people are assigned to one race or another shifts over time. Third, the salience of racial designations and the degree of differential ranking, from large to small inequalities, also changes over time, as does its relationship to power and class differences.

The term "ethnicity" is used here to more readily recognize and acknowledge what the use of the term "race" would obscure.[2] Changing references to various categories of people are illustrative. For a long time, "colored" was the generally accepted and widely agreed-upon term for Americans who were liberated from slavery or had descended from slaves. Then, for a while, a more precise-sounding term came into use, "Negro." Later, for some years, the self-identified term "Black" became generally preferred, and it remains so for many communities. The term "African American" has become commonly, but not universally,

Fighting Better. Louis Kriesberg, Oxford University Press. © Oxford University Press 2023.
DOI: 10.1093/oso/9780197674796.003.0005

the conventional term. I mostly use it to emphasize the ethnicity connotation; it conveys the sense that some dark-skinned people are Americans, with the same kind of further specification that German or Swedish Americans have.

The idea that "White" designates a race was always doubtful. For some people it might refer to descendants from particular European countries, a term that was gradually expanded to include South Italians and Jews, who had not been regarded as "White" in the past.[3] Ethnicity is sometimes treated as subsidiary to races, but I will use it broadly, as an identity based on a large number of people who believe they share a common history and fate, often constituting being a "people" or a "nationality." Ethnic membership is characterized by a shared religion, language, cuisine, and other possible cultural features. In America, ethnicities generally refer to people who immigrated or descended from immigrants from a particular country or region.

Membership in ethnicities is largely ascribed; that is, membership is socially determined by a person's parentage. It is not an achieved status that can be acquired by members' efforts or by other means. Large differences in rankings of ascribed statuses are inconsistent with full democracies, because the status is passed on between generations and overall social mobility is thereby hampered.

Clearly, differences in people's ranking in ethnicity are likely to be contentious. People of lower ranking tend to struggle to raise their status, while people of higher ranking tend to fight against that. Conflicts, nevertheless, may be dormant for various reasons. On one hand, low-ranking people may incorporate their inferiority into their self-conceptions as true and justified, or they may believe they could not better their ranking if they did fight to do so, or their dissatisfaction on one status dimension is compensated for by higher rank on another status dimension. On the other hand, high-ranking people may conduct themselves to foster one of the kinds of accommodations by low-ranked people noted above.

Furthermore, a social group of any given status ranking may also have a similar ranking in class or power rankings. They may tend to coincide in their rankings, but they may also be inconsistent for many members of the social group. Such consistencies and inconsistencies affect the emergence and course of conflicts. Discerning the relationships among the three dimensions is complicated and disputed. On the whole, the status ranking of African Americans is associated with similar class and power rankings. For particular persons, their ranking in each dimension influences their standing in the others.

Popular conceptions in America about race and ethnicity have been shaped by Americans' earlier conduct of colonialism and enslavement. Ever since humans began settling the territory of what is now the United States, there has been cultural diversity. Before Europeans began settling in North America, the various Indigenous peoples had developed distinctive languages and ways of life.[4]

Contact with Europeans decimated their population and radically transformed their cultures, but they have survived as an important "Indian" community with numerous variations across the land. Gradually, and then more swiftly, people from different parts of Europe came and settled different areas of America. Early on, people from different African societies were enslaved and brought to work, mostly in the plantations and slave camps in the southern colonies.

The American concept of race was developed by the dominant strata to explain and justify treating some humans as inferior and even treating some of them as if they were commodities. The Indigenous peoples were deemed to be wild and inherently inferior.[5] This contributed to zero-sum thinking that what one side gains must be at the expense of others. In reality, in more equal relations all parties may benefit by cooperation. Nevertheless, the terms White and Black were defined in laws in various ways over time and space in the United States.[6]

Many complexities were inevitably challenging to the idea of rigidly identifying superior and inferior kinds of humans when some people were enslaved. Recognizing and encouraging high skills among some slaves meant increasing their value, although that meant denying their immutable inferiority. Moreover, the very borders distinguishing one "race" from another were socially constructed and difficult to impose on the blended reality. For example, in the United States, unlike in most other countries, skin color is viewed as binary: White and Black. This was generally sustained, despite the obvious blending that exists to the point that some Blacks pass as Whites and some Whites as Blacks, depending upon social circumstances. With increasing intermarriages, offspring have more possible choices, including identifying themselves as biracial, mixed, or of multiple races.

The US Census Bureau has gathered statistics about race from its outset, documenting different classifications. It began with the constitutional requirement to count slaves as three-fifths of a person.[7] The number of races grew, and in 2000 the census allowed each person to mark more than one race. In the 2010 census, fourteen different boxes might be chosen, two of which allowed for more detailed specification. Two races remained: White and Black (African American or Negro). No clear rationale accounts for the list of races.

"Races," as social constructs, are subject to change and may even disappear. For example, my father left tsarist Russia and entered the United States at age seventeen, in 1910. In 1935, he filed for citizenship in the United States. In an official document, his race was stated to be Hebrew, his nationality Polish, his religion Jewish, and his color White. Jews in various places and times have been regarded as members of a single separate race. In the United States, Jews have been officially treated as Whites, but for most people, for much of the time, they were regarded as "different" and subjected to conventional prejudice and discrimination. They have become viewed simply as "White" since the 1940s.[8] The rise in the status of Jews was generally not regarded as causing a decline in the status of Christians. But, in recent years, the rise in White nationalist movements, which espouse anti-Semitism, questions that.

Scholars have come to recognize that racial designations are social constructions based on inherited physical features but are not necessarily related to social and cultural features. The term "ethnicities" is more suitable to realities; it refers to large groups of people who identify with each other as sharing a common history and fate. Members tend to share a religion and a language. However, in the United States, one's ethnicity may be based simply on the origins of one's parents or more distant ancestors. In this chapter, I focus on African Americans, who are an ethnic group that is deemed to be a race by most Americans and the US Census.

Fights for More African American Equality, 1945–1955

The circumstances of African Americans living in the United States at the end of World War II are well analyzed in Gunnar Myrdal's highly influential 1944 work *An American Dilemma: The Negro Problem and Modern Democracy*.[9] In 1937, the Carnegie Corporation of New York invited Myrdal to become the director of a comprehensive investigation of the Negro in the United States. The Carnegie trustees thought that as a non-American, this Swedish economist could conduct such an investigation in an objective and dispassionate way.

Myrdal traveled widely in the southern states and throughout America; he consulted with many leading Negro and White figures knowledgeable about the circumstances to be analyzed. He drew together a large team of social scientists to study the many facets of the Negro problem and the White problem. The product, a volume of more than 1,500 pages, was enormously influential in shaping how African Americans and other Americans viewed themselves and each other. It helped frame what those relations were and what they might be.

The book's influence arose from presenting a profound insight. In Myrdal's introduction to the book, he writes that Americans are moralistic and rationalistic, relative to other branches of Western civilization. Those qualities are the basis for the American creed of offering liberty, justice, and fair opportunity for all.[10] The creed is based upon valuations that may vary among different people but exist to some degree in all Americans, and therefore Americans are troubled by confronting even varied views of the deprivations, segregation, and discrimination suffered by Negroes. The book gave primary attention to what goes on in the minds of Whites about the Negro problem. The American dilemma persists, even as circumstances change.

Circumstances of African Americans after World War II

Some improvements in the conditions of African Americans in employment and military service during World War II raised their expectations for greater progress. In the minds of many White Americans, the victory in the war and the celebration of democracy would enhance the legitimacy of African Americans' claims for equal rights. However, the deeply rooted prejudices, widespread segregation, and structures of discrimination against African Americans persisted in many arenas.

The circumstances in the southern states were particularly harsh. After the defeat of the Confederacy, Reconstruction (1865–1877) established equal rights for the freed slaves in the former Confederate states, backed by the Thirteenth, Fourteenth, and Fifteenth Amendments to the US Constitution, as well as federal laws and Union soldiers.[11] However, in the 1870s, White Democrats began to use paramilitary groups to regain power in the state legislatures. Reconstruction was ended by a political deal between Democratic and Republican Party leaders, by which Rutherford B. Hayes (Republican) became president and the Union soldiers were withdrawn, allowing White Democrats to use terrorism to impose Jim Crow laws throughout the South. Their legislation and intimidation suppressed voting by Blacks and poor Whites.[12] They instituted laws to impose segregation in all spheres of life, and legal efforts to challenge them were overcome by the 1896 Supreme Court decision in *Plessy v. Ferguson*, which ruled that separate facilities could be constitutional.

Racial segregation and discrimination were not as blatant, widespread, and violently imposed in the rest of the country. However, governmental and nongovernmental practices across the country tended to sustain severe segregation and discrimination. Some evidence of this will be noted, before we turn to examine the struggles to increase equality of opportunities and of outcomes for African Americans to raise their status.

In the 1920s and 1930s, restrictive covenants were very widely employed by real estate developers to prohibit a buyer from reselling or transferring property to members of a specified race, religion, or ethnic group.[13] Such deeds were most commonly used to exclude African Americans from certain urban neighborhoods. This was fostered by various entities, including colleges, universities, and religious institutions. For example, the University of Chicago supported property owners' associations that used restrictive covenants and other practices to prevent African American families from living nearby.[14]

The various governmental and nongovernmental policies and actions that discriminated adversely against African Americans were not viewed as manifest conflicts. Some of them were not widely visible and others were regarded as "natural," and inevitable in the way the housing market worked. There were disputes and intense conflicts in local circumstances, such as when an African American family moved into a White neighborhood and faced threats and violence.

During the Great Depression, many people were unable to meet their mortgage payments and lost their homes, and many others could not afford to buy a house. Government assistance was made available through the Home Owners' Loan Corporation (HOLC), in order to avoid foreclosures of houses held by families unable to meet their mortgage payments. The HOLC would purchase existing mortgages and issue more favorable ones. The HOLC needed to know that a borrower would be able make the regular payments on the mortgage and that the mortgage matched the value of the house. It hired local real estate agents to make these appraisals, and they fully adhered to their national code of ethics to maintain segregation between Whites and African Americans. The HOLC created color maps of every metropolitan area, indicating which neighborhoods were risky for mortgages, which were colored red, and those that were safe, which were colored green. Red-colored neighborhoods, where African Americans lived, were deemed risky and mortgages were not granted, regardless of individual family income. This practice was known as redlining. This discrimination against African Americans as a lower-ranking status group was assigned on the basis of a lower-income class ranking associated with some, but hardly all, African Americans.

In addition, in 1934, President Roosevelt and the US Congress established the Federal Housing Administration (FHA), which would insure bank mortgages for persons whom their appraisals indicated would have low risks of default.[15] The mortgages were for twenty years and would cover 80 percent of the purchase price. The new mortgages were amortized, so that each payment included some principle of the loan and the borrower would own the home only when the loan was fully paid off. Homeownership is the major form of capital possessed by Americans.

The administration of FHA insurance, however, was highly discriminatory. When a bank had a prospective home buyer, it would apply to the FHA for insurance on the loan. The agency undertook an appraisal of the property, usually hiring a local real estate agent. It provided an *Underwriting Manual* to guide their work, which presented a rational for sustaining segregation. It included such statements as: "If a neighborhood is to retain stability it is necessary that properties shall continue to be occupied by the same social and racial classes." And "Natural or artificially established barriers will prove effective in protecting a neighborhood . . . from adverse influences . . . [including] prevention of the infiltration . . . of lower class occupancy, and inharmonious racial groups."[16]

In the aftermath of the long depression and a major war, the need for housing was great. Various government programs to make housing more available were undertaken, but the programs tended to sustain segregation and disadvantage

African Americans.[17] This was true for public housing developments during the New Deal years and during World War II, as well as later. The passage of the 1949 Housing Act, proposed by President Harry Truman, illustrates that partisan political contention was central in this process. Conservative Republicans had long strongly opposed government involvement in the housing market. To stop passage of the bill, they proposed an amendment to prohibit segregation and any racial discrimination, which would ensure that the southern Democrats would oppose the bill. The liberal congressional Democrats believed that they had to allow segregation in order to get public housing built, for African Americans and for Whites. The National Association for the Advancement of Colored People (NAACP) objected and unsuccessfully sought a bill that would not sacrifice integration.[18]

Most of the public housing consisted of large apartment buildings. Apartments for White tenants were located in White neighborhoods, while apartments for Black tenants were placed in Black neighborhoods. This further ensured that the schools would be segregated, and given the primacy of local funding for public schools, those in predominantly White school districts would have more and better resources. Soon, moreover, Black public housing projects had waiting lists to enter while White ones had vacancies.

Many local and national government policies, as well as real estate business policies, fostered segregation. Many areas were zoned for single-home residencies. As a result, suburbs and other urban areas were set aside for relatively well-to-do buyers. Rental housing was not available for lower-income people in such zoned areas. This was not framed as a way to sustain racially segregated neighborhoods, but it proved difficult to appear to fight against homeownership.

Following World War II, government benefits were directed to help veterans, but in their administration, economic and status inequality between Whites and African Americans was increased. The Veterans Administration (VA) insured mortgages to returning servicemen, but it followed the policies of the FHA, as stated in its manual.[19] Consequently, African American veterans were generally denied mortgage insurance, particularly, in all-White housing developments.

In addition, in the first years after World War II, the GI Bill for veterans was of great advantage for White veterans in going to college, but it did not help African Americans in actually getting admitted to institutions of higher education in the South, except those who were able to enter historically Black colleges.[20] Many of the government programs relating to housing, higher education, and other matters in this period did help reduce economic inequality among Whites. They were progressive in that regard. However, the way they were administered was often discriminatory, increasing the income and wealth inequality between African Americans and Whites. It is not unusual that policies that are progressive for some sets of people are regressive for some others.

In nearly all spheres of endeavor—economic, political, and cultural (including sports and entertainment)—African Americans were largely excluded from national, leading positions, even immediately after World War II. There were, nevertheless, some notable exceptions. In 1945, John H. Johnson began publication of *Ebony*. In 1946, the Nat King Cole Trio began the first African American network radio series and Fisk University appointed its first African American president, Charles Spurgeon Johnson. In 1947, Jackie Robinson became the first African American to join a major league baseball team. In 1948, Alice Coachman Davis became the first African American woman to win an Olympic Gold medal (in high jumping). In 1949, Wesley A. Brown was the first African American to graduate from the US Naval Academy. In 1950, Ralph Bunche won the Nobel Peace Prize for his mediation relating to the Arab-Israeli conflict, and Gwendolyn Brooks won the Pulitzer Prize in poetry, the first African American to do so.

In light of the American dilemma, it may be puzzling that the official and unofficial discrimination was so great and so persistent. For decades, the American government, many nongovernmental entities, and many individuals engaged in practices that enhanced inequality in the status of Whites relative to African Americans. There were uncontested economic, status, and power benefits for many Whites that served to maintain these practices. Those presumptions were often mistaken, however, which became evident through the actions of many African Americans and Whites, which I consider next.

Postwar Fights to Improve Race Relations, 1945–1954

Having participated in military and civilian efforts in waging World War II, many African Americans anticipated significant improvements in their life circumstances after the war. Furthermore, following a war waged against racist Fascism and upon entering a Cold War against totalitarian Communism, US political leaders believed that to counter Soviet ideological attacks on American racism, they needed to reduce discrimination suffered by African Americans. This is congruent particularly with two core ideas of the constructive conflict approach: first, conflicts are interconnected, and second, each contending side in a conflict is heterogeneous. During the early years after World War II and of the Cold War, increasing status equality between Whites and African Americans would come. It would come in some arenas earlier than others, and by diverse pathways, including constructively waged conflicts, legal institutions, and the self-destructive failure of strategies meant to oppose greater equality.

Presidential Actions

President Truman knew of "repeated anti-minority incidents immediately after the war in which homes were invaded, property was destroyed, and a number of innocent lives were taken."[21] Therefore, in December 1946, by executive order, he appointed a committee to investigate what was going on and deliver a report, which it did in October 1947. He had addressed the NAACP in June 1947, declaring his motives for his civil rights program to provide equal opportunity for everyone. In February 1948 he proposed legislation to establish a permanent Commission on Civil Rights in accord with the recommendations of the committee, which included establishing a Fair Employment Practices Commission to prevent unfair discrimination in employment.

Truman, running for the presidency, insisted that his recommendations be incorporated into the Democratic platform for the 1948 presidential elections. Truman anticipated that some Democrats from the South would splinter off. Indeed, some Southern Democrats, led by Strom Thurmond, governor of South Carolina, bolted the Democratic Party and ran as "Dixiecrats." Moreover, former vice president Henry A. Wallace ran for president as a Progressive. The Republican candidate, Thomas E. Dewey, governor of New York, was widely expected to win, and appeared overconfident. Truman, however, traveled the country intensively, attacking the "do-nothing" Republican Congress, and standing firm on his civil rights platform. His authenticity triumphed and he remained president, with a Democratic-held Senate and House. The Democratic Party had a mandate to advance civil rights, somewhat liberated of dependence on its southern wing.[22]

Racial segregation in the armed forces was especially embarrassing and wasteful of human resources in the eyes of America's political leaders when the United States had emerged as the primary world leader.[23] In July 1945, the US Army conducted a survey of opinions among White officers and sergeants after a colored platoon was transferred into their company. Their opinions were overwhelmingly favorable to Black soldiers, finding that the Black infantry had performed very well in combat, and that Black and White soldiers got along very well. By these accounts, there were no good reasons for segregation in the armed forces.

Some African Americans agitated and lobbied to end segregation in the military services. In 1945, the civil rights leader A. Philip Randolph had founded the National Committee to Abolish Segregation in the Armed Services. Then, in 1947, he and Grant Reynolds established the Committee against Jim Crow in Military Service and Training. They proposed civil disobedience to fight for African American rights. They argued that African Americans should avoid the draft and, as Randolph said at a Senate Armed Services Committee hearing, "refuse to fight as slaves for a democracy they cannot possess and cannot enjoy."[24]

On February 2, 1948, Truman informed Congress that he would end discrimination in the armed services. On July 26, 1948, he issued Executive Order 9981, establishing the President's Committee on Equality of Treatment and Opportunity in the Armed Services. It was accompanied by Executive Order 9980, which created the Fair Employment Board, which was charged to eliminate racial bias in federal employment. It took several years to actually end all segregation in the armed forces—it was officially ended on September 30, 1954, when the Secretary of Defense Charles Wilson announced the closing of the last all-Black unit.[25]

Social Movement Organizations' Use of Institutionalized Methods

The arenas within and the ways in which African Americans suffered discrimination and poor circumstances varied greatly. Different African Americans could reasonably choose diverse methods in different places to bring about desired improvements. These differences were long debated among African Americans and their allies. Naturally, different choices were made and acted upon by different people in different places and at different times in their struggle for justice, equal opportunities, and liberty.[26] Two dimensions are relevant in this regard. One dimension relates to the choice between pursuing integration and pursuing separation. In either case, equality in economic conditions and social respect would also be sought. The other dimension relates to the methods of struggle used; varying from relying on legal and political institutions to relying on direct coercive strategies.

The NAACP, founded in 1909, when nearly all African Americans lived in the South, tended to emphasize integration and rely on institutionalized legal methods to improve the quality of education for African American children. Walter F. White became the NAACP executive secretary in 1931 and was active in seeking to influence leading political figures. In this approach he differed from W. E. B. Du Bois, a very important analyst and interpreter of the African American experience.[27] White grew increasingly estranged from Du Bois as Du Bois's analysis and policies became increasingly militant.

In 1935, Charles Huston was appointed Special Counsel of the NAACP, and he led the legal campaign to end segregation until 1940, when he was succeeded by Thurgood Marshall, who remained with the NAACP until 1961. Marshall argued thirty-two cases before the Supreme Court, which would improve the conditions, rights, and status of African Americans, and was successful in twenty-nine of them.[28] This included *Shelley v. Kraemer* (1948), which ended state courts enforcing racially restrictive covenants in real estate transactions throughout the country.

Very consequentially, the southern states all mandated segregated schools for Whites and Blacks, in accord with the 1896 Supreme Court decision in the *Plessy v. Ferguson* case, which held that legalized separation was permissible if the conditions were equal. Of course, the school conditions generally were not equal. The NAACP, however, chose to bring a case where the schools' conditions were not highly unequal. This enabled the lawyers, led by Marshall, to argue that segregation was inherently unequal. This argument was supported by a variety of social science data, including material from *An American Dilemma*.

The class action case was based on the circumstances in Topeka, Kansas— legal segregation existed there, which was allowed, but not required, under Kansas law.[29] Oliver Brown was the named plaintiff; his daughter was in the third grade and was required to take a school bus to a distant school for Blacks and not allowed to go to a nearby school for Whites only. Other parents joining the case had similar experiences. When the case reached the Supreme Court upon appeal, it was joined with five other cases, and argued on December 9, 1952. The Department of Justice filed a friend of the court brief, which stressed how school segregation hurt the United States internationally.

The Supreme Court could not reach consensus quickly and it had the case reargued on December 8, 1953. Chief Justice Earl Warren was convinced that the Court should rule that segregation was unconstitutional, and he and others who agreed sought to convince those who did not, until consensus was reached. On May 17, 1954, the Supreme Court ruled 9–0 that public schools and educational facilities could not be equal if they were segregated.

The imposition of segregation was oppressive. Furthermore, the influential social psychological "contact hypothesis" proposes that contact between individuals of different groups will reduce prejudice and improve relations between them.[30] Considerable research confirms this, but the effects vary with circumstances and processes, such as the status equality of the groups. School integration and equal educational achievement would be difficult to realize when African Americans are subjected to prejudice and discrimination in employment, housing, and many other spheres.

It was understood that official segregation in other public spheres was also inherently unconstitutional. Implementing the decision in *Brown v. Board of Education of Topeka* was to be challenging, given the great prevalence of residential segregation, as discussed later in this chapter. It immediately was a spur to social movement actions. Social change results from the reverberations between official and unofficial actions.

Mixed Methods of Struggle, 1955–1969

In the tragic history of colonization, slavery, and ethnic discrimination in the United States, the fight for civil rights and against the Jim Crow system in the South is rightfully celebrated. Starting in the mid-1950s, many different people, across all levels and regions of the country collaborated constructively to reduce the oppressive conditions endured by African Americans in the South. Social movement campaigns and government legislation influenced each other, but also had their own trajectories.

Social Movements

Nonviolent methods were an important feature of the civil rights struggle. One organization relying on nonviolent strategies was the Congress of Racial Equality (CORE), founded in 1942 by a few students at the University of Chicago.[31] They were pacifists, social activists, and most were White; the organization grew and spread across the country. In the summer of 1947, while I was visiting Los Angeles, a friend asked me if I wanted to participate in a sit-in at Bullock's department store's Whites-only lunchroom. On the next Saturday, I joined an interracial group, which occupied the space. We sat quietly and were not served, thereby shutting down the lunchroom; there was no violence. Years later, I learned that the sit-in was attributable to Glenn E. Smiley, who was trying out nonviolent applications to fight segregation in Los Angeles, and worked with CORE and then with Martin Luther King Jr.

The bus boycott in Montgomery, Alabama, was the first major episode in the civil rights struggle that ended Jim Crow and transformed the South. It began as a response to the arrest of Rosa Parks on December 1, 1955, for refusing to give up her seat to a White passenger.[32] Parks was secretary of the local NAACP chapter and had recently participated in a program on nonviolence at the Highlander Folk School in Tennessee. Immediately after the arrest, Jo Ann Robinson helped organize the Montgomery Bus Boycott, and mimeographed over 52,000 leaflets for the boycott. The Montgomery Improvement Association was founded and the twenty-six-year-old pastor of Montgomery's Dexter Avenue Baptist Church, Martin Luther King Jr., was recruited to be president.[33] The boycott was effective in creating alternative ways to get to work and to shop, such as car pools. It was also very effective in attracting national attention and support.

The boycott finally ended after 381 days, following a series of court decisions. Five Montgomery women, represented by the NAACP, sued the city, seeking to invalidate the busing segregation laws. On June 5, 1956, a Montgomery federal court ruled that any law requiring racially segregated seating on buses violated the Fourteenth Amendment to the US Constitution. The city appealed to the US Supreme Court, which upheld the lower court's decision on December 20, 1956. The next day, Montgomery's buses were integrated.

Severe resistance, including violence, however, was used to restore segregation in city buses. Snipers fired into buses, and in January 1957, four Black churches and the homes of prominent Black leaders were bombed. In response, on January 30, 1957, the Montgomery police arrested seven bombers, all of whom were members of the Ku Klux Klan. That largely ended the busing-related violence.

Then, in January 1957, after consultations with Bayard Rustin, Ella Baker, and C. K. Steele, King invited them and about sixty ministers and other leaders to the Ebenezer Church in Atlanta with the goal of establishing an organization to coordinate nonviolent direct action to desegregate bus systems in the South. At a follow-up meeting in February, a new organization was founded, with King as president. It came to be called the Southern Christian Leadership Conference (SCLC). It had an office in Atlanta and sought to mobilize churches and fraternal orders to apply nonviolent methods to foster racial equality.[34] This was highly controversial work from less confrontational Black leaders and from more militant young Blacks. Nevertheless, it proved to be effective, and King had a close team of leaders with whom he could consider and devise alternative strategies.

After the Montgomery Bus Boycott, a series of other notable nonviolent direct-action campaigns occurred. Beginning in the summer of 1958, numerous sit-ins were made in the South and elsewhere in the country. For example, the NAACP Youth Council supported sit-ins in Wichita, Kansas, at a Dockum Drug Store lunch counter; after three weeks, the segregation policy there was ended, and soon thereafter, segregation in all the chain's drug stores in Kansas was ended. A student sit-in successfully followed in a Katz Drug Store in Oklahoma City. Sit-ins had been done earlier, but never so widely. Students planned and trained to undertake sit-ins so that they could nonviolently resist and limit the violence that they might encounter. In April 1960, sit-in leaders met at a conference at Shaw University, in Raleigh, North Carolina, and founded the Student Nonviolent Coordinating Committee (SNCC). By the end of 1960, student sit-ins to desegregate all kinds of public facilities had successfully happened throughout the South and beyond.

In 1961, SNCC undertook the Freedom Rides to desegregate interstate bus service in the South, which was protected by the Constitution's interest in interstate commerce.[35] The interracial SNCC riders were often viciously attacked at bus stops in the South. They were arrested and jailed when they used "White only" restrooms and fountains. People across the country were appalled, and more interracial groups joined the Freedom Rides. On November 1, 1961, an Interstate Commerce Commission desegregation order, which was directed by President John F. Kennedy, took effect. Signs for segregated seating on the buses ended and "White" and "Colored" signs for restrooms and water fountains in terminals for all interstate travel were taken down.

The next major civil rights event occurred in Birmingham, Alabama, in 1963. The Alabama Christian Movement for Human Rights invited Martin Luther King Jr. and the SCLC to come to Birmingham. SCLC leaders decided to set a limited goal: to desegregate Birmingham's downtown businesses, and to rely on nonviolent methods to do so.[36] The city government, however, issued an injunction against all such protests, which the campaign defied. The local leading police official was Eugene "Bull" Conner, the Commissioner of Public Safety, who was known to fiercely favor segregation. When the campaign began its protests, many of the protesters were arrested and jailed, including King on April 12, 1963. Then the SCLC decided to train high school students to join the protests, and more than a thousand students gathered at the 16th Street Baptist Church to do so on May 2. As they marched out, fifty at a time, they were arrested. But on the next day, when students again marched from the church, they encountered police dogs and city fire hoses streaming water at them. National television and newspaper front pages displayed the appalling attacks on children. People across the country were outraged, and the Kennedy administration urgently intervened in the ongoing negotiations between SCLC and the White business community.[37] On May 10, they quickly reached an agreement, which included ending segregation in public accommodations downtown, establishing a committee to end hiring discrimination, releasing jailed protestors, and establishing channels of communication between White and Black leaders.

Marches and assemblies are another frequently used form of nonviolent protest, and two examples in the civil rights struggle should be noted. A. Philip Randolph and Bayard Rustin strove to organize a March on Washington, which would have a broad set of goals relating to civil rights, employment, housing, and voting rights, and also would have a broad coalition of organizers, including progressive labor unions and other liberal organizations. The March was held on August 18, 1963, and drew great attention by its size and speeches. King delivered his "I Have a Dream" speech and President Kennedy met with the speakers that evening.[38]

The marches from Selma to Montgomery, the Alabama state capital, contributed greatly to passage of the Voting Rights Act of 1965. SNCC had begun a voter registration campaign in 1963 and sought help from SCLC in 1965. King and other SCLC leaders came and joined in marches in Selma, which met violence from police. The police killing of Jimmie Lee Jackson, at a march in nearby Marion, resulted in an escalation of the protest: a march from Selma to Montgomery, fifty-four miles away. The march on March 7, 1965 was led by John Lewis of SNCC and Hosea Williams of SCLC. Six blocks into the march, when marchers crossed the Edmund Pettus Bridge, leaving the city and entering the county, county police and state troopers, some on horses, attacked the marchers. They used clubs, whips, and tear gas to drive them back into Selma, seriously injuring many marchers. Lewis was dragged to safety after being knocked unconscious.

Again, the vicious violence against peaceful demonstrators backfired. A second march, with hundreds of people from across the country was undertaken, which King led and then turned around, in accord with a federal injunction. After further demonstrations and negotiations, the injunction was lifted, and then a peaceful march was made from Selma to Montgomery, with the protection of federal troops. Furthermore, the Voting Rights Act was passed by Congress and signed by President Lyndon Baines Johnson in 1965.

In March 1998, King went to Memphis, Tennessee, to support the striking sanitation workers there, as part of his planned Poor People's Campaign. Shockingly, King was assassinated there, and the civil rights movement he was seeking to redirect failed to arise. The turmoil of the end of the 1960s was overwhelming, as discussed in chapter 2. Nevertheless, the civil rights struggle had achieved very much that has been enduring.

Assessing the Civil Rights Movement's Actions

The strategies, which the many civil rights organizations adhered to, accomplished a great deal to constructively transform African American relations with Whites in the South and also in the entire country. On the whole, the strategies were consistent with the ideas of the constructive conflict approach. This should not be surprising, and the lessons of the civil rights struggle and similar struggles have provided data for the evolving constructive conflict approach.

The civil rights activists relied on both institutionalized and noninstitutionalized methods of conflict and tried to make each approach enhance the other. Another hallmark was the reliance on nonviolent direct action; this was a form of coercion that also conveyed powerful persuasive inducements and possible future benefits. The actions testified to the activists' strong convictions but did not convey hostility and argued for their cause in terms of presumed shared values and beliefs. They carefully tried to present the conflict as one that could be transformed so that widely enjoyed benefits would follow. They recognized that each side had diverse elements, and that they could have White allies; indeed, they welcomed and received support from sympathetic Whites. Finally, the activists understood that conflicts are interconnected, so they could shift their priorities as circumstances changed. King and other leaders in many movement organizations consulted with each other and considered alternative tactics for different localities and times.

The direct actions were dramatic and got national attention; moreover, what was objectionable was made apparent. A multitude of efforts were couched in terms of widely shared American beliefs and values about fundamental American rights: equal opportunities, justice, and liberty, for all. They were consistent with the American creed. Those who relied on nonviolent means of struggle were generally viewed as heroic and righteous. Their actions added validity to their cause. It was a source of raised self-regard, pride, and status for African Americans, who participated or watched the fight for equality and justice.

Clearly, the violent method of struggle adopted by some Whites to defend the Jim Crow system was ultimately counterproductive for their cause, as it only generated more opposition to that system across the country. The officials in the South who protected the perpetrators of violence and even murders lost their own legitimacy in the judgment of most people in the country. As often happens, repression of nonviolent actions produces a backlash. Nevertheless, violence and the threat of it could and does intimidate some African Americans in many localities. More ambiguous discrimination and other behavior that serves to repress rising African American status persists, as is discussed in later chapters.

The civil rights leaders had adopted strategies emphasizing nonviolence and fundamental American values and concerns. They gained widespread public support. All these produced direct meetings with the Kennedy and Johnson administrations and even coordinated actions. Yet, naturally, civil rights activists generally pushed for more good legislation, better implementation, and more rapid enactment. Kennedy was not comfortable with the demands for legislative action that King and other civil rights leaders sought, and in return the activists were wary about him.[39] Robert Kennedy, the attorney general in his brother's administration, became impassioned about the activists' struggle in the face of the violence they encountered. President Johnson had his own long-held convictions to increase the status and economic conditions of African Americans and others suffering discrimination and deprivations.

Throughout these years, some of the officials in the federal and many state governments tried to block and counter persons and groups fighting for and implementing more equality. Most notably, J. Edgar Hoover, director of the Federal Bureau of Investigation (FBI), conducted covert and overt actions to undermine Martin Luther King Jr. personally and the efforts of the civil rights movement organizations.[40] State and local officials also resisted school desegregation. This was drastically evident in Little Rock, Arkansas, in 1957, when the governor, Orval Faubus, directed the National Guard to block the entry of nine African American students to attend an integrated high school. President Dwight Eisenhower nationalized the Guard and ordered them back to their barracks. He then sent elements of the 101st Airborne Division to protect the students' entry.

The civil rights direct actions in the 1950s were focused in the South, but appeals for support were made and won in the rest of the country. Remarkably, the gains were institutionalized nationally by federal legislation, by Civil Rights Acts, and by court decisions. For example, moves toward reducing segregation in public schools were made throughout the country. The convergence of many social movements, political circumstances, and wise leaders made possible the remarkable progress toward justice and equal opportunity for all.

Alternatives and Implementations

Fighters in conflict about improving the status and circumstances of African Americans in the South were not limited to civil rights organizations and their opponents. Fights were waged by officials at all levels of government, and by all kinds of people who were not officials, to bring about changes and to oppose them. The changes go beyond court decisions, legislation, or agreed-upon settlements. Such endings may be implemented fully or variously ignored or even rejected.

Alternative Ways to Reduce Deprivations in African American Status

Despite improvements, particularly in the South, many African Americans were disappointed by the growing gap between rising expectations of progress and their continuing relatively high rates of poverty and discrimination. Frustrated expectations are a frequent cause for revolution.[41] Concurrently, during the years following World War II, middle-class White Americans tended to leave the cities and move to the suburbs, aided by government policies and real estate practices. Businesses that had once provided jobs and tax funding in the cities were also leaving. At the same time, more than three million job-seeking African Americans moved from the South to the cities of the North and West. Increasingly, the downtowns of large cities became home to low-income African Americans, many of them from the South.

The civil rights struggle had been focused on the circumstances in the South. The national legislation and court decisions promised less segregation and discrimination, but such change faced resistance and was slow. For many African Americans, particularly in large urban centers, the nonviolent strategies were too slow and inadequate. Consequently, often large riots erupted in the very areas in which rioters lived. Major riots erupted in Birmingham in 1963, Los Angeles in 1965, and Chicago in 1966. Then in 1967, riots broke out in Tampa, Cincinnati, and Atlanta. More than 110 cities had riots on April 4, 1968, after Martin Luther King was assassinated.[42]

In a sense, the riots were not acts of contention, but rather expressions of anger and frustration. Done collectively, they also gave some people the gratification of getting stuff from stores; sometimes particular stores were torched and others were protected. However, there were no shared demands made on particular others to make specific changes. It was not an effective way to persuade the White public that conditions for African Americans should be improved, since the riots made them worse. Indeed, Richard Nixon and other conservatives used the riots to justify their law and order policies. They also took backward steps away from integration and retreated from improving the living conditions of African Americans.[43]

The official response was given by the National Advisory Commission on Civil Disorders, chaired by Otto Kerner, governor of Illinois, and established by President Johnson. It conducted a comprehensive analysis of the causes of the riots and provided a broad set of recommendations to prevent future ones, in a report issued in 1968. It concluded that the "nation is moving toward two societies, one Black, one White—separate and unequal." And that "White institutions created it, White institutions maintain it, and White society condones it."[44] Its recommendations proposed what communities, the police, and the media should do, as well as improved policies in employment, education, the welfare system, and housing. This was not the report that President Johnson wanted, as he had become embroiled in the Vietnam war tragedy.[45] Nevertheless, the language of the report resonated, as I discuss later.

Alternative Kinds of Strategies

The drive toward African American integration into White society was not the only approach that African Americans always chose in order to achieve more equality and justice. More nationalist and separatist paths were sometimes urged, and to some degree were tried over time. This was the case in the 1960s. This is particularly relevant for status inequalities, since they entail devaluing people held to be in lower regard. One possible response is to accept those views and hold oneself in low self-esteem. That is painful, even self-hating at times, and often hampers acting effectively to improve one's status. The prevalence of skin lightening and hair straightening products in earlier decades was evidence of such sentiments.

In the 1960s, some social movements among African Americans emphasized Black beauty, power, and pride and practiced some measure of separation from Whites. This was a way of reacting to and managing lower status ranking. It took, and continues in, various forms. Many African Americans in the 1960s adopted the phrase "Black is Beautiful" to counter the negative ideas associated with being dark-skinned.[46] Afro and other natural hair styles became popular.

Relatively separatist and/or militant social movement organizations also became important. For example, the Nation of Islam (NOI), which had been founded in the 1930s, had been small and isolated, but it rapidly expanded in the 1950s and 1960s. Malcolm X, who learned about it while in prison from 1946 to 1952, joined the NOI, and is generally credited with its rapid expansion in membership. In March 1964, however, Malcolm X left the NOI due to his shift to more traditional Islam and his interest in working with other civil rights leaders.[47] On February 21, 1965, he was assassinated. In March 1966, three NOI members were convicted of the assassination. In 1965, following Malcolm X's death, Louis Farrakhan emerged as his successor.

In 1966, Bobby Seale and Huey Newton founded the Black Panthers in Oakland, California. Its activities included armed citizens' patrols intended to monitor the conduct of police officers and reduce police harassment and misconduct. It also conducted various community programs, notably the Free Breakfast for Children programs. The FBI and some local police departments conducted surveillance, infiltration, harassment, arrests and imprisonment, and even assassination to destroy the party.[48]

A particularly bloody conflict escalation was a Chicago police raid in 1969. Two years earlier, the FBI had identified Fredrick Hampton as a radical threat. Hampton was prominent in Chicago as chairman of the Illinois chapter of the Black Panther Party. On December 4, 1969, during a predawn raid at his Chicago apartment by the Chicago Police Department, Hampton was shot and killed in his bed. During the raid, Panther Mark Clark was also killed and several others were seriously wounded. In 1970, a coroner's jury held an inquest and decided the deaths to be justifiable homicides.

Briefly, the suppression produced supporters and membership grew, peaking in 1970, but membership then declined as Panther infighting grew and the group became isolated. Covert FBI operations contributed greatly to the internal disputes. By the early 1980s, the Black Panthers had dissolved. The legacy of a destructive conflict remains, however.

Whatever the good intentions of the Black Panther Party, and the sense of pride it gave its members, its militancy and separatist ideology did not win many allies. Indeed, J. Edgar Hoover, the FBI director, and local police departments viewed and depicted the Panthers as great threats and were not constrained in suppressing them. The outcome was not constructive in advancing American ideals of freedom and justice. It is not difficult to imagine the conflict being waged less destructively. The Panthers might have sought allies for steps to counter extreme incidents of police misconduct against African Americans. On the other hand, officials and nongovernmental organizations in Oakland and Sacramento might have responded more positively to the concerns raised by the Panthers. Officials in Washington might have exercised more oversight and control of Hoover and the FBI he directed.

One aftermath of the police raid should also be noted. The survivors and relatives of Hampton and Clark filed a civil lawsuit, which was resolved in 1982 by a settlement of $1.85 million. The City of Chicago, Cook County, and the federal government each paid one-third to a group of nine plaintiffs. Revelations appeared that tied the killings to the illegal FBI COINTELPRO program.[49] Documents associated with the killings suggest the Hampton's death was an assassination under the FBI's initiative, not a justifiable homicide. The killing of Hampton had many reverberations. In 1990, the Chicago City Council unanimously passed a resolution, commemorating December 4, 2004, as Fred Hampton Day in Chicago. The resolution read in part: "Fred Hampton, who was only 21 years old, made his mark in Chicago history not so much by his death as by the heroic efforts of his life and by his goals of empowering the most oppressed sector of Chicago's Black community, bringing people into political life through participation in their own freedom fighting organization."[50]

Implementation

In some arenas, agreements that were reached after direct actions, court decisions, and legislation to reduce segregation and discrimination did in fact reduce them. Sometimes, however, nothing much changed or the circumstances may even have reverted to what they were before the seeming end of the dispute. On the whole, many of the decisions won by the civil rights organizations in the South had some degree of realization, particularly those relating to the use of public accommodations, to the increase in the integration of schools, and even to participation in voting. Furthermore, the national legislation and court decisions were implemented to some degree throughout the United States.

The racial integration of public schools was a major focus of attention. The 1954 Supreme Court *Brown* decision did not immediately produce integrated education for all African American children. Indeed, many White politicians in the South initially undertook what some called "Massive Resistance," even siphoning state taxes to support private White academies.[51] The fight was on, and was largely waged within the three branches of the federal government, pressured by various nongovernmental organizations and activists.

The segregationists scored a victory by securing a provision in the 1964 Civil Rights Act prohibiting busing to overcome racial imbalance. Little action was taken to desegregate the schools until African American protests prompted desegregation actions starting in the early 1960s. Some cities and states began to introduce plans to desegregate schools. In addition, federal courts, spurred by the NAACP began to oversee school desegregation. For example, in 1962, in Syracuse, New York, the local CORE chapter and other activists conducted several actions to protest the Board of Education's failure to develop plans to desegregate the city's highly imbalanced schools.[52] In response, in August 1962, the Syracuse mayor established an Education Study Committee, under the aegis of the State Commission for Human Rights, to discuss the issue. Its members included school board members, civil rights leaders, and staff members from the State Commission. In June 1963, the New York Commissioner of Education directed all school districts to report on the degree of imbalance and their plans to end it. In July, the Education Committee reported that there was a pattern of imbalance and that it could be corrected by modifying school boundaries.

The Syracuse Board of Education promptly implemented a plan to modify school boundaries, so as to somewhat reduce segregation and overcrowding in a school that had predominately African American students. In the following academic year, 1964–1965, more extensive boundary-shifting and school closures were carried out and student transfers were implemented to reduce segregation. Nearly all students who were transferred were African Americans who entered schools that had predominately White student bodies. There was some grumbling, particularly among African American parents whose children bore the major burden of busing. The rapidity of the integration steps resulted in some problems in preparation. Helpfully, some Whites volunteered to ease the difficulties for students socializing across ethnic lines.[53]

In the following year, two schools that had predominantly African American students were closed, and their students were bused to schools that had predominantly White students. Considerable attention was given to winning public support for the changes and to assist African American students who lagged in academic attainment. The consequences of these changes were carefully monitored, and the studies found that the African American students who had been transferred into predominantly White schools had improved academic competencies compared to African American students who remained in largely African American–attended schools. There was no evidence that White students in schools with African Americans bused in were adversely affected compared to White students in schools without such transfers.

During the late 1960s, and into the 1980s, segregation substantially decreased across the country, particularly in the South.[54] Only in the Northeast, which started as relatively integrated, did segregation actually increase during those years. School integration generally increased as a result of the policies of local school boards, but often in response to court orders and to protests against segregation. Importantly, in 1968, the Supreme Court, in Green v. County School

Board of New Kent County, ruled that school districts must adopt more effective plans to reduce segregation. Court-ordered desegregation plans followed.[55]

The busing of students had increased for many years to consolidate schools into larger and better-resourced facilities. Then busing students from one neighborhood to another became a primary way to integrate schools. In some places there was resistance to this, but on the whole it became a common practice. Schools became less segregated until 1988, when segregation actually began to increase. I examine that resegregation trend in the next chapter.

Government housing policies and market processes sustained residential segregation, as discussed earlier in this chapter. The ill effects of such segregation were increasingly recognized by some political leaders in the 1960s and were noted in the 1968 Kerner Commission Report. President Johnson had introduced fair housing legislation in 1996, but passage was blocked by filibusters made by Senate opponents in 1966 and 1967. In 1968, Senator Walter Mondale successfully led the fight to overcome the filibuster and ultimately the bill was passed by both Houses of Congress and signed into law.[56] The great attention given to the Kerner Report, the assassination of King, and the subsequent riots all spurred sufficient concerns about the discrimination suffered by African Americans to win enough votes to enact the Fair Housing Act of 1968. This indicates that opponents of more equality, by choosing to kill a leading proponent of it, acted counterproductively. The legislation, among other provisions, outlawed refusing to rent a dwelling to any person because of race, and it outlawed racial discrimination in the terms, conditions or privileges of the sale or rental of a dwelling.[57] Despite passage of the Fair Housing Act of 1968, other actions of the government were conducive to increasing racial segregation and have had negative consequences for the quality of African American housing, as discussed earlier. Another such set of actions relates to the construction of new highways, which often cut through neighborhoods where African Americans lived.[58] Little attention was given to losses suffered by the displaced people in those neighborhoods. Furthermore, they were built to ease the way for suburban dwellers to live in White-inhabited areas and commute to work. These highway projects were often related to urban renewal projects.

In 1949, the federal government, and some states, began to help cities clear areas near the downtown centers that were primarily occupied by African Americans.[59] The displaced African Americans, however, received little assistance in obtaining new housing, while the cleared land was then opened for new developments that were attractive to higher-income people. The changes were portrayed as widely beneficial. These changes, however, contributed to discrimination, segregation, and inequality. Displaced African Americans who did own property lost their equity and faced difficulties in obtaining new housing in the limited areas available.[60]

No national movement of resistance arose to prevent or even mitigate the ill effects of urban renewal projects. In many localities, however, protests were mounted. CORE leaders sought to mobilize the people living in areas that were to be cleared or were already being displaced, but they had only limited success.[61] Protests were mounted in many localities across the country, but they could not claim victories that might have inspired greater resistance and win more support. For example, I outline how the conflict was waged in Syracuse, New York, in the early 1960s. The construction of Highway I-81 through the city and the nearby expansion of the Upstate Medical Center entailed the relocation of people in areas of high African American concentration (the 15th ward). The new construction had resulted in the relocation of 488 families by the summer of 1963; and the planned East Side Urban Renewal Project was anticipated to result in the removal of 838 families more.[62] In addition, 394 individuals out of the 632 who were to be relocated had already moved. At the same time, Syracuse was experiencing major demographic changes. Between 1950 and 1960, the non-White population increased from 5,068 to 12,281, while the White population decreased by 11,768.

While Syracuse officials stressed moving people from substandard housing to housing that met minimal standards, many people were concerned that African Americans, constrained by housing discrimination, would suffer severely in the relocation process. Local chapters of CORE, NAACP, and the IUE (International Union of Electrical Workers) Civil Rights Committee joined together to minimize the damages and reduce segregation. In June 1963, representatives of CORE met with the mayor, William. F. Walsh, presenting a program to improve the relocation process and threatening direct action if it was not adopted. George Wiley was influential in choosing a confrontational direct-action strategy.[63] In August, civil rights groups marched on the County Court House, and days later they began a long series of direct actions at urban renewal demolition sites. The mayor responded by proposing establishing a broad civil rights commission, but he did not lay out a program of actions. The direct actions escalated in numbers involved and in tactics used through September 1963; CORE tactics in many other cities were also becoming more radical.[64] Many Syracuse University faculty and students joined the frequent protests at demolition sites, which included chaining themselves to bulldozers.[65] Once faculty and students and other protesters were getting arrested, the local newspapers headlined the protests. To some extent the protests were the story, rather than the solutions that were being proposed to fix urgent problems.

Certainly, attention was raised about the plight of the African Americans who were being relocated to nearby neighborhoods that were an expanded ghetto. A broader range of tactics might have produced less segregation and better housing. Closer work with the people who were being relocated might have made discrimination by banks and real estate dealers more visible, more challenged, and less likely. Closer work with local churches and New York State agencies might have reduced the unfortunate consequences of the destruction of the 15th ward.

Conclusions

Constructively waged fights did contribute to significant progress toward more equality between African Americans and Whites, especially in the South. Perversely, destructively waged conflicts by supporters of segregation often backfired and actually contributed to progress. Moreover, relatively fierce contentious efforts to overcome or simply reduce the great status inequalities of African Americans relative to Whites were often ineffective or even counterproductive. Having good intentions, alone, is not enough to make progress. It is necessary to recognize that the American creed and the values asserted in the American Declaration of Independence and in the Pledge of Allegiance are not universally shared as the primary goals for social policy.

Conduct that did not rise to the level of overt conflict also contributed to making progressive change. People striving to live and work in ways that do not harm others, with due consideration of their rights, do contribute to better relations among all ethnic groups. On the other hand, harmful changes can occur without overt conflict when people act for immediate, narrow self-gains. The next chapter examines more and less constructive ways to make constructive changes in the status of African Americans since the 1960s.

Considering the progress toward greater equality for African Americans beyond the place and time of the southern civil rights struggle, the evidence presented in this chapter indicates limited progress. Government policies, market forces, and widespread prejudice sustained considerable discrimination and segregation. However, progress is multidimensional. Progress in many arenas for many people can occur alongside stagnation and even backsliding for other people in some realms. The next chapter analyzes that complexity, after 1969.

6

African American Advances and Backlashes, 1970–2022

This chapter discusses several arenas in which conflicts regarding the status of African Americans have been waged since 1970. I examine the subsequent development of the fights analyzed in the previous chapter, and also some newly salient ones. The conflicts pertain to equality in education, housing, justice, and well-being.

To begin, it is useful to mention elements of the post-1970s context for efforts to raise the status of African Americans since 1970. First, I note Whites' changing attitudes about African Americans and policies toward them. In general, Whites' attitudes have become much more tolerant and sympathetic since 1970.[1] Asked whether the country was spending too much money, too little money, or about the right amount for improving the conditions of Blacks, about 30 percent of the respondents said "too little" between 1973 and 2014, but this rose to 52 percent in 2018. The percentage responding "too much," fluctuated around 20 percent beginning in 1973 and then gradually declined to 7 percent in 2018.

The following question was asked seven times between 1988 and 2018: "On the average (Blacks/Africans-Americans) have worse jobs, income, and housing than White people. Do you think these differences are 1. Mainly due to discrimination? 2. Because most . . . have less in-born ability to learn? 3. Because most . . . don't have the chance for education that it takes to rise out of poverty? 4. Because most . . . just don't have the motivation or will power to pull themselves up out of poverty?"

Interestingly, in 2018, 50 percent of the respondents chose lack of access to education and 45 percent said discrimination—both choices having increased a little since 1988. On the other hand, 36 percent chose lack of motivation and only 8 percent said inborn ability—both choices having declined since 1988. Overwhelmingly, Whites did not attribute the poor conditions of most African Americans to their own failings. Nevertheless, in 2018, 57 percent agreed that African Americans should *not* receive special treatment to make up for past discrimination. That was a substantial decline from a little over 70 percent in 1994. The question phrasing may have implied a zero-sum circumstance, adversely affecting Whites.

Fighting Better. Louis Kriesberg, Oxford University Press. © Oxford University Press 2023.
DOI: 10.1093/oso/9780197674796.003.0006

Many important structural conditions occurred that also impact the status of African Americans. Across the country, urbanization and suburbanization continued to increase. Ethnic diversity expanded, with Hispanic and Asian immigration. The US class structure was growing ever more unequal, as the wealth and income of a small elite increased, while middle- and working-class salaries and wages were relatively stagnant. Overall, Whites and African Americans have large differences in wealth, earnings, homeownership, and rates of unemployment, rising and falling in tandem.[2] More specifically, the financial crisis of 2008 greatly reduced income and wealth of Whites and of African Americans, but African Americans did not experience the recovery that Whites did, increasing the wealth and income gap between them.

I now turn to four arenas of conflict—education, housing, justice, and well-being—and examine how well or badly they were waged. The above noted changes constitute a context for the way specific fights were conducted.

Equal Education

Education is crucial for most paths to good employment and respect. Its attainment by all Americans is a widely shared American goal. Nevertheless, many intense conflicts have been fought about how to attain more educational equity. Many of those fights have focused on some aspects of the schools, and not so much on the social conditions in which African Americans live, even though there is overwhelming evidence the social and economic factors significantly impact educational achievement. That may seem less threatening for some Whites than confronting major structural circumstances, but that limits the effectiveness of the attempted educational policies.

School Integration

The previous chapter reported that significant progress toward increasing school integration did not happen until the late 1960s. Then, segregation of African American students in public schools did begin to decline, but only until the late 1980s, when segregation began to increase. The magnitude of the changes depends upon which alternative measure of segregation is applied.[3] Two methods are most commonly used: measures of exposure and measures of unevenness. White and African American exposure measures indicate the proportion of African American students attending largely White schools. White and African American unevenness measures indicate the proportion of African American (or White) students who would have to change schools in a district so that the schools would have similar proportions of African Americans (or Whites). By both kinds of measures, segregation declined nationally from 1968 to 1987.

The southern civil rights struggle, followed by national legislation and court decisions, sought to integrate the African American and White students together, so they would have equally good educations and opportunities. Indeed, progress was made into the 1970s. Research documents that African American students' educations benefitted by attending schools that were predominantly attended by White students.[4] White students' educational benefits were not affected by the attendance of African American students.

Most of the rise in integration occurred between 1968 and the mid-1970s. Little action had been taken to desegregate public schools for many years after the 1954 Supreme Court *Brown v. Board of Education* decision. Nevertheless, in the mid-1960s the federal courts, spurred by the NAACP, began to impose school integration plans. It followed the Supreme Court's 1968 decision in *Green v. County School Board of New Kent County*, which required school districts to adopt desegregation plans that were likely to be effective. Court-ordered school integration plans became widespread, but the court orders expired, and the plans were not generally continued. Despite the 1954 *Brown* decision, federal integration plans were attacked by Ronald Reagan's and George H. W. Bush's Justice Departments and a Supreme Court with William Rehnquist as Chief Justice. White resistance and other developments tended to undermine school integration, and by 1988, segregation began to increase across the country.[5]

The degree of resegregation between 1988 and 2019 varies with the measure used. The measure of African American exposure to White students seems most important for realizing the value of school integration. Schools with many White students are likely to have more educational resources and students with backgrounds that are conducive to studying and achieving well in their education, compared to schools in predominantly or entirely low-income African American neighborhoods.

By the late 1980s, the proportion of White students in public schools was declining for a few reasons. First, it followed from the declining White birth rate. Additionally, because of the increasing income inequality, discussed in earlier chapters, more Whites could afford to move to suburban homes or send children to private school. Fundamentally, segregated housing, as described in the previous chapter, was widespread in the United States. This was sustained by federal, state, and local government laws and rules. Finally, developers and real estate salespersons often promoted and sustained racially segregated housing.

Residential segregation by ethnicity and by income have especially dire consequences for the equality of public schools. In America, public schools are largely controlled and funded by local school districts. Poor school districts lack the abundance of resources that endow schools in wealthy districts with diverse learning opportunities. Some federal and state funds are often provided to lessen those inequities, but they are too small to make significant changes, and under President Trump they were further reduced. In local instances, however, considerable improvements have been made.[6]

Another major development was a change in the ethnic composition of the United States. For much of American history, the public schools had been where immigrants from diverse European countries and their descendants all melted together to form the American nationality. America's ethnic composition was impacted by changes in immigration sources. Immigration into the United States underwent several major shifts. Beginning in the early 1900s, immigration shifted from northern and western Europe to southern and eastern Europe. In 1921 and 1924, laws were enacted to cap the total level and impose quotas favoring immigration from northern and western Europe. Restrictions began to fall in 1943, when a limited number of Chinese were allowed entry. In 1965, country quotas were ended and family reunification and skilled immigrants were favored. Immigration shifted away from Europe and grew from Asia and particularly from Latin America.[7]

Between 1970 and 2016–17, the proportion of Whites in public schools greatly declined, from 79.1 to 48.4 percent; the proportion of Latinos in public schools rose, from 5.1 to 26.3 percent; the proportion of African Americans changed little, going from 15.0 to 15.2 percent; and the small number of Asians rose, from .5 to 5.5 percent.[8] The ethnic composition of public school students by 2018 certainly makes integration of African American with White students demographically less likely. African American students too often attend schools in poor neighborhoods, which lack the educational resources of schools attended by White students.

The conflicts between supporters and opponents of school integration were largely conducted through institutionalized methods. The fights were conducted within all three branches of the federal government and also significantly at the state and local level. Various nongovernmental organizations and activists as well as officials were engaged. In addition, the conflicts were conducted within political and ideological contexts as well as major residential and immigration trends. Some White parents evaded integration by sending their children to parochial and other private schools. Many other families simply moved away from the neighborhood where schools were being integrated, moving out of cities to largely White suburban neighborhoods.

Very significantly, political leaders at the state and local level engaged in policies that had major impacts on the progress of school integration. In the 1970s, many officials took actions that supported people who resisted such progress and sustained segregated public schools. For example, the Republican governor of Michigan, William Milliken, and Detroit city officials enacted policies to increase racial segregation in schools. In August 1970, the NAACP filed suit against Governor Milliken and other state officials, and cited a close relationship between unfair housing practices and educational segregation. After a lengthy trial beginning in April 1971, District Judge Stephen J. Roth held that the State of Michigan and the school districts were accountable for the segregation. Then, the Appeals Court withheld judgment on the relationship between housing segregation and education, but specified that it was the state's responsibility to integrate across the segregated metropolitan area. The accused officials appealed to the Supreme Court.[9]

In its 1974 *Milliken v. Bradley* 5–4 decision, the Supreme Court overturned the lower courts' decisions, holding that school districts were not obligated to desegregate unless it had been proven that the lines were drawn by the districts with racist intent.[10] The majority argued that because government policies had not segregated Detroit schools, the suburbs could not be part of a government-mandated correction. This consequential decision has blocked government efforts to extend school funding across district lines, which would help equalize funding of public schools. It protected White flight to the suburbs from helping pay for schools in other districts. Yet, as the analysis in the previous chapter documents, federal government policies actually have long helped impose segregation in housing. That analysis drew heavily from Richard Rothstein's *The Color of Law*; Rothstein wrote that book to prove, by the Supreme Court's own reasoning, that the US government is obligated to support policies to reduce housing segregation.[11] Allocation of tax funds for public schools is a complex political issue and is not well suited to winner-loser court decisions.

From the outset, in many localities, Whites organized to resist school integration. Overt conflicts erupted in the 1970s against court-ordered desegregation plans. A particularly intense fight was waged in Boston between 1974 and 1988, which received much national mass-media attention, contributing to the difficulties of relying on busing to advance school integration.[12]

Boston's highly visible school integration conflict began in 1965, when the Massachusetts General Court (the state legislature) passed the Racial Imbalance Act. The act required that public schools not have more than 50 percent of their students be non-White, or risk losing state educational funding.[13] The Boston School Committee promptly refused to support implementing the legislation. In 1972, the NAACP filed a lawsuit (*Morgan v. Hennigan*) against the Boston School Committee, alleging the public schools were segregated. In June 1974, Judge W. Arthur Garrity found that the schools were unconstitutionally segregated, and the Supreme Court upheld the ruling.[14] To implement the Racial Imbalance Act, Garrity adopted a busing plan developed by the Massachusetts State Board of Education. This included busing students between a high school in a highly segregated African American neighborhood and a high school in a poor White neighborhood.

Several community organizations played significant roles in developing and implementing efforts to increase the integration of Boston schools. The depth and breadth of the actions, however, did not engage opponents in preparing for and taking graduated steps. The Citywide Education Coalition was established in January 1972 in regard to appointing a new Boston school superintendent and went on to engage in providing information related to racial imbalance in the schools. Following the June 1974 court order, it established a center, published a newsletter, and organized informational meetings about the imbalance. This was funded by a $78,000 grant by the Federal Law Enforcement Assistance Administration.

In July 1974, neighborhood groups opposing school integration plans unified in a Boston-wide militant organization, adopting the name ROAR (Restore Our Alienated Rights). Its executive board, chaired by city council member Louise Day Hicks, met weekly in the city council chambers. ROAR organized school boycotts and protest rallies throughout the city.

In December 1974, the Boston Home and School Association, which had been founded in 1906 to foster cooperation between students' homes and schools, was granted intervener status in the 1972 *Morgan v. Hennigan* lawsuit. Thereby, it was able to submit suggestions and offer modifications and proposals for the plan adopted by the Court.

The NAACP had brought the case that resulted in a court desegregation plan. It continued to advocate for school desegregation and make suggestions to further advance it. Other community organizations conducted programs that directly served to cooperatively increase desegregation. For example, in 1965, African American parents established "Operation Exodus." They transported their children from schools in the neighborhoods in which they lived to schools in predominantly White neighborhoods. To start, 400 students were transported, and in 1970 the number had risen to 1,100. However, the African American students were often met with hostility and sometime refused attendance.

Another organization was more successful in directly increasing school integration. It was the Metropolitan Council for Educational Opportunity (METCO), which began operations in 1966. It was designed to improve the educational opportunities for African American students in Boston's low-performing school districts and to increase the diversity of predominately White suburban schools. Originally funded by the US Department of Education and the Carnegie Corporation, the program continued with Massachusetts state funding. In 1966, the first 220 students, aged five to sixteen, rode buses from Boston neighborhoods to schools in seven suburbs that had volunteered to participate: Arlington, Braintree, Brookline, Lexington, Lincoln, Newton, and Wellesley. The program has greatly expanded its operations.[15]

The fight in Boston about reducing school segregation was intense, violent, and destructive, particularly from 1974 to 1976. More constructive undertakings began to be undertaken, which were aided by consultations and mediations by Charles V. Willie, who joined the faculty of Harvard University School of Education in 1974.[16] He served as a consultant and court-appointed master in major school desegregation cases in various large cities, including the landmark case of the 1988 "Controlled Choice" plan designed by Willie and Michael Alves and used in Boston for ten years.[17] The controlled-choice approach divided Boston into three zones in which parents would list their three top-choice elementary schools. In 2013, a new plan greatly reduced busing. By 2014, metropolitan Boston public school enrollment was 64 percent White, but within Boston, it was only 13 percent White, making meaningful integration difficult.

Another method to bring about school integration was to create magnet schools in low-income African American neighborhoods, which would be attractive to White parents so they would choose to send their children to them. This was undertaken widely in the 1970s, often with innovative learning. However, to some extent the schools had two curricular tracks, one preparing students for college and the other not, resulting in considerable separation of White and African American students within the schools.

Nationally, this magnet school strategy faded. The increase in specialized schools at the high school level, in the arts, sciences, and vocational subjects, has become an alternative way to draw multiethnic students from larger residential areas. The failure to equalize African American education by integration of schools prompted some people who sought greater equality to at least equalize funding for schools serving predominately African American students. Some of the benefits from integrating schools are attributable to the greater funding and the resources it can buy in predominantly White schools.

School Funding

Extensive research documents that increased school spending is causally related to improved student outcomes.[18] Moreover, as noted in chapter 4, funding for public schools hugely varies among the states and local school districts. This is attributable to the American practice to fund public schools largely with school district and state tax revenues. Funding for schools varies greatly among the school districts within each state and also among the states, which compounds possible disparities. Each state government can allocate its school funds to alleviate the disparities among the school districts in its state. That was the case in only twenty states, where high-poverty districts received at least 5 percent more combined district and state funds per capita than did affluent districts, according to a 2017 report.[19] In twenty-three states, the same amount of money was allocated to every school district. In four states, the highest poverty districts were actually allocated *less* than were the more affluent districts.

The four most progressive states, Delaware, Minnesota, Utah, and Ohio, provide their highest-poverty districts, on average, with between 27 and 81 percent more funding per student than their lowest-poverty districts. In contrast, the most regressive states gave significantly less funding to their highest-poverty districts. In Illinois and North Dakota, high-poverty districts get only about 80 cents for every dollar in low-poverty districts, while in Nevada high-poverty districts receive only 71 cents to the dollar.

Some of the difference in school funding are due to varying taxable income and wealth of districts and states. However, many of the disparities depend upon political and ideological conflicts. Public attitudes about taxation, the size of government, equality of opportunity, the value of education, and the rights of poor people and African Americans matter greatly. Vested interests of dominant persons and groups also influence the disagreements about the raising and allocation of tax revenue.

Many efforts have been made to fund public schools more equally within and among the states, sometimes effectively.[20] The NAACP led in litigation efforts, but with limited success. State legislatures have at times played important roles equalizing school funding within their states.[21] Mobilization of public pressure within regressive states might be effective. Coalitions of parents and teachers within states that underfund poor school districts could be significant change agents. Finally, national funding should be increased to give more assistance to states relatively underfunding school districts with high poverty rates. As will be noted later, however, state governments may reduce their own funding of education, nullifying the federal aid. Targeted federal funding should be considered. For example, students tend to have better educational achievement in smaller-sized classrooms than in larger class sizes. Funding specifically to reduce very large classes should be considered by foundations and the federal government.

Improving Achievement

As the battles raged on, the framing of the discussion shifted from integration to funding to achievement. In the 1980s, attention in the United States began to focus on improving student educational achievement, as measured by standardized tests. In 1981, the US Department of Education established the National Commission on Excellence in Education, which issued its report, *A Nation at*

Risk, in 1983.[22] The report stated that all American children, regardless of race or class, are entitled to a fair chance and the tools to develop their individual minds. The report presented evidence of the failures to fulfill that entitlement. For example, about 17 percent of all US seventeen-year-olds can be considered functionally illiterate. Among minority youth, it may be as high as 40 percent. Furthermore, scores, as measured by the College Board's Scholastic Aptitude Test (SAT), consistently declined in verbal, mathematics, physics, and English subjects. The report made recommendations to raise the requirements for graduation from high school, to raise the expectations about the level of knowledge to be attained, to increase the time spent on school work, and to improve teacher preparation and/or make teaching more rewarding.

The interest in advancing standards-based education and assessment grew and became manifested in the Improving America's Schools Act of 1994 (IASA). It reauthorized the Elementary and Secondary Education Act of 1965 (ESEA), which was part of President Johnson's War on Poverty. The ESEA had focused federal funding on improving education of disadvantaged children in poor areas, and IASA expanded coverage to all students and required regular assessments. But by 2000, a testing industry was flourishing, and most states had instituted regular testing in large-scale assessments.

This culminated in the No Child Left Behind Act (NCLB), signed by President George W. Bush in January 2002. It proposed that educational reform for all students be driven by public accountability, measured by standardized testing. Federal funding would be provided for the testing and for locally selected ways to improve educational outcomes. The NCLB required schools to make Adequate Yearly Progress (AYP) in standardized test scores.[23] Stiffer consequences followed if a school consecutively failed its AYP. After a fifth consecutive year of failures, a restructuring plan would need to be implemented, which might include closing the school, converting it to be a charter school, hiring a private company, or other transformation.

The many problems of relying too much on standardized testing were raised by many persons concerned with improving education.[24] Among the matters raised are the following concerns about the effects of high reliance on using such testing: it fosters teaching for the test, neglects nonverbal subjects, reduces attention to a variety of ways of teaching and learning, is vulnerable to cheating, drains money from classroom needs, and entails biases of social class, ethnic, and other cultural variations.

In any case, NCLB was a very expensive undertaking; the states spent $1.7 billion per year on testing, compared to $423 million during the year before NCLB began.[25] In December 2015, President Obama signed the Every Student Succeeds Act (ESSA), with bipartisan support. It allowed much more leeway to the states and reduced the heavy reliance on testing. The states were required to submit accountability plans to the Department of Education to become implemented in the 2017–18 school year.

Throughout the many years of legislation and public concern about primary and secondary education, lagging educational achievement of African American students was manifest. Overall, there has been some advancement in the achievement of African American students, but the gap between White and African American students has generally remained, with a little narrowing, at times, by one or another measure. For example, with the increasing importance of completing high school, the completion rate for African Americans rose from 27 percent in 1964 to 88 percent in 2015.[26] This largely closed the gap with Whites, whose completion rate rose from 51 percent in 1964 to 93 percent in 2015, but completing college had become needed for middle-class jobs. The proportion of African Americans earning a BA degree increased from 4 percent in 1964 to 23 percent in 2015. However, Whites also greatly increased their college completion rates, rising from 10 percent in 1964 to 36 percent in 2015, thus maintaining a large gap. African American attendance at elite colleges and universities did not increase; it has been flat, despite efforts at affirmative action and increasing diversity.[27]

The college experience for many college attenders, however, is disheartening. Nearly 40 percent of all college undergraduates drop out without a degree.[28] Rising tuition costs and decreasing government assistance is further increasing the obstacles for non-affluent African Americans to graduate. Furthermore, the burdens of college student debts are immense, especially for African Americans.[29]

The Historically Black Colleges and Universities (HBCU) have long been major providers of higher education for African Americans. Morehead State University and Howard University are prominent examples. Although HBCUs serve just 0.1 percent of the overall student population, they account for 20 percent of African American students who complete bachelor's degrees.[30] African American graduates of HBCUs are very well served by them, compared to African American graduates of other colleges.[31] They are much more likely to report that their professors cared about them as persons. Significantly, the tuition revenues at HBCUs are less than at other colleges.

Much attention about African American educational achievement has been focused on test scores in grade schools, in reading, mathematics, and the natural sciences. The National Assessment of Education Progress (NAEP) began conducting annual tests in 1969. It is congressionally mandated by the National Center for Education Statistics (NCES), functioning within the US Department of Education's Institute of Education Sciences. Data from national tests are the bases for much of the analysis of trends in White and African American academic progress in reading and mathematics. Overall, mathematics scores increased between 1996 and 2017, and reading scores rose only slightly. During the same period, the gap between the mathematics and reading scores of White and African American students persisted but had decreased to a small degree.[32] It is hard to attribute these changes to federal education policies or to other changes in the circumstances of African Americans' well-being. I discuss the changes in various measures of well-being later in this chapter.

Governmental educational policies are highly focused on the schools and their classrooms. However, it is well known that educational achievement is profoundly influenced by the home, neighborhood, and society at large. As the 2020 pandemic revealed, African Americans were much more likely to become sick and die from the COVID-19 virus. African Americans tended to have many underlying health conditions, live in crowded conditions, and work in dangerous settings, all of which increased their vulnerability. African American children are also more likely than their White counterparts to be frequently hungry and ill and live in crowded dwellings, which hampers study and school attendance. Furthermore, the prospects of good educations leading to good-paying jobs are hardly certain, reducing the motivation to work hard at school studies.

Many government and nonofficial programs jointly can mitigate these conditions. One such federal government policy was established by the National School Lunch Act in 1946, which continues to provide good food in public schools. In many localities there are community recreation centers, health centers, libraries, and local organizations that provide mentoring services. Much more is needed, however, including better housing and better-paying jobs.

Recognizing the multiple ways that educational achievement among African American students is hampered, multiple interconnected changes are needed. Each of these approaches may be resisted by some Whites; their prejudice, mistaken self-interest, and zero-sum thinking contribute to opposing investment in improving the circumstances of African American citizens. Countering such thinking is needed in order to improve African American students' achievement by improving the students' circumstances. Engagement by those who are aggrieved can be important and should be encouraged.

It is widely recognized that high levels of parental involvement can provide many benefits for students, and that policy programs can enhance such conduct.[33] For example, child care, summer school, and out-of-school programs have been effective in improving students' educational achievements. These are of particular value for African American students.[34] They occur in some forms at the national, state, and local level, by governmental and nongovernmental auspices. Raising the scale of such efforts would be helpful, and I discuss this after considering the importance of improving the competence of teachers.

Teachers' competence and certain teacher attitudes and beliefs are related to good test scores and grades by African American students. Research indicates that teachers' test scores are strongly associated with how much students learn.[35] Furthermore, schools with predominantly White students have a higher proportion of competent teachers than do schools that have predominantly African American students.

In addition, teachers' general ways of thinking are related to African American students' educational achievement. Thus, some teachers and parents believe that genetics are highly related to educational achievement, while others believe that effort and hard work are highly related to educational achievement. An emphasis on innate ability is likely to have negative consequences for African American students. For example, experimental research compared teachers of STEM courses (science, technology, engineering, and mathematics) with different beliefs about ability, particularly whether it was fixed or that it could be changed.[36] Students of teachers who adhered to the idea that ability could change improved in their grades more than did the students of teachers who believed in innate ability. This difference was especially strong among African American students.

Finally, I consider some possible strategies that can help overcome the conditions that hamper African American students attaining equal educational levels of achievement. I focus first on parental engagement and then on teacher competence. Preschool child care has profound effects on subsequent educational achievement; for example, talking and reading to a young child helps build the child's language skills. The Head Start program began as part of President Johnson's War on Poverty, first as a summer program, in 1965, and then as a year-round preschool program in 1996. In 1994, the Early Head Start program for children from birth to three years of age was established. These programs are designed to help communities overcome some of the problems of disadvantaged children. They help link families to healthcare providers and other community resources. Research documents that the comprehensive programs benefit preschoolers when they go on into schools.[37] A variety of research by James J. Heckman and others documents that demonstration programs for disadvantaged children have beneficial long-term benefits for education and employment.[38]

Research also finds, however, that the positive benefits of Head Start fade some after a few years in the public schools. A new federal program might adapt what has been learned from the preschool work to assist disadvantaged children in primary grades of school. This would entail working with the parents and teachers of the students, perhaps in local workshops and meetings, dealing with matters that concern the people in the community and school.

Improving the competence of teachers would be another avenue for advancing the educational achievement of African American students. Improving the training of teachers and raising their salaries and working conditions would help attract and retain excellent teachers. They are pathways to progress, but matters of ideology and short-term self-interest often obstruct improvement. I stress a particularly important matter of contention: teacher unions. For example, among the numerous Republican victories in the 2010 elections, right-wing Scott Walker became governor of Wisconsin and the Republican Party won control of the state legislature. They quickly moved to enact Act 10, which drastically reduced collective bargaining rights for teachers and other public-sector workers and slashed their benefits.[39] The fight was ferocious and the impact great, but I focus only on the effects on the teachers and analyze how the fight might have been waged more constructively.

Among the many provisions of the 2011 Act 10, it mandated annual recertification, prohibited paycheck deductions for dues, limited raises in base pay to inflationary increases, and cut benefits. This was accompanied by tax cuts to the wealthy, which were to be offset by reduced school budgets and increased revenue from private-sector growth. There was some growth, but less than in the country as a whole. The consequences of these and other Act 10 provisions on Wisconsin teachers and other public-sector workers were quick and severe.[40] Average benefits paid to Wisconsin teachers fell almost 19 percent between the 2010–11 school year and the 2011–12 school year. Voluntary teacher turnover rose. More teachers left the profession than in the previous year. More teachers left the district. More of the remaining teachers had less experience as teachers. All of these changes would be deleterious to student learning, and indeed student test scores declined. Significantly here is that the decline was particularly large among the schools that were already lagging behind.

The large Republican victory against the teacher and other public-sector unions was in part a result of the prior successes by the teacher unions in Wisconsin.[41] The unions in Wisconsin had been militant and possessed considerable political clout for many years. However, in 1993, under the Republican governor Tommy Thompson, Wisconsin set limits on collective bargaining, and in 2008, teacher salaries declined 6 percent. Then, in 2009, the Democrats regained control and ended the constraints. But public sympathy for teacher benefits was slipping.[42]

The teacher unions, in the years before 2010, might have had more success in their struggle by using more constructive strategies. The unions may have been more engaged working with other community organizations and supporting their concerns. The unions might also have framed the gains they sought as a valuable way to improve the quality of the students' learning experience. Retaining experienced teachers does result in better learning. Indeed, there is solid research that reports better student performance in schools with teachers who are organized and conduct collective bargaining.[43] In 2017, when an attempt was made to introduce Act 10 provisions in Iowa, the Iowa State Education Association was relatively successful in rebuffing the effort.[44] It provided community services and emphasized the benefits the students received from them.

In closing this section on education, I note that African American progress made in education will increase their progress in other domains. Progress in other arenas would help in closing the education gaps. Needed changes are more likely to be achieved if people do not simply decry their dire conditions and the injustice of them. It is also necessary to identify what changes should be made and what strategies can establish the desired changes. That offers hope. As I mentioned in the previous chapter, the CORE leader George Wiley sought to specify what the civil rights fighters wanted to get. Setting concrete, credible goals is vital. To wage the struggle constructively, the goals should be framed so that mutual benefits will be evident, or at least plausible. Improving the education of African American youth would have broad benefits for all of American society. It would reduce later costs, dangers, and other problems. It would enhance economic productivity, advance creativity in all fields, and move the country closer to realizing its ideals.

Housing Equality

The housing segregation described in the previous chapter basically has continued. The increasing income inequality, stagnant wages, and greater construction costs have strengthened the market forces that increased segregation by income level. This contributed to and underlay continuing segregated housing for African Americans. Their residential segregation remains greater than the segregation of Latinos or Asians, and many of the segregated African American neighborhoods are occupied by low-income residents. Significantly, research reveals that African American housing segregation has declined some since 1970.[45] That research found two developments that help account for this.[46] First, the increase in the class standing of some African Americans, at least in many localities, enabled them to move to the suburbs and other more expensive, primarily White, neighborhoods. Second, new housing built after the enactment of the 1968 Fair Housing Act no longer was built for separate White and Black residents. Several other developments also help account for the decline. Importantly, many of the segregation-inducing policies and circumstances discussed in the previous chapter were ended or curtailed. That includes redlining, restrictive covenants, and large-scale segregated public housing.

In addition, the downward trend in prejudiced White attitudes toward African Americans, noted at the beginning of this chapter, probably eased African American residence near Whites. A strategy that some African Americans have long utilized to counter segregation could be more widely applied. That is, individual families persistently seek houses to buy or apartments to rent in White neighborhoods. With effort, they find Whites willing to sell or rent housing and they brave the chance of some neighborly hostility. That is a form of nonviolent resistance. In Poland, in the mid-1980s, when the Communist government had suppressed Solidarity, the organization struggling nonviolently to make Poland democratic, many members of Solidarity simply spoke and acted as if they were free.[47]

New Policies and Actors

Very significantly, many activist individuals across the country help organize local groups to remedy problems in low-income African American neighborhoods. Sometimes those efforts are aided by foundations, activist groups, and government programs, as was the case during the War on Poverty, described in the previous chapter. One wide-ranging activist organization was influential, but for a limited time. The Association of Community Organizations for Reform Now (ACORN) was an activist association of nonprofit organizations with a broad political agenda. It was founded in 1970 by Wade Rathke and Gary Delgado to advocate for voter registration, low- and moderate-income families, healthcare, and other social issues. At its largest, ACORN had over 500,000 members and more than 1,200 neighborhood chapters. Its work was generally well regarded, holding contracts with US government agencies. In the fall of 2009, it was targeted by James O'Keefe and Hannah Giles, right-wing covert operatives. They engaged low-level staff members of ACORN, trying to solicit remarks indicating improper conduct, which they secretly recorded. They then heavily edited the videos to convey very bad practices. The edited videos were published on Andrew Breitbart's Website BigGovernment.com, and then other outlets. This generated extensive, negative publicity for ACORN. Despite several independent investigations that had begun by December 2009, which revealed that no criminal activity by ACORN staff had taken place, government agencies ended contracts and funders ended their contributions. ACORN was unable to continue its operations in the United States, and in November 2010 its US offices effectively closed. ACORN International, however, continues operations in many countries in the Western Hemisphere and elsewhere.

More broadly, beginning in the 1970s, a local community development movement, often focusing on housing issues, emerged. Local organizations became established, many using words such as "community development corporation" in their names. As in the civil rights struggle in the South, nonofficial activists often collaborated with officials in institutionalized processes.

The efforts were in part responses to the worsening housing conditions in low-income neighborhoods, particularly African American ones. As discussed in the previous chapter, various policies and practices in the 1960s and 1970s contributed to deteriorating conditions in inner-city neighborhoods. While housing in suburbs boomed, inner-city real estate values declined. African American residents who could afford to do so, moved out. In the inner cities, landlords abandoned and even burned their properties.[48] The dire circumstances stirred diverse actors to seek remedies.

Sometimes, a local crisis is the spur for major, ongoing transformations. That was the case in Bedford-Stuyvesant, an impoverished Brooklyn neighborhood of 450,000 African Americans, where, on July 16, 1964, an off-duty White police lieutenant shot and killed a fifteen-year old African American boy.[49] A few days later, rioting started and continued for three nights, resulting in many arrests and extensive property damage. No significant responsive action occurred until November 21, when the Central Brooklyn Coordinating Council hosted an all-day conference, attended by six hundred local civic, religious, and political leaders, who discussed ways to improve the area. They decided that the Pratt Institute's Planning Department would conduct a six-month survey of local challenges and potential redevelopments. The study of a small area of the neighborhood found that much of the housing was dilapidated, but 22.5 percent of buildings were owner-occupied and 9.7 percent of buildings were owned by individuals who lived close by. The Planning Department's report concluded that New York City should "mobilize all necessary antipoverty and other social welfare and educational programs" to save the neighborhood from further decline. However, no such action happened.

In February 1965, Robert F. Kennedy, the junior senator from New York, visited Bedford-Stuyvesant and then met with African American community leaders. He was taken aback and troubled by their treatment of him. Civil Court Judge Thomas Jones, the highest-ranking African American politician at the time, said to him, "I'm weary, Senator. Weary of speeches. Weary of promises that aren't kept. . . . The Negro people are angry. Senator . . . I'm angry too."

Late in 1965, Kennedy gave a speech on race and poverty, expressing his concern that America's racial crisis was shifting to the urban North, while White support for Black demands was declining. Early in 1966, Kennedy gave more speeches, which broke with Lyndon Johnson's Great Society efforts, arguing that the situation for African Americans was worsening and that their engagement as well as private-sector involvement were necessary. Kennedy decided to create his own antipoverty program and asked his speechwriter Adam Walinsky to work on that. Walinsky and Thomas Johnston, another Kennedy aide, consulted with a wide range of people throughout the country. Johnston spent much of his time in Bedford-Stuyvesant trying to sort out differences among the community's middle-class leadership. Kennedy recognized that the Johnson administration and Brooklyn's White Democrats probably felt politically challenged by his project, but Kennedy gained support from Republican leaders, including Mayor John Lindsay of New York City and the senior senator from New York, Jacob Javits. He also earned corporate support from several major business leaders.

In October, Kennedy, his staff, and community leaders decided to initiate a comprehensive, coordinated set of programs for Bedford-Stuyvesant. "Initial plans included coordinated programs for the creation of jobs, housing renovation and rehabilitation, improved health sanitation, and recreation facilities, the construction of two 'super blocks,' the conversion of the abandoned Sheffield Farms milk-bottling plant into a town hall and community center, a mortgage consortium to provide subsidized loans for homeowners, the founding of a private work-study community college for dropouts, and a public campaign to convince corporations to invest in industry in the neighborhood."[50] Isolated, no one of these programs was likely to succeed, as was the case in numerous prior undertakings across the country. The brilliance of the plan is that the programs could support each other. Announced in December 1966, initially it was to rest on two nonprofit entities. The first, the Bedford-Stuyvesant Renewal and Rehabilitation Corporation, would consist of twenty established civic and religious community leaders, with Judge Thomas Jones as leader. It was to design antipoverty programs and retain basic decision-making authority. The second, Distribution and Services, was to secure financial and logistical support for the former. It would be run by an all-White board of businessmen.

That proposed structure was quickly rebuffed by many community residents. It was widely criticized by critics from the left for its heavy emphasis on corporate business leaders. Kennedy sought to salvage the undertaking by expanding and broadening the leadership structure and African American participation. On April 1, Judge Thomas Jones announced the formation of the new structure, the Bedford Stuyvesant Restoration Corporation (BSRC). It was the first community development corporation in the United States. In 2017, The BSRC celebrated fifty years of successful restoration led by African Americans.[51] It remains focused on providing housing for poor and middle-class African Americans, and in addition to restoring residences, it created the Bed-Stuy Family Health Center and founded a supermarket, each of which became independently owned and operated.

Community Development Corporations (CDCs) have become a major factor in shaping housing policy.[52] Much policy about housing, particularly as it pertained to housing for African Americans and poor people in general, was directed from the top. These top-down policies relating to urban renewal and public housing were not widely beneficial. The field of community development emphasized and generated policies that were made formed the bottom up. Some of the social movements and social conflicts of the 1950s and 1960s, having failed to bring about large-scale transformations, prompted activists to turn toward more local, achievable projects. More attention was also given to working with the people who were to benefit from the projects.

One new form of community development entailed the appearance of financial intermediaries. These included foundations, government agencies, and local banks joining together to raise funds to provide loans and advice on ways to improve housing conditions or local businesses for low-income or minority persons. For example, in 1968, a block club in Pittsburgh was organized for such purposes, calling themselves Neighborhood Housing Services (NHS).[53] In a series of subsequent steps, the federal government became involved in fostering and supporting such local organizations. The Ford Foundation also became highly engaged in supporting local organizations, helping in developing communities in need of economic assistance. With such added resources, social banks began making loans that regular banks shunned.

Gradually, the federal government greatly increased its role with neighborhood community development organizations, in Democratic and sometimes Republican administrations.[54] Significantly, the Housing and Community Development Act of 1974 was passed by the United States Congress, after years of haggling with President Nixon, and signed by President Gerald Ford in August 1974. It allowed government block grants and required some of the funds to be for low-income families, which was reinforced in President Carter's administration. Local government agencies, so prodded, began to contract development work to neighborhood nonprofit organizations, including CDCs. Congress also created a new rental housing subsidy program that came to be known as Section 8, which has been often modified over the years in subsequent legislation to provide new ways to benefit low-income families.

In 1983, Section 8 introduced a voucher program for low-income families, which provided supplementary funds for their rental payments. This was a major shift in housing policy that had been contested for many years.[55] In trying to provide adequate housing for low-income families, the prevailing school of thought had for many years taken a supply-side point of view and supported allocating money to build new large-scale buildings for low-income families. Such buildings seemed economical to build. Another way of thinking became an alternative one: provide housing allowances, a kind of demand-side approach. The allowance is portable and subject to more consumer choice; it tended to be favored by Republicans. With many variations, usually households pay 30 percent of the household's gross income, minus deductions such as dependents under eighteen years of age. The Department of Housing and Urban Development (HUD) pays the lender a fair market rate for the balance of the rent.

This alternative program, which relied on private developers and private landlords, emerged and was widely adopted because of the recurrent failures of building very large public housing complexes for poor people. Cabrini-Green in Chicago and Pruitt-Igoe in St. Louis were celebrated when they were built in the 1950s, but the massive high-rise complexes required expensive maintenance. Funds for that were reduced, and soon the projects began to physically and socially deteriorate and became significantly uninhabited. In the early 1970s, Pruitt-Igoe was dynamited, and the land cleared. In 2000, Cabrini-Greene began to be demolished, leaving some row houses that had been built much earlier. Of course, many better-built and -maintained high-rise buildings abound, but at higher rents than low-income families can afford. However, large high-rise housing projects limited to low-income residents would require ongoing supplementary funding to be sustainable. Vouchers that would integrate low-income families into existing housing for higher-paying renters would be more sustainable.

It might be expected that a voucher system would tend to increase the integration between White and African American households, but research does not support that effect.[56] However, research indicates that voucher use does result in greater economic integration, with some increase in low-income persons residing in housing environments with higher-income neighbors. Unfortunately, housing associated with Section 8 vouchers sometimes are poorly administered by real estate owners and managers, for their profit, while the renters are trapped by fines and endure poor living conditions. The properties of Jared Kushner in Baltimore are a celebrated instance of this.[57]

In 1986, President Reagan and the US Congress created the Low-Income Housing Tax Credit.[58] It has become the longest-lasting and largest affordable housing program in US history. It is based on a public-private partnership model, offering tax incentives to attract private equity investments for both nonprofit and for-profit developers. It is a flexible and decentralized system, but also one that yields great profits for lawyers, accountants, and investors. It provides government subsidies for affordable housing, but does not entail government construction, which had so dramatically failed. The engagement of all the stakeholders in constructing and sustaining affordable housing is making a large contribution to providing affordable housing for low-income Whites and African Americans.

The final housing development strategy I mention in this section is the establishment of Purpose Built Communities. The East Lake neighborhood of Atlanta Georgia, created in the mid-1990s, was the initial model for this strategy. It was created from a deteriorating public housing project, having African American residents, with financial support from individual donors and foundations.[59] It was conceived as a holistic neighborhood, including schools and mixed kinds of housing. Some of the African American residents participated in the plans and, after the new construction, returned to live there. In addition, many White residents moved in. This new, diverse neighborhood has proved to be very attractive, and since its creation and founding, thirteen other Purpose Built Communities have been created across the country. They demonstrate how diverse neighborhoods can be successful. But the high level of funding and careful planning they require probably limits very wide expansion.

General housing policies also impact the provision of adequate, affordable housing for African Americans, particularly those having low incomes. For example, local zoning laws often restrict areas of development to single-family housing, contributing to segregation.[60] Zoning changes could reduce such segregation by generating more dwellings in areas that would be attractive to people desiring rental apartments (or condominium purchasers) in larger buildings. Such developments could provide affordable housing for a wider range of residents. Such zoning remedies would be more likely to be realized at the state level than at small local levels.

Targeted tax incentives have sometimes been offered as a way to induce banks, real estate developers, and financiers to invest in low-income areas. For example, this was the case with provisions in the 2017 Republican tax legislation. Profits from investments in federally certified opportunity zones could avoid federal taxes, while they were touted to provide housing, businesses, and jobs in distressed communities.[61] But, instead, areas within the opportunity zones were found from which untaxed profits would come from constructing expensive apartment buildings and hotels, storage facilities, and housing for college students. Again, policies to benefit the disadvantaged turned out to greatly benefit the rich.

Overall, an expanded voucher program would be the easiest and most effective way to provide decent housing for African Americans and all other Americans.[62] Presently, there are large waiting lists of people who are eligible for vouchers, but they must wait many years to ever receive one. Universal voucher programs could be advantageously adopted and carefully administered in the United States as they are in many countries. More broadly, the lack of affordable housing indicates the prevailing gap between the high costs of housing and the low level of income for many Americans. That gap might be alleviated if low incomes were raised, which might be done in various ways, including rising minimum wages, strengthening trade unions, and enhancing safety net benefits. As discussed in chapter 4, after the end of the Cold War, the possibilities of a peace dividend resulting from reduced military expenditures briefly arose.[63] Such a dividend might have contributed to financing decent housing for low-income Whites and African Americans.

As discussed in the previous chapter, African Americans did not receive the benefits Whites widely had to purchase homes, the most important source of wealth for Whites. Some of the government, banking, and real estate practices that hampered African American homeownership gradually declined. However, African American homeowners were particularly vulnerable to the effects of the housing bubble and its collapse in 2007–2008. Banks had made overly risky loans, and when the bubble burst, foreclosures were widespread. Government assistance propped up the large banks after Lehman Brothers went bankrupt, but little was done to keep people from losing their homes.[64] Houses are most Americans' major form of wealth. The collapse of the housing market was particularly devastating for African Americans, many of whom were vulnerable to losing their houses, their only wealth.[65]

Equal Justice

The quality of treatment by laws, the courts, and police are good indicators of the status of African Americans. In this section, I focus on the inequities they have experienced in the criminal justice system and the struggles to reduce the inequities since 1970. The fights for more equal justice were often particularly intense and sometimes resulted in gains, but also setbacks. I begin with discussions of major changes in laws, crimes, and punishments, noting differences for Whites and African Americans. I focus first on changes related to national legislative contentions.

National Political Policies, 1970–2020

The crime rate in the United States began a steep and steady rise after the end of World War II, peaking in 1980 and again in 1991, when it began to decline.[66] Public concern about crime grew in the 1960s and some political leaders roused the fears, promising to restore "law and order." Contentions about how to deal with crime are examined in this section of the chapter, taking into account that African Americans are more likely than Whites to be offenders and also victims of homicides and many other violent and nonviolent crimes.[67]

Riots, massive protests, assassinations, and FBI surveillance marked the late 1960s and greatly influenced the 1968 presidential election campaigns between Richard Nixon, Hubert Humphrey, and George Wallace (running as an independent). Nixon campaigned as a "law and order" candidate, using TV ads, for example, showing angry White faces and burning buildings as he said that "the first civil right of every American is to be free of domestic violence."[68] Nixon won a plurality of the popular vote by a narrow margin, but won the Electoral College by a large margin. Wallace won five states in the South, but a shift of a few thousand votes for Wallace in North Carolina and Tennessee and a tiny shift in New Jersey or Ohio, and the election would have to have been decided in the House of Representatives, controlled by the Democrats. Democratic electoral successes in the 1970 midterm elections indicated the lack of popular support for Nixon's approach to crime and law and order. A September 1970 Harris poll found only 39 percent of the public viewed it favorably.

Nevertheless, Nixon moved ahead to gain support for the Republican Party in the South and had a mixed record in policies relating to African Americans. Several progressive gestures and policies are noteworthy.[69] He signed the Voting Rights Act of 1970, nationalizing the 1965 legislation. His administration acted so that federal contracts would not be made with companies and labor unions that practiced discrimination. It set guidelines and goals for affirmative action hiring for African Americans. Nixon signed the Equal Employment Opportunity Act of 1972, which increased the power of the Equal Employment Opportunity Commission (EEOC) to enforce bans of workplace discrimination. His special assistant, Robert J. Brown, an African American business leader, arranged the promise of more than $100 million in federal funds for Black colleges and greatly increased federal purchases from Black businesses.

Nevertheless, Nixon's relationship with African Americans is widely viewed as regressive due to the "War on Drugs" that he initiated in 1970.[70] Interpreting his motivation for the legislation is influenced by a frequently cited quotation. John Ehrlichman, Nixon's assistant for domestic affairs, revealed in an interview with Dan Baum in 1994 that the Nixon administration's public enemy was not really drug abuse. Rather, the real enemies were the antiwar left and Blacks, and the War on Drugs was designed to wage a war on those two groups. In the interview he said, "The Nixon campaign in 1968, and the Nixon White House after that, had two enemies: the antiwar left and Black people. You understand what I'm saying? We knew we couldn't make it illegal to be either against the war or Black, but by getting the public to associate the hippies with marijuana and Blacks with heroin, and then criminalizing both heavily, we could disrupt those communities. We could arrest their leaders, raid their homes, break up their meetings, and vilify them night after night on the evening news. Did we know we were lying about the drugs? Of course, we did."[71] This judgment was expressed in 1994 and may have been influenced by the developments that followed the 1970 legislation, but it rings true in light of what happened over that period of time.

The Comprehensive Drug Abuse Prevention and Control Act of 1970 actually included some provisions of a public health nature to treat drug addiction and for education to reduce demand. Primarily, it ranked substances by balancing potential for abuse against medical usefulness. Heroin, marijuana, and LSD were deemed to have the highest potential for abuse.[72] Penalties were also ranked, including that possession of a controlled substance for one's own use, without an intent to distribute, was deemed a misdemeanor.

The major increase in measures directed at stopping drug abuse by punishing providers and users of prohibited drugs occurred during Ronald Reagan's presidency (1981–1989). The Comprehensive Crime Control Act of 1984 established mandatory minimum sentences. Enforcement actions greatly increased as well; for example, the annual budget of the FBI's drug enforcement units rose from $8 million in 1980 to $95 million in 1984. Also significant for contributing to the disparity in African American incarceration rates, Congress, in 1986, enacted a 100 to 1 sentencing difference in penalties for the trafficking or possession of crack cocaine compared to trafficking powder cocaine; presumably, the greater severity of the problem warranted more severe penalties.[73] However, this was widely seen as discriminatory against African Americans, who were more likely to use crack than powder cocaine. The strong efforts to suppress drug abuse, particularly of crack cocaine, contributed to the great disparities between African Americans and Whites in overall rates of criminal arrests, convictions, and incarcerations.

Political leaders generally seemed to assume that being "tough" on crime was popular, despite evidence about the high monetary and social costs of such punitive policies and their failure to actually reduce drug abuse. Several RAND Corporation studies were made of the effectiveness of diverse government programs, and they found the drug interdiction and supply efforts were highly expensive and not effective, while the treatment programs to reduce demand were more effective at much less cost.[74]

Nevertheless, the next major crime bill, the Violent Crime Control and Law Enforcement Act of 1994, signed by President Clinton, did not embody any such radical shift. Indeed, Clinton wanted to demonstrate that the Democrats were tough on crime, as well as adopting innovative strategies.[75] Trade-offs with Republican were necessary to obtain passage of this 356-page bill. The provisions included banning assault weapons, covering nineteen specific semi-automatic weapons, which was resisted fiercely by the National Rifle Association (NRA).[76] The bill also provided funding for the COPS Office, which fosters community policing practices, to hire 100,000 more police officers. The bill included the Violence Against Women Act and the Jacob Wetterling Crimes Against Children and Sexually Violent Offender Registration Act. Among many other provisions, funding for more prison beds was included.

The trends of decreasing crimes and rising incarcerations continued after the 1994 crime bill, with debates about how various provisions in the bill contributed to increasing the magnitude of the trends, and various other provisions contributed to reducing the magnitude of the trends. Given the diverse orientations of the bill, it is possible that all those consequences occurred to some degree from different provisions. At the time, the bill's shortcomings as regards equal justice for African Americans was not given much attention, and many Black political figures, reflecting African Americans' concern about high crime rates in their communities, supported the bill. The NAACP did oppose it. However, in 2020, the bill's ill effects for African Americans was more apparent, and in the Democratic contest for the Party's presidential nomination, Joe Biden was criticized for his leadership in the Senate in passing the bill.

Some correction of the prejudicial provisions in the 1994 crime bill was accomplished in Barack Obama's presidency with passage of the Fair Sentencing Act of 2010. There had been many efforts to reduce or end the great disparities in penalties relating to powder and crack cocaine, which failed. Research studies had found that the alleged differences in the effects of using the two kinds of cocaine were not well founded.[77] Sometimes good evidence can win victories. Provisions of this bill included ending the remaining disparity between penalties for crack and powder cocaine. It eliminated the five-year mandatory minimum for first-time possession of crack cocaine. Also, it raised the amount of crack cocaine in someone's possession that would result in minimum prison terms. These changes would reduce the number of federal prisoners.

In the first decades of the new century, Republican as well as Democratic leaders were becoming concerned about the shame and great cost of the extremely large and increasing number of prison inmates in the United States. Congressional attention to this concern isolated one relevant element in this problem, recidivism. In 2008, President George W. Bush had signed the Second Chance Act, which would assist states and communities so that former prisoners would not return to prison. Small grants would be made to help them in four areas: jobs, housing, families, and mental health. This interest in reducing recidivism was continued in the First Step Act, which President Trump signed in December 2018.[78] Its focus was on improving prison programs before prisoners were released.[79] It included provisions reducing the sentence for crack cocaine offenses, and by July 2020, more than three thousand persons, mostly African Americans, had been released from prison.[80] However, in the administration of the law by the Department of Justice headed by Bill Barr, and by the actions of judges, the number of releases was lower than it could have been, congruent with the law.[81]

Overall, legislation about crime has had a mixed history in keeping all Americans safe from violent and nonviolent injuries by other persons. The definitions of what is criminal and the policies to minimize the crimes are matters of contention. Generally, the contentions are about the severity of punitive ways of dealing with crime. More constructive noncoercive alternatives were always available and promoted by some of those suffering the penalties, and by many other people as well.[82] The established crime control industry and the political value of appearing to be tough counters scholarly explanations for criminal conduct and evidence-bounded policies that would minimize criminal conduct.[83] Relevantly, as discussed in chapter 3, stress and social problems are related to economic inequality. That is the case for crimes' association with inequality. There is evidence that, internationally, homicide rates are positively related to income inequality among countries, and also among the fifty American states.[84] Also, higher-income inequality is related internationally to higher rates of imprisonment. The same is true for the fifty US states. African Americans, insofar as they have less income than Whites, would tend to have life experiences that contribute to resorting to street crimes, not white-collar crimes. Moreover, many African Americans often suffer indignities due to their lower status as African Americans, which generates more stress.

Another theoretical approach focuses on how criminal behavior is often a response to living in circumstances that block opportunities for earning an income that enables at least a modest lifestyle.[85] Blockages can occur at many life stages, including early poor educational experiences, lack of models or expectations of success, lack of helpful connections, and discrimination by employers and lenders. Social disorganization is emphasized by other analysts.[86] The idea is that in addition to the circumstances noted, impoverished neighborhoods lack the strong social bonds that would control and inhibit many kinds of criminal conduct.

Clearly, various forms of deprivation are conducive to many forms of criminal behavior. More resources devoted to greatly reducing the deprivations would enable more people to be productive contributors to American society. It would also enable substantial reductions in federal, state, and local expenditures in policing and incarceration. Large-scale social movements making that case have not emerged. However, resistance and demands for improvement in specific arenas have occurred, as discussed next.

Personal and Organizational Interactions with Officials

The lived reality is that individuals interact with particular police officers, judges, and prison guards. Those interactions are the immediate cause of conflicts at the interpersonal and intergroup levels. Many African Americans have regularly complained and protested about mishandling by police officers, including unwarranted stops, interrogations, and arrests. Indeed, disparities occur at each stage of the criminal justice process in the treatment of African Americans compared to Whites. A comprehensive survey of the process as of 2008 provides the base for examining variations over time due to policies and conflicts.[87] The process begins with the first encounter between a police officer and a possible offender. There is research that confirms that police officers are more likely to initiate traffic stops of African Americans than of Whites, and also more likely to search African American autos, but no more likely to discover any contraband.[88] These interactions attest to the risks of "driving while Black," which too often escalate into detentions or even worse results, such as shooting or choking deaths.[89]

Possible detention is the next step; that is, arrested persons are locked up while awaiting trial or another hearing. This is widely practiced in the United States: 65 percent of jailed persons in 2019 were unconvicted defendants. Again, African Americans were about 29 percent of the people in local jails who were not convicted in 2002. African Americans were 43 percent of these detainees, although they constituted less than 15 percent of the American population.[90] A study of sentences in Delaware courts from 2012 to 2014 examined African American–White disparities in criminal adjudications. It found that almost a third of the cases involve pretrial detention. The analysis controlled for differences in the severity of charges and criminal histories, and it found that cash-only bail and pretrial detention increased a defendant's chances of being convicted, pleading guilty, being incarcerated, and receiving a longer incarceration sentence. Bail and pretrial detention also contributed 30 to 47 percent of the explained African American–White disparity in court dispositions.[91] Other research has documented that judges and juries sentence African Americans more harshly than Whites when the circumstances of the allegations are similar.[92]

Given the severe disparities experienced by African Americans at each step of the criminal process, the relatively high rates of incarceration for African Americans are not surprising. Thirty-eight percent of prison and jail inmates are African American, though they constitute less than 15 percent of the overall population. An African American male born in 2001 has a 1 in 3 likelihood of imprisonment, compared to a White male's 1 in 17 chance.[93]

Fighting Criminal Injustices

Many nongovernmental organizations strive to overcome the unjust treatment of African Americans in various stages of the criminal justice process. For many decades, the NAACP has been a leading organization, working on a broad range of issues in four criminal justice areas: sentencing reform, effective law enforcement, eliminating barriers for the formerly incarcerated, and survivor victims of crime. It works in these areas utilizing legal channels, lobbying for legislative measures, and joining campaigns for particular policies advancing justice for African Americans.[94]

The American Civil Liberties Union (ACLU) was founded in 1920, "to defend and preserve the individual rights and liberties guaranteed to every person in the country by the constitution and laws of the United States." During the civil rights struggle in the South, the ACLU engaged in some court proceedings to obtain equal treatment for African Americans.[95] For example, in 1967, in a case that reached the Supreme Court, the Court overturned a murder conviction of an African American because African Americans were systematically excluded from serving on juries, ruling out that practice. The ACLU also prevailed in cases that strengthened the rights of all citizens relative to police investigations and court proceedings. The victories tended to protect relatively vulnerable persons, whether White or African American. For example, in 1961 the case of *Mapp v. Ohio* reached the Supreme Court, which decided that evidence obtained by searches and seizures that violated the Constitution would be inadmissible in a state court. In 1996, the federal decision in the *Miranda v. Arizona* case required police to inform suspects of their right to have an attorney present during questioning. Significantly, the Supreme Court, taking up the *Gideon v. Wainwright* case, ruled in 1963 that indigent individuals being tried must be provided with attorneys.

The Southern Poverty Law Center (SPLC) is a nonprofit advocacy organization, founded by Morris Dees, Joseph J. Levin Jr., and Julian Bond in 1971. Based in Montgomery, Alabama, it focuses on pursuing legal cases against White supremacist groups, on gathering information about far-right extremist organizations, and on promoting tolerance education programs. Beginning in 1979, it has been particularly effective in suing Ku Klux Klan chapters and similar organizations on behalf of persons injured by them. The large sums awarded by favorable verdicts have bankrupted or reduced the operations of sued organizations.[96] The SPLC also has initiated or joined suits challenging institutional racial segregation and discrimination. In 1981, the SPLC started its Klanwatch to monitor Klan activities. This has expanded to gather information about hate groups, organizations that attack or malign an entire class of people. This information is widely accepted and cited in academic and media coverage of such groups and related issues. In addition, the SPLC has provided information about hate groups to the FBI and other law enforcement agencies. It has been successful in raising funds from small and large contributors, amassing total assets of $492.3 million in 2018.[97] As a consequence of reactions to President Donald Trump's conduct, public contributions in support to the SPLC increased greatly, and two hundred additional lawyers were hired.[98] In February 2020, Margaret Huang, former chief executive officer at Amnesty International USA, became president and CEO of the SPLC.

Sometimes individuals have acted in ways that have opposed and reduced injustices suffered by African Americans and created lasting structures that raise the status of African Americans. Bryan Stevenson is such a person. Born in 1959, in a rural town in southern Delaware, he experienced the informal segregation of the time.[99] He won a scholarship to attend Eastern University in St. Davids, Pennsylvania, and went on to earn a JD from Harvard Law School. After graduating in 1985, he joined the government-supported Southern Center for Human Rights. In 1989, he was appointed to run the Montgomery operation and its death-penalty defense work, a cause born when he was a law student. Alabama had the highest death penalty rate per capita in the country, and it was disproportionally imposed on African Americans.[100] After the US Congress cut off funding, he established the nonprofit Equal Justice Initiative (EJI), which prioritized death penalty concerns. One of the first cases Stevenson undertook was that of Walter McMillian, who was imprisoned and awaiting his execution. Stevenson's intervention discredited the prosecution's case and McMillian was found not guilty and released from prison in 1993. Stevenson has particularly litigated to end severe punishments imposed on children. He successfully argued before the Supreme Court in *Miller v. Alabama* (2012) that mandatory sentences of life without parole for children seventeen and younger are unconstitutional.

Stevenson recounts many painful injustices and his own legal failures and triumphs in his moving and widely acclaimed book *Just Mercy*.[101] With the visibility he so gained, he went on to develop the National Memorial for Peace and Justice and, associated with it, the Legacy Museum: From Enslavement to Mass Incarceration. They opened in 2018 in Montgomery, dramatically commemorating the about four thousand persons who were lynched in the South between 1877 and 1950. In addition, he and others fostered placing plaques across the country where lynching occurred. All this calls attention to the terrorism that crushed the short-lived period of Reconstruction, during which African Americans began to flourish in freedom. Perhaps the enduring mass incarceration of African Americans is a legacy lynching. Perhaps, too, drawing attention to the terrorist subjection of African Americans induces contemporary Whites to feel more sympathy for poor African Americans now.[102]

Black Lives Matter

Long-smoldering dissatisfaction with deprivations tend to erupt into collective action when a particular grievance is manifested in an extreme event or when a path forward becomes visible and attainable, and especially when the two circumstances coincide. A long-standing grievance of many African Americans has been the highly unjust police and judicial conduct they either experience directly or know about through the experiences of others. Particularly worrisome are encounters of police or White men with unarmed, nonthreatening African

Americans that escalate into the killing of an African American. In 2012, an un-punished killing triggered a local protest that spurred a social movement to correct such events. On an evening in February 2012, in Sanders, Florida, Trayvon Martin, a seventeen-year-old African American, was walking back with snacks to the house where he was a guest, when George Zimmerman, a local volunteer watch coordinator, saw him and thought he looked "suspicious."[103] Zimmerman called the police, as he often did as a watch coordinator, and reported this "suspicious guy." The police dispatcher told him not to follow the young man. Despite this, Zimmerman stopped his car and chased after Martin, got into a fight with him, and shot and killed the unarmed Martin.

The local police questioned Zimmerman for five hours, but because of Florida's Stand Your Ground law, they did not prosecute Zimmerman for murder. This decision was fiercely protested, in Sanders and nationally; the largest demonstration was held in New York City on March 21, 2012.[104] Ultimately, Florida's governor, Rick Scott, appointed a special prosecutor. Consequently, Zimmerman was tried for murder, but he was acquitted on July 13, 2013, after a jury trial. Protests erupted across the country, as the NAACP and many other old and new organizations expressed their outrage. An Oakland-based activist, Alicia Garza, wrote on Facebook: "Black people. I love you. I love us. Our lives Matter. Black Lives Matter." An activist friend, Patrisse Cullors, converted it into #BlackLivesMatter, which went viral.

Federal conflict resolution interventions helped avert violent escalations in the past. The Civil Rights Act of 1964 had created the Community Relations Service (CRS) within the Department of Justice. The CRS began and continues to provide mediation, dialogue training, and other consultation services to communities experiencing identity-based disputes.[105] Thomas Battles, a regional director of CRS, went to Sanders and brought together local faith leaders and city officials to discuss ways to de-escalate community tensions. Following the acquittal of Zimmerman, Battles facilitated a public meeting of church, city, and law enforcement leaders and community members to discuss the verdict and local concerns. Peaceful demonstrations continued, but no arrests were made.

The Black Lives Matter (BLM) movement erupted nationally, fueled by a large-scale uprising in Ferguson, Missouri, following the killing of Michael Brown, an eighteen-year old African American, by a White police officer, on August 9, 2014. Brown's body remained in the street for hours, as stories about what had happened spread. Apparently, there had been an altercation in which Brown had talked back and disobeyed an order by the officer. Vigils and protests by African American religious and civic leaders and by African American youths soon followed.

Some protests escalated to include violent acts and looting as heavy-handed police conduct increased. African American and White activists from other cities joined the fight and formed numerous activist groups. The police resorted to bringing in and using military equipment and tactics. On August 14, Missouri governor Jay Nixon sent the Missouri State Highway Patrol to help the local police. Long-standing grievances by African Americans were driven by police harassment and widely imposed fines.

Fortunately, the ongoing violent conflict drew President Obama's attention, and on August 23, he ordered a review of the distribution of military hardware to state and local police. The US attorney general, Eric Holder, asserted that displays of force in reacting to mostly peaceful demonstrations could be counterproductive. Furthermore, on September 4, he announced that the Department of Justice (DOJ) would investigate possible misconduct of the Ferguson Police Department (FPD). In March 2015, the DOJ announced finding that the FPD had engaged in misconduct against the citizenry of Ferguson, including discriminating against African Americans and applying racial stereotypes, as well as determining that Ferguson depended on fines and other charges generated by the police.

The uprising and federal intervention produced some progress in the conditions and status of African Americans in Ferguson.[106] Three years after Michael Brown's death, Delrish Moss, an African American police officer experienced in community outreach, was police chief. The proportion of African American police officers had increased. The seven-member city council had three African American members, when it had only one earlier. A store that had been looted and burned had been replaced by the Urban League Empowerment Center to offer job training and placement. The progress did not entail a fundamental restructuring of African Americans' circumstances.

A more constructively waged fight by the protesters entailing winning broader collaboration with local community organizations might have achieved more progress. Sometimes, various organizations, in pursuit of their specific goals, compete with similar organizations for influence rather than joining in collaboration with them.[107] The repressive methods of the Whites who opposed the changes were indeed counterproductive for them. More constructive methods of struggle by all sides might well have achieved greater progress for all the citizens of Ferguson. Mediation might have been helpful at various stages of the conflict. A staff of nineteen Community Relations Service conciliators in 2018 is certainly much too few for the numerous community conflicts in America.[108]

The BLM social movement, however, continued to grow, becoming a wide network of local chapters, led by local members.[109] The Movement for Black Lives soon emerged to aid connecting local organizations, and so helped provide resources for rapid mobilization at particular localities when an unarmed African American was killed by police officers. Local leaders were often young

African American women activists, creating a new vision, bypassing past charismatic leaders.

A chronicle of African Americans killed by police continued, including Walter Scott on April 4, 2015, in North Charlestown, South Carolina; Alton Sterling, on July 5, 2016, in Baton Rouge, Louisiana; Philando Castile on July 6, 2016, in St. Paul, Minnesota; Stephen Clark, on March 18, 2018; and Breonna Taylor on March 13, 2020. Then, on May 25, 2020, the murder of George Floyd in Minneapolis, Minnesota, ignited a most remarkable social movement in the United States and around the world. Three years earlier, Black Visions had been established in Minneapolis as a base for community mobilization and actions. It held rallies and educational sessions and it built relations with City Council members, often focusing on police violence within the context of systemic racism.[110] Given the national attention to the killing of Floyd, the calls of some young activists to defund or even abolish police departments flowed out of Minneapolis.

Several factors converged to produce the sudden massive and diverse protest marches around the country, which persisted for weeks and had significant consequences. First, videos of Floyd's death were stunning and were rapidly spread by social and mass media. A White police officer was pressing his knee on the neck of a prone African American man, whose head is twisted on the street, saying, "I can't breathe." The police officer looks casually at the cameras, with one hand in his pocket, as three other policemen stand by and onlookers cry out, "you're killing him." This persisted for over eight minutes, as the struggling man on the street became inert. That shocking video was played over and over again.

Media attention across the country, and abroad, focused on the death and responses to it. Floyds' relatives and friends expressed their grief and outrage; the police chief apologized; and city, state, and national officials decried what had happened, as did former presidents and White and African American civic leaders. Some official actions were speedily taken. On May 26, the Minneapolis mayor and police chief announced the firing of the four police officers who were responsible for Floyd's death.[111] Protest demonstrations quickly spread to over two thousand cities and towns across the country. Surprisingly, the protesters were diverse in ethnicity, age, and experience. The demonstrations were generally nonviolent and peaceful; in some localities, police expressed solidarity with demonstrators, while in other localities confrontations escalated into forceful actions. In a few cities, opportunistic persons looted and burned businesses, while gangs looted items to sell later. In some instances, armed outside vigilante groups appeared to confound the protest.[112]

Various specific demands for reforming police departments were made in cities, towns, and some states. Many reforms had been discussed over many previous decades, and some had been implemented, with beneficial results in many, but not all, cases. In many cities, community policing policies were implemented, training relating to diversity and implicit bias were instituted, and recruitment of officers was more diverse. For many decades, civilian review boards have been instituted to oversee and restrict bad police conduct, but often the police departments and police unions have effectively nullified their efforts.[113]

The magnitude of the social movement to end extreme police violence against African Americans and related matters was extraordinary. Many developments and circumstances converged to bring about this major phenomenon. The prevailing COVID-19 pandemic that had been underway for almost three months might have been expected to dampen mass demonstrations. Yet, on the contrary, school-age persons were not in school and many persons were not employed; they could march with masks, socially distanced and outdoors.

Furthermore, African Americans had been experiencing some improvement in their class, status, and power standing in prior years and expected more improvements. Yet the pandemic had revealed that their living circumstances made them more vulnerable to becoming sick and dying from COVID-19. Moreover, their employment circumstances deteriorated even more than did those of Whites as the economy drastically declined. Rising expectations interrupted by declining conditions is the well-established formula for revolt.[114]

Another striking feature of the BLM protest marches was the participation of Whites as well as people of color. As I noted at the beginning of the chapter, White public opinion had become less and less prejudicial since 1988, and in 2018 Whites generally did not attribute most African Americans' poor conditions to their own failings. More specifically, in late June, 44 percent of White voters viewed the Black Lives Matter movement "very favorably" and 52 percent believed that there was a pattern of excessive police violence against Blacks.[115]

The focus on police killing African Americans does not necessarily deal with the systematic injustices in the criminal justice system, outlined above. African Americans are generally sensitive to those injustices at each stage of the prosecution of crimes and of incarceration. Greater class, status, and power equality for African Americans are needed to fully overcome criminal injustice. Unfortunately, the energy and power of the BLM movement arose at the time that President Trump was seeking re-election during a mishandled response to a pandemic. He chose to escalate confrontations with protesters, proclaiming he was a law-and-order president. White supremacists saw an opportunity to gain influence. On the other hand, an increasing number of Americans worried that the opportunity to transform race relations in the United States was being lost and American democracy itself was becoming endangered. Political polarization was deepening.

Equal Well-Being

The well-being of African Americans is a broad area of concern, and has many indicators, including life expectancy, economic standing, and social-cultural respectability. Trends in each are noted and actions helping account for them are discussed.

Life Expectancy

African American life expectancy is shorter than White life expectancy.[116] Thus, in 2017, expectation of years of life after birth was 78.8, 76.4, and 81.2 years for all Whites, for White males, and for White females, respectively. The number of years for African Americans was 75.3, 71.9, and 78.5 respectively. The gap was greater for males than for females, whose life expectancy was very close to that for White females. The gap between Whites and Blacks had been hugely different many years ago. Even in 1970 they had been larger and very slowly drew closer.[117] The comparable number of years of life for Whites in 1970 was: 71.7, 68.0, and 75.3; for African Americans: 64.1, 60.0, and 68.3.

The discrepancies between African Americans and Whites vary greatly among the states, as do living conditions and healthcare provisions.[118] The discrepancies indicate that social, medical, and political policies impact on changes in life expectations sometimes similarly and sometime differentially between the two communities. For example, New York City had substantial decreases in homicide and HIV/AIDS mortality during the 1980s and 1990s. These changes particularly benefitted African Americans and help explain the increase in their life expectancy in New York. These improvements are attributable to local policies, such as the aggressive identification of and treatment for people with HIV/AIDS. Policies that reduced drug- and alcohol-related deaths also contributed substantially to increasing overall life expectancy in New York City during the same period.

COVID-19 had a much greater deadly impact on African Americans than on Whites in all age categories.[119] The crude death rate for African Americans is twice the rate for Whites. But two conditions should be taken into account: older persons are more likely to die from the disease, and the White population is older than the African American population. Consequently, calculating age-adjusted COVID-19 death rates reveals that the death rate for Black people is 3.6 times that for Whites.

Three circumstances of many African Americans, which are related to their lower status and class rankings, help explain their relatively high likelihood of dying from COVID-19: exposure, vulnerability, and medical care. First, African

Americans are disproportionately represented in essential work settings such as healthcare facilities, farms, factories, grocery stores, and public transportation. Some people working in these settings are more likely to be exposed to the COVID-19 virus because they entail matters such as close contact with the public or other workers, are unable to work from home, and lack paid sick days. In addition, many African Americans experience exposure because they live in crowded housing that make it difficult to follow prevention strategies. Disproportionate unemployment rates for African Americans during the COVID-19 pandemic can cause greater risks of eviction and homelessness or sharing of housing.

Second, the COVID-19 virus tends to be more deadly for people who have underlying health problems. African Americans are more vulnerable to the COVID-19 virus since they are more likely than Whites to have such illnesses, such as hypertension, obesity, diabetes, and lung disease. Moreover, lower-income African Americans are more likely than Whites to live in neighborhoods that are environmentally injurious to their health.

Third, access to healthcare can be limited for African Americans because of factors such as lack of transportation, child care, or ability to take time off from work. Furthermore, historical and current discrimination in healthcare systems can interfere with African Americans obtaining healthcare. Some people from racial and ethnic minority groups may hesitate to seek care because they distrust the government and healthcare systems.

COVID-19 revealed the vulnerability of low-paid, essential workers. The workers demanded safer working conditions and better pay to meet greater costs for many to do their work safely. They held strikes, protests, and walkouts, which won considerable public sympathy. Soon, legislation was passed at all levels of government to provide extra pay for persons doing hazardous work. Recognition of the important work that was generally not shown great respect and was paid low wages abruptly rose. Better compensation would certainly help increase income equality in the United States and secure more effectiveness in raising public health conditions for all Americans. These circumstances provide opportunities for new coalitions to wage constructive conflicts, with communal broad benefits.

Economic Standing

The relationship between the status and the class standing of African Americans must be discussed. In chapters 2–4, I traced rising income and wealth inequalities in America. I stressed conflicts between, on one side, advocates of policies to reduce poverty and improve the economic conditions of the working class, and on the other side, advocates of policies to reduce such governmental policies and maximize the freedom of markets. African Americans tended to adhere to the first camp. Advocates in the other camp often suggested that the beneficiaries of government social policies were disproportionally undeserving non-Whites.

In the previous chapter and this one, I focus on conflicts in arenas where race differences have been salient: education, housing, criminal justice, and well-being. One reason for doing so is that conflicts in the United States have increasingly been conducted in terms of identity politics, which involves, on one side, social groups struggling to free themselves from exploitation, marginalization, or subjugation.[120] Such social groups include women, ethnic minorities, and the LGBTQ community. Opposing protagonists often also define themselves in status identity terms; indeed, Whites generally have tried to impose their definitions of White and non-White identities. White identity itself exists and will be discussed in the next chapter.

These status differences are viewed as ways to gain self-empowerment and legitimacy in the larger society. The importance of social identity became more widely used in the 1980s, retroactively including earlier social conflicts, such as with the Black Panthers, discussed in chapter 5. Alliances were not readily built with other groups in terms of income or wealth. To some degree, identity politics refers to conflicts about status differences rather than class differences. These are matters of debate. Antiracist activists can disagree about the analytics and applied correctness of emphasizing class or identity. For purposes of this book, it is well to recognize that in conflicts engaging African Americans, both status and class are in play. African American insistence on respect and dignity, celebration of Black culture, and interest in self-segregation are important, even if they can hamper cross-racial coalitions and framing of issues. Actually, some developments in the conception of what the United States is, and in the many achievements of African Americans, has resulted in a wider range of policies along with increasing solidarity of Whites and African Americans. Before discussing these developments, I recognize the continuities in the struggle of African Americans and Whites against poverty. In chapter 5, I mentioned that Dr. Martin Luther King Jr. went to Memphis, Tennessee, to support the striking sanitation workers there, as part of his planned Poor People's Campaign. He was assassinated there, but the mass demonstration was held nevertheless, on May 12, 1968, in Washington, DC. A week later, protestors erected a settlement of tents and shacks on the National Mall, which was called Resurrection City. They held out for six weeks, living in miserable conditions.

Many years later, a few steps were taken that led to a reinvigorated Poor People's Campaign, led by co-chairs the Reverend Dr. William J. Barber II and the Reverend Dr. Liz Theoharis. In 2013, William Barber (head of the North Carolina Chapter of the NAACP) joined with other progressive religious leaders in conducting civil disobedience protests, organized in what they called "Moral Mondays." They arose to protest the numerous right-wing Republican laws enacted after the 2012 elections in North Carolina, which gave the Republicans control of the governorship and both houses of the legislature.

Soon after Trump became President in January 2017, and continuing Martin Luther King's vision, Barber issued a National Call for Moral Revival to "confront the interlocking evils of systemic racism, poverty, ecological devastation, militarism and the war economy, and the distorted moral narrative of religious nationalism."[121] During the summer of 2018, poor people and moral witnesses in forty states launched the Campaign with a season of nonviolent civil disobedience, and a new organism of state-based movements was created. In June 2019, over a thousand community leaders gathered in Washington, DC, for the Poor People's Moral Action Congress, which included the release of *Poor People's Moral Budget*, and a hearing before the House Budget Committee on the issues facing the 140 million poor and low-income people in the nation. Leaders of the Poor People's Campaign toured twenty-five states to mobilize people for the campaign. This culminated in the Digital Mass Poor People's Assembly and the Moral March on Washington, on June 20, 2020. The national context for the march was not opportune for influencing the federal government, but it contributed to sustaining pressure to care for the poor by a new administration in Washington.

Official and Organizational Actions Fostering Equal Social Recognition

Many institutional and official developments and policies since 1970 have helped raise the status of African Americans. As discussed earlier in this chapter, after immigration country quotas were ended in 1965, immigration from Asia and particularly from Latin America greatly increased.[122] Also, identity politics began to be more salient in the 1960s. Gradually, the notion that the United States was a melting pot faded. The caste-like nature of African Americans had always contradicted the melting pot characterization of the country.[123] Even as the caste-like nature of African Americans' position in the United States decreased, so did the idea that America was a melting pot. Other designations appeared and disappeared, such as "salad" or "mosaic." America was diverse. African Americans being one of many components of diverse America enhanced their respectability.

Additionally, diverse actions institutionalized the new understanding. Significantly, African American studies programs and degree-granting departments were established in the 1960s and 1970s, initially responding to student and faculty activism.[124] The Black Power movement and the Black Panthers, discussed in chapter 5, were especially strong in California, as were protesting university students. San Francisco State University (SFSU) students were able to create Black studies courses in the Experimental College at SFSU, and in 1968 that led to America's first department in Black studies there. Since then, programs and departments have opened across the country. Consequently, many White and African American college students have learned more than students did previously about the history and culture of African Americans. In addition, the programs generate research and publications about the past and present circumstances of African Americans, which fosters recognition and respect for them.

The establishment of Black History Month is another expression of and promoter of importance and respect. It was first proposed by African American educators and the Black United Students at Kent State University in 1969 and became officially instituted one year later. In six years, it was being celebrated throughout the country in educational institutions and various other settings. Black History Month has expanded widely, aside from educational institutions. For example, in 2018, Instagram created a Black History Month program, which then spread to other social media platforms.

The national celebration of the birthday of Martin Luther King Jr. is another national recognition and honoring of an African American person, and of the movement he led. Reaching official agreement for all fifty states, however, was contentious. After King's death, US Representative John Conyers (a Democrat from Michigan) and US Senator Edward Brooke (a Republican from Massachusetts) introduced a bill in Congress to make King's birthday a national holiday. In a close vote, in 1979, it was defeated in the House of Representatives.[125] It took popular mobilization to win passage of the bill. The King Center sought and won strong support from the corporate world and the general public. The campaign was aided in 1980 by Stevie Wonder's recording of "Happy Birthday." Then, at a Rally for Peace press conference in 1981, it was announced that six million signatures were collected for a petition to Congress to pass a law, proposed by Representative Katie Hall of Indiana, to create a federal holiday honoring King.

Senators Jesse Helms and John Porter East (both North Carolina Republicans) led the opposition to the bill, including a filibuster to prevent its passage. Despite the opposition, the bill became law. The House of Representatives passed the bill on August 2, 1983, 338–90 (with 5 members voting present or abstaining). The Senate vote was 78–22, on October 19, 1983. President Reagan signed the bill on November 2, 1983. The holiday was first observed on January 20, 1986. Several states named the holiday in different ways, referring to civil rights or, as some southern states sardonically did, adding Robert E. Lee to Martin Luther King. Gradually, over the years, such gestures ceased. Ultimately, starting in 2000, all the states simply referred to the federal holiday only with King's name.

Finally, among official and organizational actions, I want to mention the opening of the spectacular National Museum of African American History and Culture. It was established by Act of Congress in 2003, and opened to the public on September 24, 2016, as the newest museum of the Smithsonian Institution, in Washington, DC. As observed by Lonnie G. Bunch III, founding director of the Museum, "there are few things as powerful and as important as a people, as a nation that is steeped in its history."[126]

Direct Actions

It is important to recognize that status ranking does not change only by way of large-scale public contention and conflict or top-down imposed policies. The esteem of and respect for African Americans by other Americans occurs through a variety of channels. A primary way is by African Americans directly insisting upon being treated equally, brushing aside discrimination. Laws against discrimination often are initiated when such actions assuming fairness are undertaken, and equal treatment is denied. Furthermore, laws, to be effective, require that people who might be unfairly treated act as if they will not be discriminated against. Beyond laws or practices to avoid unequal practices, policies may be undertaken to enhance diversity through affirmative action measures. I note some spheres of American life in which African Americans have made contributions that have been highly recognized since the 1960s.

Engagement in prominent, highly visible, and prestigious spheres of life by some African Americans tends to reduce prejudice and increase esteem for African Americans generally. One way this happens is from collegial interactions and bonds, in accord with the social contact hypothesis, discussed in chapter 5. This can also extend to solidarity across ethnic and racial lines. Another pathway is appreciating and respecting persons performing excellently within shared fields of endeavor. Increasingly, African Americans have included persons who make outstanding careers in politics, business, sports, music, entertainment, and literature, fields with large public audiences. In many arenas, barriers had to be broken through.

In politics, obviously, African Americans have won more and more major official positions in local, state and national levels of government. In business as well, entrepreneurial successes and corporate leaders are increasing. In boxing, Muhammad Ali was dominant for very many years. Team sports were entered more slowly, and then, in some sports quite prominently, notably in baseball, basketball, and football. Some African American men and women excelled in nearly every sport, contributing to America's successes in the Olympic Games. Some athletes employed their visibility and significance to advance the rights of African Americans in their fields of endeavor and also in support of advancing the rights of African Americans more broadly. This includes making public actions of protest.[127]

African Americans have long produced and performed popular music in the United States, going back to spirituals and to jazz. Some of them have been leaders in the creation of the ever-new American popular music fashions. Sometimes, however, White performers adopt and succeed in attracting the mass White audience to the style, as occurred in the early days of rock and roll.[128] African Americans have become increasingly important performers in many of the ever-changing popular fields of entertainment, on stage, television, and social media. Sometimes persons are highly successful in many entertainment arenas. For example, Oprah Winfrey succeeded as a talk show host, television producer, actress, author, and philanthropist. Her program, *The Oprah Winfrey Show*, was the highest-rated television program in history and ran for twenty-five years, from 1986 to 2011.

Finally, I mention African American contributions to American literature. Many books, both fiction and nonfiction, convey insights into the experiences of African American men and women in the United States. For many Whites, reading these books was a mind-changing experience. Three well-known writers are James Baldwin, Maya Angelou, and Toni Morrison, who won the Nobel Prize for Literature in 1993. They were not only writers, but public figures as well.

Overall, there has been some progress toward increased equality in well-being between African Americans and Whites since the 1960s, but not for all African Americans. Moreover, for some Whites, all African Americans are regarded as lower than they regard themselves to be. Despite the progress, the inequalities are still great, and to some degree they are experienced as even more unacceptable. Moreover, the resources that even a little higher status brings raises the desire and capacity to make further advances to greater equality.

Opposition to Increasing Equality

Some Whites view the progress made by non-Whites as won at their expense and as a threat that further declines in their status will be coming. There is evidence from interviewing a national panel of respondents in 2012 and 2016 that White men who felt that their status had decreased tended to feel resentful and adopt right-wing ideas and supported Donald J. Trump in the 2016 election.[129] Beliefs about growing racial diversity and globalization tended to contribute to some Whites' sense of siege. Many White males without college educations have suffered stagnant wages, falling back on lower-paid jobs or unemployment. This is associated with the hyper income inequality discussed in chapter 4. Consequently, there may be feelings of distress and a feeling that they are disrespected. Some political leaders try to exploit these fears and identify scapegoats who they claim cause the losses. Whether they do this due to ignorance or for more malevolent reasons, they are having some success.

White supremacist groups have never completely disappeared from the United States, remaining particularly significantly in the South. Their numbers have waxed and waned, as have their open expressions of hostility and even violence against African Americans. There are other affinities that radical White supremacist groups often have. One important link is to people holding anti-Semitic sentiments and beliefs. Another important link is with antigovernment militia groups. In the chapters on class inequality, I noted the ideological struggles in America about what role government should play in fostering or moderating class inequality. Some persons favored reducing taxes in order to keep government services and regulations down. Furthermore, that meshed with opposition to government rules about gun controls. Candidate and President Donald Trump spoke and acted against gun control legislation, reduced government regulations of corporate actions, lowered taxes for corporations and the very rich, and reduced the safety net for the poor and for non-Whites. He went so far as to embolden White nationalist militia groups. Finally, for personal and political reasons, Trump's policy guide was to overturn every policy that President Barak Obama had instituted.

Most African Americans naturally would view the events associated with Trump's coming to power and many subsequent developments as setbacks to the small steps of progress that many of them believed they had been making. With their gains in status and political power, they fought back. Many Whites, who had their own reasons to oppose Trumpian policies, became strong allies in escalating political fights. Partisans who were inclined to follow norms of civility and constructive ways of contending faced an adversary in Trump who denied the utility of such ways of contending. He celebrated and practiced quickly and consistently resorting to coercion and intimidation.[130] The political polarization that intensified greatly since the 1990s, which I discuss in chapter 8, made cooperation in mitigating differences less and less likely.

Nevertheless, many Whites, sometimes in collaboration with African Americans, undertook actions to ameliorate the difficulties of advancing the equality of African Americans. One path for making progress relates to changing the way Whites view their relations with African Americans. For example, the idea and reality of White privilege became more widely acknowledged. "White privilege" refers to the advantages and benefits that White people have by virtue of their race status. Some Whites may still suffer from other challenges—poverty, poor health, addiction—but they always retain some advantage based on race status. Whites may not recognize that they have privileges, as distinguished from bias or prejudice. The advantages occur in professional, educational, and personal contexts. The concept of White privilege also implies that Whites assume the universality of their own experience to be normal. Training in racial sensitivity that includes education about White privilege has become widespread in educational institutions, business corporations, and government offices. Distressfully, in September 2020, President Trump ordered the ending of any racial sensitivity training in federal agencies.[131]

Despite setbacks and opposition, White acceptance of the movement toward greater equality for African Americans is profound, as was indicated by the report on changing public opinion noted in the outset of this chapter. It is indicated too in popular attention to books such as *Fragility: Why It's So Hard for People to Talk about Racism.*[132] Finally, attention to what America owes to African Americans and how restitution should be made has risen.[133] In July 2008, the US House of Representatives passed a resolution apologizing for American slavery and for later discriminatory laws. In January 2017, Representative John Conyers Jr. introduced a bill "to establish a commission to study and consider a national apology and proposal for reparations for the institution of slavery, its subsequent de jure and de facto racial and economic discrimination against African Americans, and the impact of these forces on living African Americans, to make recommendations to the Congress on appropriate remedies, and for other purposes."[134]

Conclusions

Certainly, in the period since the end of World War II, some reductions in the inequities between Whites and African Americans have been achieved. But considerable inequities remain, and the consequence of prior inequities have not disappeared. A profound cultural shift is underway, which is welcomed by many people in the United States but viewed as a threat by some others. During the great wave of immigrants from Europe in the later 1800s and early 1900s, in conventional thinking, the United States was a melting pot, and immigrants were to assimilate and take on US ways and manners. People who were not from Europe, however, did not "look American." In the post–World War II years discussed in this book, however, that conventional view of the US has been shifting. Increasingly, the US is now generally viewed as a diverse country. The virtues of its diversity are increasingly celebrated; they include more information and greater creativity. Diversity contributes to a more expansive economy in many spheres. The direct actions of African Americans in conducting themselves as equals have been crucial and effective.

Too often, in education, housing, justice, and well-being, conflicts have been obscured and handled within judicial, legislative, and administrative channels. Within those channels, persons with relatively high economic, status, and power ranking enacted and implemented policies that helped sustain inequalities for African Americans. Having little regard for or engagement by African Americans impacted by the policies tended to injure the long-term interests of those fashioning the policies. On the other hand, protest outbursts by African Americans that become violent can be turned against them. Broad coalitions among ethnicities and across status lines can be more effective. Resorting to mutually beneficial framings of conflicts also can be effective. These possibilities will be discussed further in chapter 8, on power. The next chapter examines other struggles to reduce additional kinds of status inequalities.

7

Conflicts about Gender and Other Collective Identities, 1945–2022

Many collective identities are prone to conflicts.[1] This is likely to be true when there is high societal consensus about the relative status of a set of identities and varying prerogatives are associated with the ranking. In the preceding two chapters, this was seen to be the case, in varying degrees over time, for African Americans relative to Whites. Certainly, collective identities may exist with little consensus of their relative standing and with few prerogatives, as is the case with fans of baseball teams or musical preferences. However, differences in ethnicities, genders, or other identities can be subject to intense contentions, depending on the rights claimed by and over members of different collective identities. In some periods and countries, they were and are matters of terrible conflicts, even wars and genocides.

In this chapter, I note only a few collective identities that have been the locus of major American conflicts since 1945. One major conflict has pertained to the struggle for equal rights for women. Another contention relates to the rights of persons identified as LGBTQ persons. Finally, I will also refer to ethnic differences, particularly discussing White identity. Obviously, everyone has many identities, with some being more salient than others, and varying over time. The multiplicity of identities results in collective identities having overlapping memberships, which makes amelioration of severe conflicts possible, but also can be aligned and intensify conflicts underway.

Movement toward Female and Male Equality

Although based on biology, human cultures vary greatly in how females and males conduct themselves and relate to each other. Cultures even shape whether or not women and men are a binary matter or if there may be humans with some other identifications. That will be discussed later in this chapter. Now we start with two primary collective identities: women and men. In all societies, humans are nearly all born and live in close relations with each other in families most of their lives. Nevertheless, men and women tend to have different sets of rights and prerogatives. The experience of close relations between men and women in every

Fighting Better. Louis Kriesberg, Oxford University Press. © Oxford University Press 2023.
DOI: 10.1093/oso/9780197674796.003.0007

society should tend to moderate intense broad contentions between women and men. On the other hand, at the interpersonal level, there are very many opportunities for intense conflicts.

During World War II, many women took over the occupational and other roles men vacated to serve in the armed forces. Generally, women were celebrated for their new activities. That quickly changed when the war ended. Families were reunited and many families were newly established. A baby boom followed. At the time, women and men had significant differences in their roles and in their rights and prerogatives. Men's dominant position relative to women was evident in countless laws.[2] The difference in gender roles were sustained by all-encompassing customs. Men were formally addressed with Mr. and their first and last names. Women were formally addressed as Miss if unmarried and Mrs. if married, and then with their husband's first and last name. A woman's personal identity disappeared into her husband's.[3] The degree of inequality in the rights of men and women at the end of World War II may be difficult to recognize now. I trace the transformation in gender roles by discussing particular changes resulting from numerous contentions, nationally, in states, organizations, and families. Much of the struggle was conducted within political and judicial institutions, but within contexts of ideological contentions, technological changes, and shifts in social relations.

Emergence and Establishment of the Women's Movement

The burgeoning of the women's liberation movement following the end of World War II is generally dated by the publication of Betty Friedan's book *The Feminine Mystic* in 1963.[4] The book documented the widespread disaffection of American women, as their lives were focused entirely on their roles as wives and mothers. Although celebrated in the years after the war, the housewife life was not enough for many women. The book was a bestseller and Betty Friedan lectured around the country. At the same time, many women wanted to join the civil rights struggle and then the later anti–Vietnam War campaign, but felt they were brushed aside by the male activists. Indeed, women who ventured out of the housewife confinements experienced great discrimination in education, entry into professions, engagement in sports, and most other spheres of endeavor.

Women sought redress to some of these many grievances through various channels. One was the national legislative process. With Democrats in control of both houses in 1963, Congress passed, and President John F. Kennedy signed, the Equal Pay Act, which prohibited sex discrimination in the payment of wages. In 1964, Congress began considering a broad Civil Rights Act. Feminists lobbied hard for adding an amendment prohibiting sex discrimination in employment.

They were successful. The act passed, including Title VII, prohibiting discrimination in employment on the basis of sex. That covered the full range of employment decisions, including recruitment, selections, terminations, and other terms and conditions of employment.

Implementation of those acts, however, would have to depend on the capacity and courage of women employees and job seekers. As is widely the case, collaboration between official policymakers and nongovernmental actors is important to bring about changes. The Equal Employment Opportunity Commission (EEOC) was officially established in 1965 to implement Title VII of the Civil Rights Act of 1964.[5] It quickly demonstrated its inadequacy. By a vote of 3 to 2, the EEOC decided in September of 1965 that sex segregation in job advertising was permissible. In response, the "dissidents" gathered together and began planning a new organization; it would be NOW, the National Organization for Women. By October, about three hundred women and men were charter members and an organizing conference was held on October 29–30, in Washington, DC. Officers were elected, including Kathryn (Kay) Clarenbach as chair of the board, Betty Friedan as president, Aileen Hernandez as executive vice president, and Richard Graham as vice president. A Statement of Purpose was adopted, addressing all women and all facets of a woman's life.

NOW became the major organization representing much of the broad social movement for women's rights, which was rising and quickly expanding. Gains won incentivized further goals. The movement bore various names with different connotations. Thus, the term "second-wave feminism" built on the first-wave emphasis on legal and political rights. The term "feminism" itself is an overarching reference to diverse ways of thinking about women and about the social relations and structures that restricts, injures, and disrespects them. Men as well as women have been and are feminists. The term "women's liberation" connotes freedom from strict roles constraining women's lives. I will generally use broad inclusive terms such as women's movement and feminism.

An early component of the women's movement was consciousness-raising groups. These were small local groups of women meeting to discuss their struggles and discontent. Comparing notes, they could discern that their troubles were not personal and attributable to particular husbands or children. They were living within a patriarchal structure favoring men. These local groups reinforced each other and were linked together by national figures who travelled the country. NOW took actions that made changes. On August 30, 1967, eleven NOW members picketed the *New York Times*'s classified advertisement office, charging discrimination against women by segregating help wanted ad columns by gender. Within a few years that practice had vanished in the country.[6]

Important direct actions were taken by individuals and local NOW chapters across the country. For example, Syracuse, New York, was the locus for local chapter activities and the home for a few national leaders in the women's movement.[7] In the spring of 1967, Betty Friedan visited Syracuse for a few days. Karen DeCrow, Rev. Betty Bone Schiess, and a few other activists met with her. Shortly afterward, a local chapter was organized, with ten charter members.[8] Individuals advanced the movement from the vantage point of their experience. Dr. Robert Seidenberg, a clinical professor of psychiatry at the New York Upstate Medical Center, while going through medical magazines, noticed that women were targets for being tranquilized. "They were depicted as demeaning, as silly things that had to be drugged."[9] He made slides of the ads and showed them to drug companies, he testified before Congress, and he spoke at psychiatric meetings and in public settings throughout the country. The implications of undue tranquilizing of women helped curb the practice.

Another widespread 1950s practice injurious to women was frequently required management job moves, with a trailing wife. Seidenberg wrote the first book analyzing the many severe damaging consequences of this practice.[10] Recognition of the ill effects upon wives and children and women's increasing claims for equality have greatly reduced the practice and made job moves more of a joint decision. It has come to be more recognized that fewer moves and full consideration by all concerned parties when a move is contemplated have widespread benefits for the many parties impacted by a possible move.

The Syracuse NOW chapter engaged in many interventions to open up organizations and localities to end their exclusion of women. For example, taverns generally refused to serve women at a bar if they were unaccompanied by men. When a woman alone was refused service at a Hotel Syracuse bar, NOW members picketed the hotel.[11] Faith Seidenberg, a lawyer, sued the hotel and lost, twice. Then Seidenberg and DeCrow, when in New York City, went to McSorley's Ale House, which famously had not served women for 114 years up until then. They were thrown out. Seidenberg sued and won, which resulted in a state law forbidding discrimination on the basis of gender in places of public accommodation. Chapter members also strove to equalize pay in Syracuse area schools and universities, which is a long-enduring, widespread issue. They won admittance to male-only institutions and organizations. Such integration would help equalize career advancement and civic engagement opportunities. These conflicts were waged by NOW chapters and other organizations countrywide.

A great many books and articles began to be written and read that expressed feminist experiences and perspectives.[12] Some works reported and analyzed the unfair conditions of women's lives and their low status.[13] Many such works proposed changes that would rectify and redress the grievances. Other works presented historical accounts that presented information about the notable roles of women, which had previously been untold.[14] The extensive feminist literature expressed different ideologies and viewpoints, including ethnically specific approaches to feminism, notably black feminism and intersectional feminism.[15] Numerous women's studies courses as well as undergraduate and graduate programs were established and widely proliferated. This fostered a substantial market for books. In December 1972, the influential *Ms* magazine was launched, with Gloria Steinem as editor.

Much of the outpouring of literature served to recruit and mobilize women and men to support and actively join the women's movement. Some of the literature served to persuade men as well to support the policies favored by the women's movement. One stream of argument is to emphasize the ongoing unfairness of male dominance and discrimination. Another stream of persuasion is to assert the superiority of feminine qualities over various elements associated with masculinity.

A prominent line of argument that has probably been significantly effective is that rigid gender roles injure men as well as women. Many men are sometimes stressed by not feeling free to express some emotions or actions that seem too feminine. Many men believe they must be tough and assertive, even when thoughtful consideration of others would be more productive. Fathers engaging in more child-care time would likely have better relations with their children and have more fun. The benefits for men of women's liberation is an important factor in the large and speedy cultural transformation in the relations between men and women in the United States since the 1960s.

Confronting Opposition

NOW and the feminist movement encountered large-scale opposition as well as small-scale, local disputes. Some women thought the women's liberation movement demeaned the traditional housewife. They were satisfied, even proud, of raising their children and supporting their husbands' careers. Various more elaborate critiques of aspects of feminism were raised.[16] At times, such sentiments and concerns were mobilized to constrain and avert the efforts of the women's movement and of NOW to expand women's rights and options. The intense and long-lasting struggle about adding an Equal Rights Amendment (ERA) to the US constitution is one example. Conflicts about banning or allowing abortions is another.

The Equal Rights Amendment
The ERA had been discussed and favored among many political leaders after World War II; indeed, the Republican Party's platform supported it in 1940. Concern that it would end laws providing benefits and protections to women workers resulted in opposition from trade unions and the Democratic Party.

However, the Democrats shifted and supported it in the 1944 platform, but support for the necessary legislation was weak. As discussed earlier, several bills that banned sex discrimination had been enacted in the early 1960s. Frustration at the lack of enforcement, however, had contributed to the establishment of NOW. In 1967, NOW endorsed the ERA, over the opposition of some unionists and social conservatives, who left NOW.[17] In 1970, NOW began a series of direct actions to win the congressional legislation needed for ERA. In February it picketed and disrupted a US Senate subcommittee hearing. In August, over 200,000 American women conducted a national Women's Strike for Equality, demanding full social, economic, and political equality. Protesters in Washington, DC, presented a petition supporting ERA to Senate leaders. Congressional hearings on the ERA began in 1970 and it was passed by the US Congress and signed by President Richard Nixon in March 1972. The ERA would declare "Equality of rights under the law shall not be denied or abridged by the United States or by any State on account of sex."

Ratification of the ERA by three-fourths (38) of the state legislatures, within seven years, was required for it become an amendment to the US Constitution. In the first year, 22 state legislatures had ratified it, but only 8 more ratified it in the second year. Five states revoked their ratifications, and in 1978 Congress voted to extend the deadline for ratifications to June 30, 1982. Many legal disputes about these actions remain in contention. In 2020, Virginia became the 38th state to ratify the ERA. In January 2022, President Biden called on Congress to pass a resolution recognizing that the ERA is properly ratified.[18]

The widespread strong support for the ERA in the early 1970s was countered by a politically effective opposition, which did not have to reflect the national majority opinion in order to win. Ratification within a set time period was not achieved. Many persons in NOW and others in the women's movement felt defeated by the failure to ratify ERA. The arousal and mobilization of opposition to the ERA is generally attributed to the vigorous campaign led by the socially conservative Republican activist Phyllis Schlafly.[19] She, and other opponents, stressed traditional gender roles and used traditional symbols of American housewives. They also stressed the threats that the ERA posed, particularly that women could be subjected to military conscription. Other threats included that women would lose protective legal and customary benefits relating to work conditions and to alimony and child custody, in the event of a divorce. Furthermore, it was purported that the ERA could allow same-sex couples to marry, and single-sex public bathrooms would be eliminated. Finally, they argued that the Equal Pay Act of 1963 and the Civil Rights Act of 1964 already provided adequate protection of equal rights for women.

It should be noted, moreover, that some preexisting organizations and interest groups were inclined to oppose the ERA and did so.[20] Right-wing organizations had furnished funds for mobilization efforts. Evangelical Christians, Roman Catholics, and other religious conservatives tended to be opposed, as were the Whites in the southern states.

The proponents of the ERA argued against the alleged threats and stressed that the ERA opposition ignored evidence of ongoing unequal treatment of women relative to men, and that the criticism was based on gender myths. The effort was led by NOW and ERAmerica, a coalition of over eighty organizations. Supporters held rallies, petitioned, picketed, undertook hunger strikes, and committed nonviolent actions.[21] Karen DeCrow, president of NOW (1974–1977) and Phyllis Schlafly carried out over fifty debates across the country about the ERA.[22] More generally, DeCrow sought to broaden NOW's membership and collaboration with blue-collar workers, the gay movement, and men.[23] Some NOW members dissented and thought NOW should have been more focused on gaining support for the ERA.

The fight about the ERA has implications for subsequent contentious issues. Many supporters of ERA tended to blame right-wing special interests for subverting the democratic process. The defeat called for greater political engagement to gain election to office and achieve desired legislation. Many amendment resisters, however, viewed their victory as the result of arousing the average person, especially women, to reject a maneuver by a political elite, the feminist organizations, and their political supporters in Washington, DC. The rejection of elites grew with the rise of Donald J. Trump. The battle about the ERA increased the antagonism and mobilization of feminists and counter-feminists for later battles.

Perhaps more constructive ways of handling the conflict about the ERA were possible.

One possibility emerged in the early 1950s, when the ERA began to be considered in Congress. Arizona senator Carl Hayden proposed inserting a sentence into the original amendment to preserve special protections for women: "The provisions of this article shall not be construed to impair any rights, benefits, or exemptions now or hereafter conferred by law upon persons of the female sex." This was attractive to persons who were concerned about the loss of protections with ERA and would vote against ERA without those protections. Proponents of the ERA regarded that as unacceptable. In retrospect, one might reflect on possible words and understandings that would provide essential equality in the rights of men and women and the protections of women in particular vulnerable circumstances. Negotiations in good will might have reduced antagonism and fear.

Another possible strategy is to introduce additional items into the contention and create possible trade-offs. Thus, proponents of the ERA might offer support on a matter of concern to some anti-ERA groups, such as parental rights,

anticipating reduced opposition to the ERA. Or, anti-ERA groups might offer
consent, with a small modification of ERA wording that would lesson fear of it.
Such conflict expansions and resulting trade-offs are not necessarily crude and
cynical exchanges. They might arise from better mutual appreciation of each
side's concerns and preferences.

Who Decides about Having an Abortion?

In the 1960s, the legality of abortion emerged as a major enduring issue of contention. Earlier, abortion was illegal, but in reality, a woman's decision whether
or not to have an abortion was a difficult private matter. The emerging women's
movement and the loosening spirit of the 1960s were making it a more public
matter, and some states liberalized their abortion laws. In January 1973, the
Supreme Court decided, 7–2, that the US Constitution protects a pregnant
woman's right to have an abortion, within the first two trimesters, without excessive government regulations. That decision in the *Roe v. Wade* case came to divide
many Americans for decades. Initially, leaders of NOW and the general feminist
movement had not taken a clear stance about the right of a woman to choose to
have an abortion, fearing to appear too radical.[24] However, support for a woman's
legal right to decide to have an abortion had become widespread, among both
Republicans and Democrats, and certainly among feminists. NOW leaders then
declared their strong support for the Supreme Court decision and many other
reproductive rights, such as contraception. Demonstrations followed, including
the 1.1 million March for Freedom of Choice in Washington, DC, in 2004.

In opposition, a religious New Right movement soon arose, countering both
feminism and abortion.[25] Abortions have always occurred and will not be entirely eliminated. The key issue is access to legal and safe abortions. Many religious and conservative people increasingly came to regard opposing access to
legal abortion as central to their identity and agenda. Social movements often
generate opposing movements, with each side incorporating multiple social
movement organizations. That was the case with the women's movement; while
NOW was the primary organization in the feminist movement, other organizations focused on particular matters. For example, the Women's Equity Action
League (WEAL) was founded in 1968 to counter discrimination against women
in employment and education spheres. Feminists for Life (FFL) was founded
in 1972 as a feminist anti-choice organization. As the salience of legal access to
abortion access grew, these organizations had to move clearly to be for or against
a woman's right to have an abortion. WEAL moved close to NOW and FFL
moved close to the organizations that would ban abortions and away from the
feminist ones. Bridging organizations declined and polarization intensified.

Planned Parenthood is a major nongovernmental actor in the conflict about
access to abortion in America. It originated in 1916, when Margaret Sanger, her

sister, Ethel Byrne, and Fania Mindell opened America's first birth control clinic. In 1921, it was transformed into a national organization, the American Birth Control League, and into a few other organization and name changes, until 1942, when the name became the Planned Parenthood Federation of America (PPFA). Its size and breadth of services expanded greatly, varying among the affiliates. Abortions were being provided in only a little over a half of the centers.[26]

By the time of its 2018–19 annual report, Planned Parenthood had grown to over six hundred health centers, with 2.4 million patients, and was providing 9.8 million services.[27] Of the services provided, 50 percent are testing and treating of sexually transmitted diseases; birth control information and services account for 26 percent; breast exams and Pap tests are 6 percent, and abortion services constitute only 4 percent of the services provided. Its revenues are 37 percent from government health service reimbursements and grants, 36 percent from private contributions and benefits, and 23 percent nongovernment health services revenue.

Despite the widespread services Planned Parenthood provides, from medical service to education, particularly for low-income women, it became a target for political and violent attacks. It has responded with great energy. This includes organizing and mobilizing 13 million supporters to take actions in local communities and educational campuses. Storytellers have been engaged to participate in rallies and press interviews. Demonstrations were held in Washington and across the country to oppose the confirmation of nominees to the Supreme Court with anti-choice inclinations. Finally, it has been highly active in the courts, challenging and sometimes overturning new state laws that would restrict or ban the medical service it provides. Next, I will discuss how the Republican Party became very hostile to abortions under all or most conditions, and then the extent of violence directed at providers of abortions, before turning to examine some constructive ways that the conflict about abortions is being conducted.

It is not obvious that the Republican Party would be intensely opposed to abortions and would support government banning of them, with varying exceptions. After all, the 1973 Supreme Court decision was framed in terms of protecting citizens' right to privacy, independent of government control, an idea that is generally consistent with the Republican Party's belief in limited government and personal freedom. Actually, many Republican and Democratic presidential candidates often spoke similarly of the difficult moral issues in deciding to carry out an abortion and determining who should make that decision. Yet the strong statements and actions to deny access to abortions have been made more often by Republicans.[28]

In 1969, in Richard Nixon's first year as president, he expanded the federal government's family planning initiatives and spoke of its importance.[29] He was silent about abortion and advised congressional Republicans to be so as well. In 1972, however, he spoke of his belief in the sanctity of human life—including the sanctity of the yet unborn. He did this in response to the statements of the Democratic senator Edmund Muskie, a Catholic whom he expected to face in his re-election campaign. It also fit into his belief that the Republican Party could improve its electoral prospects in the industrial Midwest by attracting conservative Catholics to the Republican Party.

The 1973 Supreme Court decision that women had a constitutional right to have an abortion posed a new situation. Republican Party leaders initially offered mixed responses. Barry Goldwater and Nelson Rockefeller lauded the decision. President Nixon did not comment directly, but the White House released a statement that abortion should not be used for population control. Six senators from both parties sponsored a constitutional amendment prohibiting all abortions except ones needed to save a woman's life. The House majority leader, Gerald Ford, less drastically, called for a constitutional amendment to rescind the decision and return the provisions about abortion to the state legislatures. However, no congressional actions were taken.

Anti-choice activists, however, sought to mobilize a broad social movement to oppose the decision. They tend to use the term "pro-life" for the movement, referring only to unborn life. I adopt the term "anti-choice" to keep the focus on the policy that is in dispute: whether or not a woman may choose to have an abortion, without excessive government regulation. The National Right to Life Committee (NRLC), which started under Catholic auspices, expanded to include Protestants, and elected a Methodist physician as its president. Many evangelical Protestants believed that the *Roe* decision was too permissive about abortion and denounced it. In addition, some conservative Republicans who had become activated in the campaign against the ERA joined the fight. As this movement against abortions grew, some Republican political figures spoke out to win its support. Notably, Senator Bob Dole, in a tough re-election campaign in 1974, waged a strong pro-choice campaign and won.

A pivotal event in the linking of the anti-choice movement and the Republican Party occurred in the 1976 presidential race. In his re-election campaign, President Gerald Ford turned away from his earlier support for a constitutional amendment to counter the *Roe* decision and tried to ignore the issue. However, he was pushed to take anti-choice positions by Ronald Reagan and other conservative Republicans. Ultimately, Ford accepted a strong anti-choice plank in the party platform, a call for a constitutional amendment to restore protection of life for unborn children. Ford was defeated by Jimmy Carter, but presumably for other matters than abortion. Reagan was in a good position to win the GOP nomination and he went on to defeat Carter and to affirm an anti-choice agenda for eight years.

In 1989, the US Supreme Court upheld the constitutionality of a Missouri law regulating abortion care, in the *Webster v. Reproductive Health Services* case.[30] Then, in 1992, the Court, in the *Planned Parenthood v. Casey* case, used the *Webster* decision to rule that states could regulate abortion care, even in the first trimester, in order to protect fetal life. The *Webster* and *Casey* decisions both gave states the power to apply stringent regulations to abortion care. The anti-choice movement then focused on passing state laws to slowly but steadily restrict access to abortions. This accelerated following the June 24, 2022 Supreme Court 5-4 decision to overturn the *Roe v. Wade* decision.[31] Consequently, political polarization was reinforced by geographic divisions. From the 1980s to the current day, the Republican party became increasingly anti-choice, so much so that even President Trump, who had been pro-choice before he became a politician, adopted a strong anti-choice posture and boasted of making abortion less available as a result of his appointments of Supreme Court justices.

Consequently, political polarization anti-choice coercive acts began to occur in the late 1970s and early 1980s, particularly targeting Planned Parenthood centers. Violence included clinic fires, bombings, other property damage, and murders. Between 1977 and 2015, there were 11 murders, 26 attempted murders, 185 arsons, and thousands of criminal acts against abortion providers.[32] Three of the murders were committed by Robert Dear in 2015 at a Planned Parenthood facility in Colorado Springs. The facility had been featured in a highly edited, inflammatory video, referenced by Dear when the police arrested him. The attack followed a great increase in hate speech and threats after the release of the misleading video. Violent acts since 2015 have escalated.[33] For example, death threats and threats of harm rose from 57 in 2018 to 92 in 2019. Trespassing incidents reached 1,507, the highest since such tabulations began in 1999. President Trump's words and actions against abortions and against Planned Parenthood were associated with increased violence and threats of violence experienced against persons working or being served at Planned Parenthood.[34]

Some acts of violence may be perpetrated by individuals acting alone, but small groups and even organizations are responsible for the acts committed so extensively. The Army of God is one such organization, identified as an underground terrorist organization by the Department of Justice and the Department of Homeland Security's joint Terrorism Knowledge Base. The group published a "Defensive Action Statement," stating that "whatever force is legitimate to defend the life of a born child is legitimate to defend the life of an unborn child."[35]

The violent anti-choice actions of such groups and individuals certainly are not constructive ways to wage a conflict. They may be effective in terrifying some people from having or providing abortions in some degree. However, they do not draw sympathy for the cause of abolishing abortion. Rather, such zealous behavior appalls many people and can undermine the cause they are intended to advance. Of course, many leaders of the anti-choice movement decry a murderous attack when it occurs, acknowledging that the consequences are counterproductive.

In 1994, the US Congress enacted the Freedom of Access to Clinic Entrances (FACE) Act. It establishes that it is a federal crime to injure, intimidate, or interfere with those seeking to obtain or provide reproductive health care services— including through assault, murder, burglary, physical blockade, and threats. This law also prohibits damaging or destroying any facility because it provides reproductive health services.[36]

FACE may have helped contain the most egregious acts of violence, but violent threats and many forms of harassment continue to occur and have even escalated. An ongoing, whole-hearted repudiation of such harassment by members of the anti-choice movement would be helpful. The unchecked practice of terrifying harassment in any circumstances weakens norms against it. It can contribute to such conduct by other groups, such as some of the right-wing groups during Trump's presidency.

A national political transformation of the bitter conflict about access to abortion may not seem likely, but it is possible. Actually, according to a 2019 national survey, only a very small minority of Americans (12 percent) favor making abortion illegal in all cases,[37] while 26 percent said abortion should be illegal in most cases. On the other side, 27 percent said abortion should be legal in all cases, and 34 percent favor legality in most cases. Furthermore, 70 percent did not want to see *Roe v. Wade* overturned, an increase from 60 percent in 1992. Of course, there are differences between Republicans and Democrats and among different states, but there certainly was no consensus for a changed legal solution.

A transformation can arise from the areas of agreement that exist and upon which some people are building. Abortion is generally viewed as undesirable. Understanding the reasons women have abortions can help develop effective policies to reduce them. Several studies in recent years have asked women who had abortions what their reasons were.[38] Interestingly, each woman offered a few interrelated reasons, phrased differently as reported in different studies. Several different themes often were stressed: their inability to properly care for raising a child and/or their inability to pursue their work, education, or other concerns and take care of a new child, and/or their lack of a stable relationship to help raise a child adequately. Some women simply felt they were too young to care for a child or old enough to have borne as many children as they wanted to care for. Carrying the pregnancy to term and giving the child up for adoption seemed less attractive than keeping the child or having an abortion.[39]

These reasons suggest many policies that could help avoid unwanted pregnancies, beginning with contraception by men and women. One frequently noted policy is to provide information and services related to family planning and birth control methods. Other policies are less often considered and pursued. They include the provision of good medical care and healthy living conditions for low-income persons. Another matter is employment benefits and/or public assistance for pregnant women and for months after childbirth. Finally, the availability of good child care at sustainable cost would also reduce the pressure for an abortion. Some members of the opposing sides of the abortion issue might collaborate to advance some of these policies. They might join with other groups striving to improve the living conditions of women who risk having unwanted pregnancies. An obstacle to this may be that many anti-choice persons are associated with the Republican Party, and are thereby disinclined to strengthen the safety net to protect and provide the care babies need.

As it is, there are networks of people from opposing sides who do cooperate to help women with unwanted pregnancies.[40] Unfortunately, leaders of opposing sides sometimes act as if they must sustain a fight without compromise in order to remain leaders. If the goal is reduced, the organization may wither away. Indeed, abortions have been decreasing. They rose in the 1970s and remained high in 1980s, but began a steady decline starting in the 1990s. Most Americans do not hold extreme positions regarding abortion, and it does not have high salience in national elections.[41] Trump tried to make it a major issue in order to obstruct abortions and win evangelical Christian support. But he lost the 2020 presidential election.

National Governmental Actions

The federal government took numerous measures that would raise the status of women and increase their equality with men. An early action was the Equal Pay Act (EPA) of 1963 to prohibit discrimination in the payment of wages on account of sex, signed by President John F. Kennedy. Because it was an amendment of the Fair Labor Standards Act of 1938, it had limitations that were overcome in subsequent legislation, including the Civil Rights Act of 1964 and the Education Amendments of 1972. Significantly, Title IX of the Education Amendments of 1972 prohibited sex discrimination in any education program or activity receiving federal financial assistance.[42] The provisions remained dormant, however, until the 1990s, when several sexual harassment cases in school athletic programs resulted in legal cases and Supreme Court decisions that revolutionized school sports activities. School funding for sport programs for women had to be equal to the funding for men's programs. As a result, female students, like males, had the experience and benefits of actively engaging in team sports and also had access to college sports scholarships. It also won attention to women doing sports, rivaling the attention given to men in sports.[43]

Other federal government actions advanced women's status and well-being. In 1987, Congress established Women's History Month. In 1993, the Family and Medical Leave Act (FMLA) was passed. It provided particular employees with up to twelve weeks of job-protected unpaid leave per year and also required that group health benefits be maintained while employees are on leave. It applies to all public agencies, all elementary and secondary schools, and companies with fifty or more employees. Employees are eligible primarily for the birth and care of the newborn child of an employee or immediate family member, or when the employee is unable to work due to a serious health matter. The bill is helpful but does not provide any funding for loss of wages or child care.

In 1994, the Violence Against Women Act (VAWA) was passed with bipartisan support and signed by President Bill Clinton. It provided funding for investigating and prosecuting violent crimes against women and allowed civil redress when prosecutors decide not to prosecute. It is intended to "strengthen responses at the local, state, tribal, and federal levels to domestic violence, dating violence, sexual assault, and stalking."[44] In 2000, the Supreme Court struck down the VAWA provision allowing women the right to sue the accused in federal court (*United States v. Morrison*), Regular renewals of appropriations have included legislative fights about adding or reducing provisions.

Finally, I note that President Obama signed the Lilly Ledbetter Fair Pay Act in 2009. This overcame the Supreme Court's 2007 decision in *Ledbetter v. Goodyear Tire & Rubber Co., Inc.*, which required that a charge for compensation for discrimination must be filed within 180 days of a discriminatory pay-setting decision (or 300 days in some jurisdictions).[45] The act restored the pre-*Ledbetter* position of the US Equal Employment Opportunity Commission that each discriminatory compensation is a wrong actionable under the federal EEO statutes, regardless of when the discrimination began. It restarted the 180-day clock every time a discriminatory paycheck was issued. The name of the act reflects that it corrected the denial of compensation due to Lilly Ledbetter for her discriminatory compensation, which she discovered many years after it had begun.

Women's Expansion into the Labor Market

A crucial aspect of increasing women's equality with men has been their movement into paid employment. Women's labor force participation rose from 34 percent in 1950 to 57 percent in 2017.[46] Men's participation declined during the same period, from 86 percent to 69 percent. Consequently, women moved from 30 percent of the labor force to 46 percent, about where it remains. Of course, most women have always been productively employed. On family farms, the wife not only was usually engaged in some gardening and raising small animals, but also in preparing food, making some clothing, and caring for a few children. With urbanization, technological advances, market changes, and increases in governmental social services, some family functions were not filled within and the family structures changed.

For many years, women's employment outside the home occurred in occupations that were deemed appropriate: teaching children, nursing, making clothing, and secretarial work. These occupations were relatively low paid. A few women did venture a profession, an academic field, or another male-dominated field. They often suffered indignities in getting the required training and then barriers in getting employed in their chosen field. The account of Supreme Court justice Ruth Bader Ginsburg's early experience is often cited, including graduating at the top of her Columbia Law School class and struggling to be employed as a lawyer.[47]

In retrospect, the transformation was massive and speedy. It was the result of the convergence of several developments, and despite prejudice, discrimination, and patriarchy.

First, major reasons included changes in occupational distribution. One change was the expansion of some occupations traditionally filled by women. For example, healthcare occupations began a strong, steady increase in 1950, particularly for professional nurses and attendants in hospitals and physicians' and dentists' offices, but not for physicians.[48] Similarly, primary and secondary teachers sharply increased between 1950 and 1970 and remained elevated. On the other hand, occupations in which males predominated shrank, such as laborers and craftsmen. Some occupations, including new ones, were relatively open to women as well as men, and they were expanding in this time period. This was true for service workers (except private households) since 1950, faculty and administrators in colleges and universities after 1960, and lawyers and judges after 1970.

A few other processes eased the entry and success of women in the labor market. First, pioneer feminists wishing to enter paid employment would see opportunities in the expanding occupations. As they prepared for them and entered them, other women could see the pioneers as models in the women's movement. Many women had a sense of solidarity and assisted each other to endure and overcome obstacles. In addition, women's status was traditionally determined by their husbands', and therefore they did not face stigma for having been unemployed and without personal income. Because a relatively high proportion of African Americans were poor, improving their status was hampered.[49] Moreover, individual women often have valuable male allies, fathers and maybe brothers, husbands, and sons, who would enjoy benefits from their success, unless the men were blinded by jealousy, little self-confidence, or misogyny.

Of course, the entry of women into occupations that were occupied entirely or largely by men was not without contentions and hardships. I will note the transitions in college and university teaching and in the practice of law. The participation of women in the faculty and administration within higher education varied greatly by discipline before the women's movement emerged. In schools preparing students entering female-dominated occupations, women faculty and administrators were common. In schools and departments preparing students for careers in the natural sciences and engineering disciplines, in which there were few women, the faculty were nearly all males. The circumstances were different in some of the humanities and to a lesser degree in the social science fields.

The discipline of political science was peculiarly masculine, as the founding fathers of the field from 1890 through the 1920s were focused on the state, ignoring women's engagement in public affairs. This was despite the activism of women to win the right to vote, impose prohibition, and advance social welfare. Indeed, a father of the discipline in the United States, John Burgess, opposed women's suffrage.[50] The analytical perspective in political science did not incorporate women. Coincidently, very few women earned doctorates in political science: the percentage had risen in the 1930s to 10 percent but declined to less than 6 percent in the 1950s. Graduates would find it difficult to obtain teaching jobs, except for women's colleges.

Change began in the late 1960s, both in women's engagement and in analytical thinking in the field. Early in 1969, the American Political Science Association (APSA) established the Committee on the Status of Women in the Profession (CSWP). Later in the year, further direct action was initiated by women at the national conference; they organized the Women's Caucus for Political Science (WCPS) to press for greater change sooner. During the 1970s and 1980s, generally in collaboration, the two groups pressured the APSA to have women on all committees and to provide child care at APSA conferences. The analytic scope of the field was broadened by research and publications on women and politics. The percentage of women earning PhDs rose substantially, reaching 42 percent in 2002, and women became as likely as men to be hired at major research institutions. In 2019, eight of the twelve new editors of the *American Political Science Review* were women.[51] Women are not equal in all regards, however, and some issues remain in contention, even about research methods.[52]

The expansion of women in the practice of law has been particularly noteworthy. One contribution to the change was the unanimous 1963 Supreme Court ruling in *Gideon v. Wainwright*, which stated that under the Sixth Amendment of the US Constitution, states must provide an attorney for defendants in criminal cases who cannot afford their own attorneys. This certainly resulted in the demand for more lawyers. Law school enrollment quickly rose from 20,776 in 1963 to 34,289 in 1970 and continued to rise until it peaked in 2010 at 52,404.[53]

Women's enrollment in law schools in 1963 was only 3.6 percent, and in 1971 it still was only 9.5 percent.[54] Some women viewed being a lawyer as a vehicle to advance women's equality. Women increasingly entered law schools and in 2017 made up 51.3 percent of all law students.

Women practicing law followed the same pattern. Before the 1970s, the rare female law school graduate, even ones at the top of the class at elite law schools, faced discrimination and had great difficulties in being hired to practice law.[55] Many outstanding women who were rejected later went on to make important contributions to American society. Moreover, women in law firms were systematically treated unfairly regarding promotions, contact with clients, and participation in law firm functions. Concerted action by women began to bring about changes for women lawyers. In 1971, a women's rights group at New York University joined with the Columbia Employment Rights Project in Columbia Law School to investigate such discrimination, and then filed a complaint against ten major law firms with the New York Commission on Human Rights. This resulted in a class action against the firms and, by 1977, settlements with all firms for guidelines assuring the hiring of women associates. Subsequently, the percentage of women in large law firms rose from 14.4 percent in 1975 to 40.3 percent in 2002.[56]

Women's engagement in military service also warrants attention as a sign of rising status equality. After the US military ended conscription and established an all-volunteer force, the number of women serving on active duty rose quickly for twenty years and then leveled off. The number of women serving as enlisted personnel grew from 42,278 in 1973 to 166,729 in 2010. The percentage of women in the enlisted ranks increased from 2 percent in 1973 to 14 percent in 2010.[57]

Attention should be given to the increased participation of women in electoral politics at the local, state, and national levels. Improvements in their status helped provide the capacity to enter higher and higher political positions. For example, the proportion of women in the US House of Representatives surged in the 1990s among the Democrats. This matter is examined further in the next chapter.

The great expansion of women into the labor market has not attained full equality for all women in all employment locations. Issues of equity in pay and advancement arise in many cases. As there is greater equality at all levels of hierarchy within all institutions, those issues will lessen. If rights are to be gained and sustained, they must be exercised. Some concerns about the increased participation of women in the labor force may foster contention and opposition to policies that foster and enhance it.

Another enduring women's movement and organization targeted employers of secretaries and clerical workers due to low wages, lack of benefits, sexual harassment, and other kinds of demeaning treatment they experienced.[58] This is the 9to5 movement, which started in 1972 as a newsletter. A year later, the newsletter publishers formed a local collective: Boston 9to5. It soon won a class-action suit against several Boston publishing companies that awarded $1.5 million for back pay to the female plaintiffs. It sought to form a union and ultimately became Local 925 of the Service Employees International Union (SEIU), with considerable autonomy. The movement gained much visibility from the 1980 film *9 to 5*, starring Jane Fonda, Dolly Parton, and Lily Tomlin, which benefitted from interviews with 9to5 members. The movement organization's current name is: "9to5, National Association of Working Women." It functions nationally, providing a wide range of benefits, services, and education for its members and the American public. It has been creative in embracing a very wide range of activities, as well as those of traditional trade unions.[59] The good will of such policies tends to contribute to more constructive consequences.

Destructive contentions often arise and persist due to zero-sum ways of thinking, such as thinking that women's gains must be at the cost of men's losses. In some circumscribed setting that may be so, but from a broader perspective it does not hold. When more women entered the labor force, the job market actually grew.[60] Clearly, when women left the family household, some of the work they did as unpaid family members began to be done by other people outside the family, who were paid. In addition, an analysis of 250 metropolitan places from 1980 to 2010 found that "as more women joined the work force, they helped make cities more productive and increased wages."[61] This would be expected insofar as the entry into the labor market is attributable to an expanding job market. The earlier discussions of the shifting occupational distribution and of jobs being created to replace unpaid family labor suggest that a growing job market was occurring in certain metropolitan areas.

Domestic Violence and Sexual Harassment

Women in America, as elsewhere, have long suffered from violent abuse by intimate partners with whom they were living.[62] The abuse has gone by many names, such as domestic violence, wife beating, and violence against women. At the beginning of the period examined in this book, there were no national laws prohibiting such conduct, and the state laws, where they existed, were not very effective. Domestic violence began to become a recognized social problem in America in the 1970s.[63] The organized undertaking to reduce domestic violence has been long, marked by progress and setbacks. It began with the establishment of shelters for battered women and crisis services. The first US shelter opened in Phoenix, Arizona, in 1973, and many other shelters soon opened across the country. Local and national organizations began to give attention to such domestic violence and public awareness grew. Women's service organizations, such as the YWCA, and federal agencies became involved in establishing shelters. Local chapters of national women's organizations such as NOW took actions to arouse public concern.

It was not an increase in violence against women that generated the growing attention, but rather the growing engagement in responding to the ongoing domestic violence.[64] In the early 1970s, the *New York Times* hardly had any stories about domestic violence, but the coverage increased steadily in the late 1970s. The same was true in popular magazines and television. Sociological research has found that social movements to improve coping with a social problem do not expand simply because a social problem has intensified. The resources of the people in the relevant social movements are very important in the movements' growth.[65] Efforts to obtain supportive legislation were successful in some localities and states. National bills were proposed in 1977, 1978, and 1980, but they failed to overcome the opposition of some members of Congress. The social movement organizations moved on at the local, state, and national levels, which was a basis for collaborative action at different levels at different times. This was noted in the previous chapter regarding the Black Lives Matter movement.

The success in the passage of VAWA in 1994 is an example of the convergence of many different actors and events that brought a success. The fight was waged largely using institutionalized procedures, assisted by external actions and interpersonal relations. An NPR podcast reported the convergence of diverse actors that produced VAWA.[66] In 1989, Joseph Biden, then chair of the Senate Judiciary Committee, read an op-ed about a man in Montreal who had entered an engineering classroom, separated the women from the men, and began shooting, killing fourteen women students. Biden had wanted to do something to help women, and he then decided to get some legislation for women. He tasked the only female lawyer on the committee staff, Victoria Nourse, to write something. She was not a feminist and had no background or knowledge about legal protections for females. So she started to read about violence against women in American history. Nourse studied numerous rape trials and found that these cases were handled differently than other crimes. She found that to claim rape in some jurisdictions, you would have to take a lie detector test. There were places where prosecutors refused to take the cases. There were judges who would act as if wearing a short skirt meant assuming the risk of rape. Nourse realized that the federal protections she thought she had did not exist. The criminal law said some people are just not protected, yet the Constitution says all Americans have equal protection under the law. Something occurred to her: a constitutional basis for such a law was in the Fourteenth Amendment, which guarantees that all Americans had equal protection of the laws. The chief counsel, Ron Klain, agreed. The next task was to win enough votes in Congress to pass the legislation, which would require "muscle."

Nourse called the NOW Legal Defense Fund and connected with Sally Goldfarb, a law professor at Rutgers Law School, who was working on civil remedies for domestic and sexual violence. Goldfarb was excited to learn that Biden was excited about the VAWA. Earlier attempts to win congressional approval for actions against wife beating had failed. For example, a decade earlier, Senator Alan Cranston of California had tried to get federal funding for battered women's shelters. He brought a measure to the floor, and it was filibustered. Gordon Humphrey, a senator from New Hampshire, argued that battered women shelters were indoctrination centers. They would indoctrinate people into the idea of feminism. Senator Jesse Helms of North Carolina chimed in that battered women's shelters would encourage the disintegration of the family.

Nourse and Goldfarb needed someone who knew how to lobby. They knew that Pat Reuss was a great lobbyist and organizer on women's issues and asked her to move the VAWA forward. She said yes and went to work traveling the country, preaching the VAWA gospel. She met with a wide variety of groups, including Girl Scouts, civil rights groups, labor unions, religious denominations, and women's rights groups. She would say, "I'm here today to talk about violence against women and Joe Biden's bill. And I know he's a guy—and everybody would laugh, you know—but he really is earnest and we have a chance. And in the bill, there'll be money for shelters. And then I said—and in the bill, we have a provision that gives people a right to sue their rapist or their batterer."[67]

Now they needed to convince members of Congress directly that this was important. Reuss had a separate strategy for the politicians and coached Goldfarb on how to lobby. For example, once when Goldfarb was flying back from Washington to New York after a day of meetings, she found out that she was on a plane with Chuck Schumer, who was then in the House of Representatives. She checked on the phone with Reuss, who told her, "You approach him. You love him. You are equals right now. And you maybe drop that you chair this giant coalition."[68] They did have a conversation about the VAWA. It turned out they lived in the same Brooklyn neighborhood and Schumer offered to drive her home from the airport, which resulted in Schumer becoming a major supporter of the act.

Leaders of the coalition recognized that hearing from survivors was necessary to raise public awareness and get bipartisan congressional support. They organized a series of public hearings. Hearings were held by relevant committees of the Senate and House in Washington, DC, and elsewhere in the country. A woman in the first hearing, Marla Hanson, was a former fashion model whose face was horribly slashed in 1986 outside a Midtown Manhattan bar. Hanson had rejected sexual advances by her landlord, and he had hired two men to slash her face in retaliation. Another witness had been left partially paralyzed after she was attacked by her estranged husband. Members of Congress heard many such stories of desperation and lack of support, and they were appalled. The hearings even impacted Senator Orrin Hatch, an ultraconservative from Utah. He had a staff woman who had been cruelly beaten by her then ex-husband and heard her story at a public hearing in Salt Lake City. After that, he told his staff to come up with some compromise.

Opposition, nevertheless, remained, and some came from surprising places. First, traditional liberal allies were not very interested; even within the women's rights movement, some people believed that other issues, like family leave, abortion rights, and workplace equality, needed full attention. Furthermore, the VAWA was attached to the criminal justice system, which concerned some feminists. Simultaneously, some members of the justice system were themselves disturbed by the legislation. The biggest enemy was a group of federal judges, who are not supposed to lobby, but they thought the act's civil rights provision would enable women to use the statute to get higher settlements in divorce cases. The chief justice of the Supreme Court, William Rehnquist, led this resistance, warning that if the bill passed, the courts would be flooded with family squabbles. The judges ultimately backed down and agreed to abandon their resistance to the civil rights remedy to violence against women.

In 1994, nearly four years after work started on the bill, the political context had changed. In 1991, during Justice Clarence Thomas's confirmation hearings to the US Supreme Court, law professor Anita Hill testified that she had been sexually harassed by Thomas, her former boss. She testified before the Senate Judiciary Committee, chaired by Joe Biden. The interrogation of Hill was often hostile, and Thomas was confirmed. Biden afterward was apologetic.[69] The hearings inspired women across the country to run for political office, and twenty-four women were elected to the House and five were elected to the Senate in 1992.[70] On September 13, 1994, President Clinton signed the Violent Crime Control and Law Enforcement Act, which included the VAWA.

Sadly, VAWA did not wholly survive for long. Christy Brzonkala, a student at Virginia Tech, alleged that she had been attacked by two football players. One of the players, Antonio Morrison, was suspended. But ultimately, he was reinstated. Meanwhile, Brzonkala was traumatized and dropped out of school. She then brought a suit under VAWA. Brzonkala lost in the lower courts. In 2000, she brought her case to the Supreme Court. Chief Justice William Rehnquist wrote the opinion for the 5 to 4 decision, holding that Congress had no power to enact the civil rights remedy. Under our federal system, Rehnquist said, that remedy must be provided by the Commonwealth of Virginia.

Clearly, Supreme Court activism occurred to block progressive national legislation. The VAWA was passed with strong bipartisan support and with backing from a broad citizen coalition. When the bill expired in 2019, the House passed a new version that was stalled in the Senate when Mitch McConnell was Senate Majority Leader.

The COVID-19 pandemic contributed greatly to increasing domestic violence.[71] Families were confined in close quarters and lived in very stressful conditions, resulting from a loss of jobs and income. Disputes were more likely to arise. Violent conduct is a more likely result when good conflict management skills are absent. Biden's awareness of the harms of the increase in domestic violence with COVID is indicated by the inclusion of tens of millions of dollars in the $1.9 trillion American Rescue Plan of 2021 to help the victims of domestic violence recover from the burdens of that violence.[72]

Domestic violence is a serious social problem, and it is hard to reduce. It is often hidden and regarded with shame. Yet it has severe long-lasting intergenerational societal harms. It is dangerous for police as well; it may not be the most dangerous circumstance for police intervention, but about 5 to 10 percent of domestic violent calls do result in an assault on officers.[73] Also, it results in severe injuries, broken families, and increased poor households.

More attention is certainly needed to forge possible preventive policies. Reducing the stress many families experience might be accomplished with enhanced, sustainable child care. Good, affordable housing would, of course, be helpful. Workshops and counseling on domestic relations and how to disagree constructively could be provided to individuals, couples, and children by churches, community centers, and social work agencies.

Sexual harassment is a much more frequent phenomenon than domestic violence. The Pew Research Center examined what Americans see as needed work to do for advancing gender equality (among those who believe more needs to be done).[74] Interestingly, sexual harassment was most frequently cited: by 77 percent of all adults, 72 percent of men, and 82 percent of women.

Significantly, another social movement, #MeToo, arose in 2006 and then grew quickly after 2017 to afford some security for women, and some men, against sexual harassment in business and other institutions. Tarana Burke, a survivor of sexual harassment and an activist, first used the phrase "Me Too" on social media in 2006.[75] She intended it to help empower particularly young and vulnerable women subjected to sexual harassment and abuse by providing some empathy and a sense of solidarity. The phrase went viral in 2017, when it became related to the numerous allegations of sexual abuse, dating back to the late 1970s, by the powerful movie mogul Harvey Weinstein. Weinstein was a major film producer with many highly successful films to his credit. Celebrities joined the discussion and he was expelled from the Academy of Motion Picture Arts and Sciences. He was arrested on May 25, 2018, in New York and charged with rape and related crimes against two women. A trial date was set for January 6, 2020, and he was found guilty on February 24, 2020, of two of the five criminal charges. On March 11, he was sentenced to twenty-three years in prison.

Much discussion flourished on social media about and by celebrities, men as well as women. Allegations of harassment and abuse in academia, politics, churches, as well as the music and television industries appeared, and consequently many individuals, often with multiple charges of sexual abuse, lost their positions. Discussions flourished about changing the norms of conduct. In addition, rules were established within a wide variety of institutions to improve conduct and enhance the security and effectiveness of making charges of sexual harassment and abuse. For example, in November 2017, Jackie Speier proposed the Member and Employee Training and Oversight On Congress Act (aka the ME TOO Congress Act). In January 2018 it became an Amendment to the Congressional Accountability Act of 1995, and in December 2018, Congress passed it and President Trump signed it.[76] The provisions included a requirement that any payment by a member of Congress for any harassment settlement would be paid by the member (rather than with any federal tax funds). Also, details of the matter would be made public and not be confidential.

The widespread #MeToo publicity made conduct that had been hidden, shameful. The consensus and the speed of the congressional response was remarkable. Timing can be crucial in making changes. The congressional action was well targeted. However, no advantage was taken of the widespread condemnation of sexual harassment and abuse in order to pass legislation to prevent or criminalize domestic violence or to mitigate its consequences. That still seemed too hard to forge a winning coalition.

In summary, the improvements in the status of women in American society since 1945 have been transformative. Major steps have been taken toward greater status equality. As a result, women—and men—have more options than they previously had in choosing how to live their lives. Certainly, conflicts were waged to make advances nationally, within states and cities, within and among organizations, and within families. Often, advances were made by women simply acting directly to enact advances. Women were also persuasive about what they sought. Their words were persuasive because many men could listen and heard what their mothers, wives, sisters, and daughters had been saying. Since men and women generally live in families together, recognition of mutual benefits of greater status equality for women occurs. Thoughtful framing of specific gains to be won were pursued through political and judicial channels.

Nevertheless, there was and continues to be resistance to moves toward greater status equality for women. This is the case for some men, and even some women, who believe they will lose some privileges and status as a result of the changes brought about by the women's movement and feminism. Such fears are sometimes enflamed by extremist ideologues.

Overall, according to the American people, greater equality for women is still needed. More needs to be done. A 2020 Pew Research report found that when it comes to giving women equal rights to men, 57 percent say it has not gone far enough, 32 percent say it has been about right, and 10 percent say it has gone too far. There was some gender difference: women responded at levels of 64 percent, 27 percent, and 8 percent, respectively, and men at 49 percent, 37 percent, and 12 percent. The difference by political party is much larger. Among Republicans/lean Republicans, the responses were 33 percent, 48 percent, and 17 percent, while among Democrats/lean Democrats, the responses were 76 percent, 19 percent, and 4 percent.[77] That political party difference was evident in the earlier discussions here about conflicts regarding government policies related to gender issues.

In closing, it must be noted that the COVID-19 pandemic produced particularly ill effects for women's circumstances. Many of the service occupations in which women were employed shrank very much in response to the pandemic. Reduced household income sometimes pushed women to leave careers and employment to care for their children. The 2021 American Rescue Plan, with its provisions related to children, helped ameliorate these difficulties.

Other Collective Identities

Collective identities have increased in number and importance since the 1960s. They have become the basis for collective action as members of some groups seek to rise and become more equal, while others struggle to rise and not lose or sacrifice any entitlement. For this book, I have focused on African Americans and women. I briefly discuss only two others next, LGBTQ individuals and White supremacists. Many other identities might be discussed, including religions, ages, sections of the country, education levels, and cultural preferences. These can be important societal dividers parties in conflicts, but I will not do that in this book. It could take many volumes to do so. I will refer to some of those other identities in relation to conflicts in the next chapter, dealing with changes in the distribution of American political power.

Gay Liberation and LGBTQ

Some persons have neither male nor female identities; instead, some of them regard themselves as being gay, lesbian, bisexual, transgender, or queer (hence "LGBTQ"). Such people have long suffered great prejudice, and many of them lived hiding that identity. It was regarded as shameful and immoral, but that began to change in the late 1960s, and within a few decades a remarkable transformation and normalization had occurred. The term "gay liberation" is sometimes used to incorporate the LGBTQ struggle for the rights of persons with lesbian, gay, bisexual, transgender, and queer identities. The 1960s were marked by the claims made by people in many communities for recognition. Many organizations and publications proliferated in the spirit of a new Left. All this was the context for LGBTQ people to fight for their rights.

One important campaign was to confront and end the designation by the American Psychiatric Association (APA) that homosexuality was a mental illness. For several years counter-evidence was presented at professional meetings. Finally, in 1973, that designation was corrected and ended by a vote of APA members.[78]

Also in the 1960s, the New York City Police Department harassed bars and clubs frequented by homosexuals (the term at the time) to rid the city of undesirables. Resistance to such harassment at the Stonewall Inn in Greenwich Village, in June 1969, became the Stonewall riots, which are now commemorated by a parade every year in June. The idea of holding a yearly gay *pride* parade resonated with the movement and spread around the world.[79]

Gay men and others with LGBTQ identities were certainly becoming more visible, and as that happened, more people acknowledged and accepted their friends and relatives so identified. That helped to lower levels of homophobia in society and make the public more ready to extend rights to them.

However, beginning in the 1980s, HIV/AIDS was spreading, particularly among gay men. That tended to increase homophobia from 1986 to 1991, but it also accelerated the liberation movement because it became a life and death issue and made more gay men visible.[80] In later periods, the negative impact on views of persons with HIV/AIDS had declined from 1995 to 2011, as indicated in a report based on the Kaiser Family Foundation's 2011 Survey of Americans on HIV/AIDS.[81] Specifically, reported levels of discomfort decreased; for example, the share saying they'd be "very comfortable" working with someone who has HIV increased from about a third in 1997 to roughly half in 2011. There have also been striking declines since the early years of the epidemic in the share expressing the view that AIDS is a punishment (from 43 percent in 1987 to 16 percent in 2011), or that it is people's own fault if they contract the disease (from 51 percent to 29 percent). Very significantly, in 1995, new medications had begun to sharply decrease AIDS-related deaths and hospitalizations.

In general, the status of LGBTQ persons is raised by having good class standing. I think it is safe to say that many gay men and other LGBTQ persons are successful persons. During the height of the AIDS endemic, the early deaths of leading figures in the arts and other realms raised the regard the public had for LGBTQ persons. As their participation and engagement in American society has increased, homophobic prejudice has shrunk. President Trump's actions to reduce some civil rights of persons having LGBTQ identities may have been popular for many of his base supporters but it was not in tune with the public at large.

The idea of same-sex marriages becoming acceptable in America was a sign of growing acceptance. Pew Research Center polling reports that in 2004, Americans opposed same-sex marriage by a margin of 60 percent to 31 percent. Fifteen years later, though, the results were reversed: in 2019, a majority of Americans, 61 percent, accepted same-sex marriage, while 31 percent opposed it.[82] As in other identity matters, political party differences are marked: 75 percent of Democrats and Democratic-leaning independents favor same-sex marriage, while only 44 percent of Republicans and Republican leaners favor it .

Indeed, the change to approving same-sex marriage is a remarkable achievement by the LGBTQ community. Community members framed the issue as two people loving each other and wanting the sanctity of marriage, which was brilliant. The old, sordid image of gay men meeting each other for sexual encounters disappeared. That was replaced by images of old same-sex couples, loving partners for many decades. The 2018 elections included electoral firsts for LGBTQ candidates: the first openly gay governor and the first openly bisexual US senator.

The struggle for people with LGBTQ identities for equal status with other Americans made progress through a convergence of factors.[83] Mormon opposition spurred LGBTQ mobilization for electoral campaigns allowing same-sex marriage. It has been waged by those people taking direct action, openly living their identities. It has been waged largely through persuasive efforts. Advocates were able to raise money and to shame donors to opposition campaigns.[84] It has been opposed by some people on religious and ideological grounds and taken up by right-wing Republicans. But for conservative Republicans, generally favoring small government and few governmental regulations, it would seem inconsistent to invite the government to invade the privacy of homes.

White Identity

Finally, I turn to consider White ethnicity, the other side in most of the African American conflicts previously discussed in this book. A comprehensive analysis of any conflict should include a full treatment of all major parties in the conflict.[85] My analyses of African American conflicts in chapters 5 and 6 lacked such treatment, assuming my observations and the reader's familiarity would suffice. My discussion of White identity that follows will provide some relevant additions. The very idea of a White identity needs to be examined. In a society so much dominated by White people, their distinctive identity may be vague in content and boundaries. Its size is so broad that many other ethnic or other identities may be much more intense.

As discussed in chapter 5, races are socially constructed, and therefore I usually use the term "ethnic" to encompass what popularly might be referred to as races. These ethnic identities are largely determined by parental lines but are recognized as loosely bounded and to some degree self-selected. Americans have always given great attention to the multiplicity of ethnicities that exist in America. Ethnicities are socially constructed, mostly by those claiming the ethnicity, but also sometimes imposed by the claims and actions of nonmembers of the ethnicity. American Whites largely succeeded in homogenizing slaves of African descent from their various ethnicities into being Blacks. They also made Black and White into two peoples: in slavery, any person with one-eighth or more of African ancestry was Black.

The meaning of White in America became problematic in the period between the 1840s and 1850s, when three million Irish Catholics fled the famine imposed on them in Ireland. The nativist "Know-Nothing" movement rose up to oppose the immigration of these foreigners. Soon, many more immigrants from southern and eastern Europe, fleeing poverty and hardships, entered the country of great opportunities. Such poor people were not White enough for many of the established Whites who had come earlier. In 1924, a new American Immigration Act was passed, which set quotas favoring immigration from northwestern European countries and setting very low quotas for some others.

The people who had gotten to America were expected to assimilate into the preexisting White culture, as in a melting pot, and be White Americans with shared identities as Whites and as Americans. It never quite happened. Various national identities, religions, and ideological beliefs were generally recognized to have other collective identities within the overarching White American and American identities. This book's account begins in 1945, and the American celebration of diversity within White and American identities was gaining ground. Moreover, the overlap between White and American was decreasing. People would stop saying to someone of non-White appearance, "You don't look like an American."

The celebration of diversity was increasing due to moral and ethical changes and demographic changes. The legitimacy of the demands by Indigenous and Black people in the United States for equal rights and living conditions was difficult to deny. Indeed, the recognition by Whites that their policies had imposed dire circumstances upon non-Whites was growing. The 1924 Immigration Act was widely seen in America, and also internationally, as racially discriminatory. In 1965 Senator Philip Hart and Congressman Emanuel Celler introduced a bill that would overturn the 1924 immigration system. Both northern Democratic and Republican members of Congress strongly supported the bill, which was opposed mostly by southern Democrats. It was signed into law by President Lyndon B. Johnson.[86] The new bill prioritized relatives of US citizens and legal permanent residents, as well as refugees and individuals with specialized skills. The act maintained per-country and total immigration limits, but it also opened limited immigration from the Western Hemisphere. The act greatly increased the total number of immigrants as well as the share of immigrants from Asia and Africa.

After the 1960s, the American population of non-Whites was increasing faster than the population of Whites. The reality of the great increase in Hispanics, mostly from Mexico, in the United States was noted in chapter 5. I only note here that "Hispanics/Latinos" are allowed by the US Census to self-identify as White or as non-White. I focus my discussion in this chapter on the variety of attitudes among Whites, a highly consequential collective identity among Americans. The changes in the salience of certain views among Whites was especially critical in the appeal of Trump, as discussed further in the next chapter, on power.

The distinctive attitudes that a category of people share is a basis for them having a collective identity. Sets of attitudes may be conducive to waging constructive or destructive conflicts. For example, claims of supremacy or of defining one's self in terms of facing enemies are conducive to hostile relations. On the other hand, taking pride and pleasure from one's history, achievements, or lifestyle can be conducive to amiable relations. Consider the often-recognized difference between nationalism and patriotism. Nationalism connotes mobilization against antagonistic others. Patriotism is more internally centered and more cultural than political. Nevertheless, any recognized differences in identities are susceptible to the creation of "us" versus "them" sentiments.[87] Ethnocentrism is a very widespread phenomenon.[88]

What might seem puzzling is that collective identities can be splintered. The splintered groups can be competitive and even intensely antagonistic with each other. This has been notoriously the case among splinter groups among people who thought of themselves as Marxists in past decades. In the chapter on African Americans, we noted great collaboration among several civil rights organizations, but also intense rivalry among some other African American organizations seeking to improve the circumstances of African Americans. Among African Americans, many sub-identities waxed and waned. There were, for example, various Black separatists and also Black integrationists.

Furthermore, some "racial" designations are created by outsiders. "Hispanic" incorporates people from many different countries and territories who have strong identifications with the places they or their parents came from. That is even more clearly the case for "Asians," a term that incorporates people from countries that have waged wars against each other. Refugees from civil wars may come to America from opposing sides in the fighting, holding the identity of their side in the civil strife more important than any shared identity.

As discussed earlier, the American definition of White people has been problematic, sometimes requiring also being Protestant or, in later periods, Christian. So, even presently, White Americans can also incorporate adversarial sub-identities, by religion, ideology, or national origins.

Some White Americans claim that people of their identity are supreme over African Americans and have a right to subordinate African Americans. Some claim similar rights against Jews and persons of "Mexican" or other non-European descent. Some such White supremacists advocate keeping such hateful people from entering the United States, wishing to expel them, and even seeking to terrorize them. Some such people gather in groups and organizations to share their ideas and feelings, discuss possible actions, and sometimes act in ways they believe will implement their goals. During the civil rights struggle discussed in chapter 5, many Whites wanted to maintain Jim Crow rules subordinating African Americans. Some White supremacists used their official positions or acted unofficially to impose their views. The recourse to violence to retain African American subordination was counterproductive, however, and Jim Crow was largely overturned. That result was in large measure the result of so many Whites who condemned Jim Crow segregation.

Much research has been done about Whites' and non-Whites' views on race relations. Researchers usually simply regard people who claim to be White or appear to be White to be so. As discussed in chapter 6, most Whites presently do not express negative prejudicial views toward African Americans in answering survey questions. Indeed, a Pew Research Center study found that 55 percent of White Americans think ethnic diversity is very good for the country, and 20 percent think it is somewhat good.[89] Furthermore, 64 percent think racial and ethnic diversity has a positive impact on the country's culture, and only 14 percent think it has a negative effect, while 22 percent believe it doesn't make much difference.

Despite the acceptance of integration and even satisfaction with diversity, there is less readiness to support policies to advance interethnic equity. For example, in the same Pew Research Center report, respondents were asked how important it was for employers to promote racial and ethnic diversity in their workplaces. Among Whites, 43 percent replied it was very important and 30 percent said it was somewhat important. However, the responses were different when more action was in question. When asked if employers should only take a person's qualifications into account even if it results in less diversity, or take a person's race and ethnicity into account in addition to qualifications in order to increase diversity, 78 percent of the Whites choose the first option and only 21 percent choose the second. This suggests that goodwill alone is not enough to make advances in equity; some people need to take actions that will move others to actually act differently.

I turn now to consider the variations among Whites in their sentiments about their White identity. Ashley Jardina has integrated her and other researchers' work about White identity, identifying many degrees and contents of White persons' White identity.[90] One often used indicator of the importance of White identity is simply to ask Whites how important that identity is for them. In eight surveys done between 2010 and 2016, conducted face to face and in two cases by web, the question posed asked how important being White was for them. Five responses were offered the respondents: (1) Not at all important, (2) A little important (or Not very important), (3) Moderately important (or Somewhat important), (4) Very important, and (5) Extremely important. The middle option, "moderately important," was consistently the most frequently given. Otherwise, the responses were generally spread out about equally in the quintiles. The proportion answering "extremely important" varied between 11 percent and 21 percent. There does not appear to be any change trending over the period between 2010 and 2016.

The results are somewhat different for two other indicators of White identity: pride in being White, and degree of commonality with other Whites. Very few say "none at all" or "a little," and the most extreme choices are more likely to be chosen. Similarly to the importance question, the middle choice is the most chosen for pride (33 percent) and commonalty (42 percent). Overall, Whites possess a moderate degree of identity as Whites, and 30–40 percent have much sense of White identity. Sometimes these two indicators are combined with importance of the identity to constitute an index

A related concept to White identity is White consciousness; the concept is that various ideas are shared by people holding the same identity. That is to be expected, and the finding that people identifying as White vary greatly in the importance or strength of their identity would correlate with varying strengths of associated ideas. Survey data have been gathered about political and social views that may be correlated with varying degrees of White identity importance. I will simply identify six variables that were found to be statistically correlated with strength of White identity.[91] They are varying degrees of the following: (1) ideological identification (conservative), (2) racial resentment, (3) anti-black stereotype index, (4) Hispanic feeling thermometer, (5) Asian feeling thermometer, and (6) Muslim feeling thermometer. This indicates that Whites having very strong and important White identities also tend to have stronger conservative commitments and more oppositional ideas about non-White minority groups.

There are interesting implications of the finding that there are correlations between the strength of White identity and various political and social ideas. For example, Whites who feel extremely identified as being White and believe that Whites will soon not be the majority in America may tend to feel threatened and oppose non-White immigration. However, the many other Whites whose identity is held less strongly would not feel so threatened. Their identity as American or as Christian may be very important. Or they may believe that immigrants inherently make important contributions life in America. Of course, there is some tendency for various points of view to be clustered together. People with strong ethnic identifications are likely to have various views that sustain their identity. Nevertheless, people are able to hold views that might seem contradictory to others. After all, views are acquired from many different sources over changing times.

Conflicts are waged about these matters and conflicts change peoples' minds. Many beliefs, preferences, and other attitudes are correlated with each other and to some extent reaffirm each other. However, that does not necessarily indicate any causal direction of one view determining another. They may both be held because of some similar circumstance, allegiance, or ideology. It may also indicate that changing any one belief or preference can result in changes in others. This is crucial in waging conflicts as leaders in each side in a fight try to mobilize supporters for their cause and undermine the views of people on the opposing side. In the next chapter on conflicts related to political power, I will consider such efforts pertaining to class and to status contentions.

Conclusions

Status differences are important sources of conflicts, many of them seemingly destructively intractable. There is much commentary about the difficulties in resolving culture conflicts and bewailing the prevalence of "culture wars." Nevertheless, we have seen in this chapter that there have been remarkable transformations in the ranking of members of some collective identities that historically were devalued and oppressed. Often, an important component of such movements toward greater equity is the activism of members of the low-regarded collective identity community. They begin acting as if they deserve and are getting better regard from higher-ranking communities. They seek and require greater respect and the perquisites of higher status.

The response of higher-status communities is not always simple opposition and repression. The solidarity of the opponents is often fractured. The costs of acknowledging a rise in the status of the previously low-ranked identity are recognized as small. Indeed, there may be benefits in rising shared gains. Indeed, some people feel better for acting better to and with other humans or simply other Americans.

I share an extreme instance, with an anecdote. I visited South Africa in June 1998, when the hearings of the Truth and Reconciliation Commission were closing down. In my conversations with many Afrikaners, I was struck that most were accepting of the radical transformation that was taking place in the status and power of Whites relative to Blacks. Indeed, many expressed their pride that apartheid had ended nonviolently, and Afrikaners acceded to its ending. When I left South Africa, I happened to sit on the plane alongside a young man, an Afrikaner, who was leaving South Africa to live in Europe. He explained that he anticipated that his professional career would be capped now in South Africa, as Blacks moved up. However, he went on, that was right. Apartheid was wrong, and he was gratified by the change.

Of course, the linkages of conflicts are often quite different. A conflict that results in a step forward is followed by a backlash from the side pushed back, forcing the previous advancers to move one or two steps backward. Perhaps the fighters who had made a sizable step forward are viewed as having overreached, arousing fear and anger that energized a fierce reaction. Or the step forward had been modest in the eyes of the opposing side and therefore perhaps a sign of weakness and vulnerability, thereby prompting a strong counterattack. Often, conflicts result in one side moving two steps forward, which is followed by a conflict that results in one step back. Insofar as mutual gains are made and recognized, backlashes will be minimized.

There are no formulas predicting the way one conflict will be linked to the next outbreak. It depends very much on the resources, aspirations, and perceptions of the opponents. Using a constructive conflict approach would entail considering the alternative likely responses from the other side to reaching for one goal or another. This suggests the utility of some exploratory undertakings to learn about outcomes that would be stable. These matters will be examined further in the next chapter, which traces the ongoing struggles for power.

8
Changing Political Power Equality, 1765–2022

This chapter examines changing power inequalities from several perspectives. First, I discuss the concept of power and the forms of political power in their US historical context. This will indicate the importance of considerable equality in political power in US values and ideals. Then I consider power inequalities in connection to class and social inequalities and relations between political parties and nonofficial actors. I also examine and assess several important conflicts that relate to inequalities in political power after 1945.

Concepts of Power and Forms of Political Power

Power is a much-used term with many meanings. Many social scientists base their usage on the definition that Max Weber provided: "Power (Macht) is the probability that one actor within a social relationship will be in a position to carry out his own will despite resistance, regardless of the basis on which this probability rests."[1] Weber analyzed how that power may be viewed as legitimate and obeyed when the power wielder is accorded the authority, on one or another ground, to give orders. When authority is not accorded, orders may be resisted, and conflicts arise.

Many social scientists regard the ability to impose one's will on others as *distributive power*. Actors may be individuals or large organized entities. Distributive power may be viewed from two perspectives: from the command over others or from the capacity to be free from the command of others. There is a further implication: that the command over others is for the benefit of the commander.

Another kind of power that many social scientists recognize is *collective power*, or the capacity of a group to realize its common goals.[2] Exercising collective power entails combining resources, organizations, cooperation, and technology that enables a group, political entity, or country to prosper. The advancement of a group or a society sometimes is sought at the cost of external groups or societies. Again, that often is the source of a serious conflict. Of course, in actuality, exercises of power are often some blend of distributive and collective

Fighting Better. Louis Kriesberg, Oxford University Press. © Oxford University Press 2023.
DOI: 10.1093/oso/9780197674796.003.0008

power. Nevertheless, I will simplify the matter by referring to particular power exercises as predominantly collective or distributive.

Distributive power and collective power are connected in many ways. Within a government agency, different individuals or factions may contend among each other to determine whose policy position will prevail. Their fight may be viewed as primarily between them and to determine who defeats whom. That would seem to be a distributive power contention. But sometimes the conflict is presented and viewed as a choice about the best policy to advance the shared collectivity relative to an external one. Which designation is correct or even salient is a matter that is more or less clear and is often debated. How the actors view the power struggle influences the means of struggle. If the partisans think their fight is distributive, within a shared system, the methods of fighting may be relatively controlled. If the fight is over collective power between different collectivities, the methods of struggle may be relatively unconstrained.

The long struggle that resulted in the independence of the United States of America can be viewed as a shift from a struggle within the British Empire between colonists and the British government for distributive power to a struggle of one country against another. Starting in 1765, many people in the thirteen North American colonies of the British Empire began to nonviolently resist the increasingly objectionable actions of the British government. The British Parliament imposed ever more duties on imported goods to pay for British soldiers remaining in the colonies after the French and Indian War. The Stamp Act of 1765 particularly infuriated many colonists; it imposed a tax on documents needed in trade, legal proceedings, and civil life. With each act by Parliament, colonial boycotts of British goods and other forms of opposition grew, with ever more impact.[3] In 1773–74, cities and towns began organizing themselves outside of British rule, and British power throughout the colonies was disappearing. Resistance leaders met at the First Continental Congress in 1774 and they adopted a plan for further nonviolent struggle. Gene Sharp observed that if the plan had been followed, the colonies might have been liberated sooner and with fewer deaths than by the US War of Independence.[4] The eruption of the battles of Lexington and Concord in 1775 turned the conflict into an armed struggle. The preceding ten years of boycotts and many other nonviolent methods loosened the bonds the colonies had with Britain. The nonviolent struggle generated an independent economy, alternative forms of governance, and a sense of US identity.

The Declaration of Independence, issued on July 4, 1776, set forth the fundamental rights that all people have and the requirement that governments provide for their attainment:

> We hold these truths to be self-evident, that all men are created equal, that they
> are endowed by their Creator with certain unalienable Rights, that among these
> are Life, Liberty and the pursuit of Happiness.—That to secure these rights,
> Governments are instituted among Men, deriving their just powers from the
> consent of the governed.

These words have resounded around the world, defining democracy and its ul-
timate legitimacy. The power of the governments derives from the people and is
legitimate when the government protects and fosters life, liberty, and the pursuit
of happiness for the people.

The US War of Independence was harsh and long, even with the aid of Britain's
imperial rivals, France and Spain. Finally, King George III agreed to US independence and preliminary articles were signed in November 1782. On September
3, 1783, the Treaty of Paris was signed between Great Britain and the United
States of America and was ratified the following spring. The first constitution
of the United States was set forth in the Articles of Confederation on November
15, 1777, and ratified on March 1, 1781. It was an agreement among the thirteen
original states of the United States of America and provided rules for how the
new government would function. Its collective power, however, soon proved to
be inadequate. It lacked an executive branch to execute laws passed by Congress
and a judicial system to adjudicate disputes. Congress did not have the power to
tax nor to regulate foreign or domestic commerce.

On May 25, 1787, a Constitutional Convention began meeting in Philadelphia
to amend the Articles of Confederation.[5] George Washington of Virginia was
elected as president of the convention. Very soon, the fifty-five White, male
delegates began writing a new constitution with much greater collective powers.
They managed to balance many contradictory desired elements. For example,
they wanted a powerful national capacity and also wanted to avoid any future
tyranny, and they wished to protect important interests of the former colonies
and now states. The solution was to create a complex national government with
three branches of power so that they could check each other from becoming too
powerful. Furthermore, some matters were reserved to governance by the states.
Contentious conference disputes related to many issues, including the composition and election of the Senate, whether to vest the power of the executive
branch in three persons or a single chief executive, how to elect a president, the
nature of a fugitive slave clause, and whether to abolish the slave trade. The US
Constitution was completed and signed by thirty-nine of the fifty-five delegates
on September 17, 1787, and became effective March 4, 1789. Soon afterward, ten
amendments, known as the Bill of Rights, were added, offering protections for
individuals' liberty and justice.

The US Constitution has survived, with amendments, as a framework that is widely honored by the people, even when interpreting it is often disputed. It did not meet the aspirations of the Declaration of Independence, but it has been a vehicle toward a more perfect union. Progress has been embedded in many amendments. As frequently noted in the analysis of conflicts thus far in this book, conflicts have been waged and transformed within and against the confines of the Constitution. Undoubtedly, also, many conflicts will continue to be waged to modify the constitution by interpretation and by amendment. The text of the constitution begins with *We the People*, which recognizes the premise and purpose of the Constitution—to serve the US people. Abraham Lincoln, in dedicating a cemetery for the dead from the fighting in Gettysburg, provided a memorable definition of US democracy: He called all Americans to give:

> increased devotion to that cause for which they gave the last full measure of devotion—that we here highly resolve that these dead shall not have died in vain—that this nation, under God, shall have a new birth of freedom—and that government of the people, by the people, for the people, shall not perish from the earth."[6]

Most significantly, Lincoln proclaimed the Emancipation Proclamation, on January 1, 1863, breaking with the Constitution.[7] Amendments transformed the Constitution, ensuring equal rights for all citizens. Congress passed the Thirteenth Amendment, outlawing slavery, before the Civil War had ended. After the war ended, White southerners and organizations like the Ku Klux Klan targeted freed persons with violence. To protect African Americans' and all citizens' equal rights, Congress passed the Fourteenth and Fifteenth Amendments. The necessary ratifications of the amendments by the states were made.

Different forms of power inequalities exist in many arenas of social relations, including corporations, families, churches, universities, cities, states, and countries. In this work, I focus not only on class and status conflicts, but also on power conflicts, particularly in US political arenas—institutions of governance at the local, state, and national levels. First, I discuss changes in the relative power of different people in controlling their own and others' lives (distributive power), and in shaping policy for the country as a whole (collective power). The changes often entail conflicts, which tend to be conducted within or about the context of a political system. Thus, great collective power inequality exists when a few people at the pinnacle of power make decisions about collective choices with little regard to lower-ranking people. This would be an authoritarian system. It may be viewed as directed by legitimate authority and followed without challenge. For example, during the COVID-19 pandemic, people who believed in the expertise of medical officials readily obeyed the directives about masks and vaccines.

Persons who did not accord legitimacy to the commands tended not to obey the directives. Given different political party sentiments and state variations in party strength among the states, the level of compliance and noncompliance varied.

Collective power relates to making decisions about policy for the collectivity as a whole. In the matter of responding to rising hospitalization rates for COVID-19 illnesses, government officials may be more or less decisive about ordering lockdowns. The opinions of the public about economic or medical concerns weighs more or less, depending largely on the different concerns' strength and links with political leaders.

Simplified characterizations of the basic US collective and distributive power systems abound. At one extreme, there is a relatively egalitarian democratic model in which major decisions about societal policies are determined by the people, reflecting majority views. Even distributive power is egalitarian, with all adult individuals possessing great liberty from directives from others. Attractive as that may be to some in the United States, it is hard to imagine that this was ever fully the case in America nationally or even in any substantial communities. At another extreme, collective power and distributive power are largely controlled by a single cohesive elite, which is also hard to imagine.

There are many gradations between such extremes, varying over time, arena, and sphere. The perceptions of those gradations have also always been varied. In the 1950s–1970s, several influential social scientists stressed the great power of a cohesive ruling class.[8] Numerous books drew widespread attention to the power that a small elite had. For example, C. Wright Mills, a Columbia University sociology professor, published *The Power Elite* in 1956, in which he argued that "institutions are the necessary bases of power, of wealth, and of prestige, and at the same time the chief means of exercising power, of acquiring and retaining wealth, and cashing in the higher claims for prestige.[9] Moreover, the men at the pinnacle of these institutions associated with each other, and sometimes some of them moved from one pinnacle to another. Mills further argued that the public was not a real check on the power exercised by the elite, the members of the public having become significantly members of a mass society. Many of them were feeling alienated, isolated, and powerless. Other analysts made similar arguments about the relative strength of a unified ruling class, including G. William Domhoff and Floyd Hunter.[10]

Concerns about major decisions being made without public attention or control were widespread. President Dwight Eisenhower, in his Farewell Address to the Nation, on January 17, 1961, warned the people. He famously said:

> In the councils of government, we must guard against the acquisition of unwarranted influence, whether sought or unsought, by the military-industrial complex. The potential for the disastrous rise of misplaced power exists and will

persist. We must never let the weight of this combination endanger our liberties or democratic processes.[11]

This was said as a warning, which received much attention, but it did not have any significant consequence. Military budgets continued to expand, despite many cries for their curtailment.[12]

Numerous authors did challenge the conclusions that the United States was ruled by a united ruling class, arguing that power was more disbursed and equal.[13] They pointed to the powers of people in communities and states and to countervailing organizations such as trade unions, churches, and other voluntary associations. Although recognizing some reality to mass society qualities, voluntary associations and other phenomena countered the ill effects of those self-alienating qualities that undermine democratic engagement.

In any case, popular concern about powerful elites ruling the country to serve their own interests seemed to begin to decline in the 1970s. It is striking that popular attention to any powerful ruling class in the United States relatively dissipated as class inequality began its spectacular rise. Indeed, not only was income and wealth inequality greatly increasing, as discussed in chapter 4, but corporate profits shot up, while wages were almost stagnant.[14] Moreover, corporate lobbying and influence in formulating legislation began to rise greatly. By 2015, large corporations and their associations spent $34 for lobbying for every dollar spent by labor unions and public interest groups combined.[15] The lobbying became not only directed at minimizing government regulations, but also intended to gain positive benefits. For example, pharmaceutical companies opposed adding prescription drugs to Medicare, fearing increased government bargaining power through bulk purchasing. However, around 2000, industry lobbyists thought of supporting prescription drugs as a Medicare benefit along with the provision forbidding bulk purchasing. There is another kind of linkage between corporate and political high positions, the "revolving door." Lobbyists are often well-paid, especially when they serve an industry they got to know while working in the government. Anticipating such moves can influence the conduct of persons in government service.

Finally, politicians and political parties, needing ever more money in order to win elections, raised their dependence on monetary contributions by large corporations and wealthy persons. Moreover, the wealthiest contributors provide the greatest proportion of money to candidates' campaigns.[16] In 2010, the Supreme Court decided the *Citizens United v. Federal Election Commission* case, ruling that the government could not restrict independent expenditures for political communications by corporations. This 5–4 decision further increases the influence of major corporations in political affairs. As discussed at the beginning of this book, inducements in conflicts include coercion, persuasion, and benefits. Money is an often-used benefit to purchase desired behavior. That is often regarded as corruption, except when it is legally allowed, as it often is in the United States.

International Context

Despite increasing class inequalities and increased political power gained by the very rich, public attention to ruling elites' domination of national policies declined. Understanding any conflict is improved by considering interlocking conflicts. A major matter that helps explain the lack of public attention to class conflicts was the high salience of foreign conflicts. Following the end of World War II, the Cold War quickly erupted in crises related to Soviet-aligned governments in eastern and central Europe. The threat of Communist expansion was heightened by the Soviet Union's first nuclear weapon test on August 29, 1949. Then, on October 1, 1949, Chairman Mao Zedong proclaimed victory in China's civil war and the founding of the People's Republic of China. Chiang Kai-shek, leader of the defeated government, retreated to the Chinese island of Taiwan.

Other wars followed. On June 25, 1950, military forces of the Democratic People's Republic of Korea (DPRK), seeking to unify Korea, attacked the Republic of Korea (ROK) to the south of the 38th parallel, the line of division between the two governments claiming sovereignty over the entire peninsula.[17] DPRK forces moved quickly down the peninsula. President Harry Truman, in accord with United Nation resolutions, dispatched military forces to aid the ROK. Under the command of General Douglas MacArthur, allied forces forced the DPRK forces north to the previous border along the 38th parallel. In an over-reaching action, the US and associated forces continued to move close to the border with China. Chinese troops entered the war and forced the US and allied forces back to a line close to the 38th Parallel, and a stalemate resulted. That continued until an agreement to cease all hostilities was signed on July 27, 1953, by representatives of the United Nations Command, the Korean People's Army, and the Chinese People's Volunteer Army.

The Vietnam War emerged slowly, but became a large, lengthy, and highly divisive war that absorbed US attention. It originated in the broader Indochina wars of the 1940s and 1950s, when Ho Chi Minh's Viet Minh, inspired by Chinese and Soviet Communism, fought the colonial rule first of Japan and then of France. The eight-year French war effort was largely supplied by the United States. Nevertheless, finally, it was defeated at the Battle of Dien Bien Phu in May 1954. Consequently, in July 1954, Vietnam was divided at the 17th parallel as a temporary demarcation line. North of the line was the Democratic Republic of Vietnam, controlled by the Vietnamese Communist Party. In the South the French transferred most of their authority to the State of Vietnam. Nationwide elections to decide the future of Vietnam, North and South, were agreed to be held in 1956.

President Dwight D. Eisenhower began providing assistance to South Vietnam. By late 1955, Ngo Dinh Diem had consolidated his power in the South. Diem called for a referendum only in the South, and in October 1955 he declared himself president of the entire Republic of Vietnam. It is worth considering what would have happened if a referendum had been held in all of Vietnam. Even if the Viet Minh had won, the US well might have achieved the good relations that were achieved after a terrible war. The number of people killed during the Vietnam War includes as many as 2,000,000 civilians on both sides and some 1,100,000 North Vietnamese and Viet Cong fighters.[18] Between 200,000 and 250,000 South Vietnamese soldiers died. The Vietnam Veterans Memorial lists more than 58,300 names of members of the US armed forces who were killed or went missing in action.

As it happened, the Diem regime had many inadequacies. During 1963, US dissatisfaction with it grew. Diem's brutal crackdown against Buddhist demonstrators convinced US leaders that Diem could not effectively oppose the Communist threat. As a result, in November 1963, they did not avert several South Vietnamese generals overthrowing Diem's government, and Diem's assassination. The stage was set for greater and greater US military engagement in Vietnam, as North Vietnamese forces won over more and more of South Vietnam until they conquered all of it in 1975.

I discuss domestic US conflicts relating to Vietnam and other foreign conflicts later in this chapter. Suffice it to say here that public attention to foreign wars and threats was high until the end of the Cold War in 1992, and then renewed on September 11, 2001, by the Global War on Terrorism (GWOT) and subsequent "endless wars." The importance of military activities did enhance the military budget and increase the military-industrial-congressional ties.[19] Leaders of the military forces and associated industries themselves had increased influence.[20]

Social Status Changes

Other matters also drew public attention away from concerns about ruling elites. The efforts of collective identity communities to improve their status, discussed in chapters 5–7, were in the forefront of much attention. The struggles of women and of African Americans for equal rights and opportunities and for dignity and respect entail a power dimension, by resisting domination by men and Whites, respectively. These struggles manifested distributive power applications from the perspective of those who would deny the dominative distributive power of men and Whites. From a broader, longer perspective, however, the struggles enhanced collective power for America. The enhanced capacities of people whose contributions were previously constrained came to benefit the whole of US society. In making these gains, they reduced the primacy that class inequality would otherwise have on power inequality.

Actually, status inequalities also significantly contribute to power inequalities and to differences in political power. Status politics, as well as class politics, have long been contentious regarding US power inequalities.[21] Class politics have generally been divided between left and right sides, those striving for redistribution of income and wealth and those striving to maintain or increase their income and wealth. Status politics tend to focus on resentments by persons and groups who fear declining status or those who strive for improving their status.[22] Class movements tend to arise in periods of economic depression while status movements tend to arise in periods of economic prosperity. Since no clear political processes exist to deal with status anxieties, irrational scapegoating can flourish. For example, anti-Catholic and anti-immigrant sentiments were relatively high and expressed by the Know-Nothing Party during the prosperous decade before the Civil War. The KKK provides another example when it flourished as it attacked minority rights during the prosperous 1920s.

Finally, status politics at times has distracted attention from class politics in the national political parties, and their electoral processes constitute the arena within which power inequalities are mostly contested. The US Constitution with its amendments provides a framework for America's unique form of popular democracy. It has evolved and serves as the primary way for political power to be widely shared. Its salience and the degree of elite or popular influence in it have varied over time.

Changes in Political Parties

I begin examining the US political party system by considering the distributive and collective power of the US public at large. Voting in elections is a minimal form of engagement in exercising political power. Eligibility was expanded when it was extended to African Americans who had been enslaved, and then to women. Notable expansions in eligibility occurred with the passage of the Civil Rights Act of 1964, which banned tests and other restrictions on registering and voting in the southern states. This was expanded to the entire country in 1965. Furthermore, in the Twenty-Sixth Amendment, the age to be eligible to vote was lowered from twenty-one to eighteen. This happened in 1971, during the

Vietnam War, and public sentiment was that if eighteen-year-old men could be drafted to be soldiers, then they were old enough to vote.

The proportion of the population of voting age who actually vote is remarkably low in the United States. Between 1948 and 2020, in presidential election years, the percentage of the population of voting age that voted varied in the 50s to low 60s. In elections between presidential elections, the percentages are even lower.[23] The low turnout is attributable in some degree to obstacles to registering to vote and to casting a ballot.[24] Politicians generally would try to rouse potential voters for them to actually vote. They would try to make it easier for them to vote. On the other hand, some politicians, fearing votes against them, would seek to discourage and obstruct voting. Unfortunately for democracy, as discussed in the next chapter, in recent years, after widespread efforts to facilitate registration and casting ballots, the Republican Party has undertaken to legislate state and city laws that would restrict voting.[25]

In addition, many potential voters are apathetic. One reason for apathy among some voters has been that the two parties were regarded as very similar, and such potential voters believed the consequences of one or the other candidate winning would not matter. The percentage of voting-age Americans who voted in the 2020 presidential was about 7 percent more than in the 2016 election.[26] Due to the COVID-19 pandemic, access to voting was greatly expanded. In addition, the contrasts between the Republican and Democratic presidential candidates were extreme.

Traditionally, the Republican and the Democratic Parties had broad and overlapping kinds of adherents, each being to some degree coalitions. Back in 1950, many analysts were critical of the diversity of opinion within the parties, which made cooperation between the parties too easy for the politicians and also trivializing party differences for the voters.[27] Some observers disparaged this, believing that if they were more sharply different there would be greater popular engagement. Of course, the parties soon became much less diverse, and their relations became more ideological and antagonistic. The political parties, particularly the Republican Party, seemed to seek distributive power for themselves. Efforts to build and exert collective power for everyone in the country seemed to diminish. This contributed to the increasing antagonism between the two parties. The changes in the composition of the parties and in the ways that they competed against each other and fought each other needs to be recognized and corrected.

As a heritage of the destruction of Reconstruction, which had been created in the South after the Civil War, the White supremacists in the South were Democrats. As noted in chapter 5, the southern Democrats were able to block even national government policies that would have made racially integrated housing feasible. They also succeeded in blocking national legislation banning lynching. When the Democratic Party moved to advance civil rights, the southern Democrats began to break away from the Party. In 1948, Strom Thurmond, senator from South Carolina, led a new party, the States' Rights Democratic Party, known as the "Dixiecrats," which was a precursor of further separation. In 1968, former Alabama governor George Wallace ran as a third-party candidate for president. Soon the southern Democrats had become conservative Republicans. The Democratic Party still retained ideological diversity, with centrists and some progressives.

Changes in the Republican Party began to become increasingly drastic, narrowing its orientation and acting with greater animosity against the Democratic Party. Early on, in the 1940s and 1950s, the Republican Party had an important moderate component, including President Dwight Eisenhower (1953–61). After Eisenhower, Nelson Rockefeller, the governor of New York, became the leader of the moderate wing of the Republican Party. The moderates, however, were defeated in 1964 as conservatives captured control of the Republican Party and nominated Senator Barry Goldwater of Arizona for president. Goldwater articulated a radically conservative orientation and set of policies for an enduring faction of Republicans.

Nixon ran as a conservative, but as president (1969–74), he adopted several Rockefeller Republican policies, for instance setting up the Environmental Protection Agency (EPA) and fostering legislation to reduce pollutants in the air, surface water, and groundwater. The acts exhibited collective power, strengthening the country as a whole. On the other hand, Nixon also engaged in distributional power fights against Democrats. He went so far as to resort to illegal actions to search for dirt against political opponents and then tried to cover it up.

Ronald Reagan's presidency (1981–89) firmly established that the Republican Party was highly conservative. As discussed in chapter 3, he devotedly, if amiably, pursued policies that increased income and wealth inequality and reduced social welfare provisions. He strove to lessen the power of trade unions and reduce government regulations. These actions were congruent with the views of America's corporate and wealth elites and were advocated by some members of those elites in the form of lobbying and making contributions to election campaigns. Some of the elite members also supported think tanks and other channels to influence the public at large.

Many Republican Party leaders couched their efforts as measures that are good for the country. That may be viewed as acting to increase collective power. In the context of the evidence presented in chapter 3, and in this book as a whole, however, that is not correct. Policies that have raised class inequality have generated immense problems for most people not in the class elite. For the class elite to capture control of political power would be highly damaging to America as a whole. In actuality, the right-wing Republican Party was engaging in distributive power fights. This became brutally clear as the Republican Party leaders focused more and more on attacking the Democratic Party leaders and obstructing their actions when they hold dominant government offices.

An important contribution to the increasingly antagonistic relations between the Republican and Democratic parties was made by Newt Gingrich's leadership in the House of Representatives.[28] Gingrich was Speaker of the House from 1995 to 1999 and he energetically fostered a combative approach by the Republican Party. Thus, intense partisanship, tribalism, and crude name-calling became pervasive and democratic norms were ignored by other Republicans. For example, Gingrich often charged that Democrats were corrupt and seeking to destroy the United States. Furthermore, he oversaw major government shutdowns, providing evidence that the government was not working. On the other hand, President Bill Clinton adhered to a policy orientation that approached traditional Republican positions. He declared the end of big government policies and the end of welfare as it had been known. Despite this, Republican Party leaders vilified Bill and Hillary Clinton personally during Bill Clinton's presidencies and afterward.[29] This extended to false conspiracy theories, for example about Clinton's murder of Deputy White House counsel Vince Foster.[30] Such false information was spread by right-wing commentators through Fox News and many other media channels. Wealthy donors such as Robert Mercer and his daughter helped fund anti-Clinton attacks.[31]

The two terms of President George W. Bush (2001–9) were not marked by Democratic obstructionism or personal attacks based on false conspiracy theories. In the aftermath of the 9/11 terror attacks, there was general support of Bush. Indeed, a more critical oversight might have avoided the overreach in the response. When some of Bush's policies contributed to the recession at the end of his term, there was bipartisan cooperation to avoid even worse consequences. However, the radical right-wing Republicans were disappointed in what they considered the liberal-leaning aspects of the administrations of George H. W. Bush and George W. Bush.[32]

The grave international mistakes made by President George W. Bush, Vice President Richard Cheney, Secretary of Defense Donald Rumsfeld, and a few others in responding to the al-Qaeda attack by launching a Global War on Terrorism (GWOT) had enduring bad consequences. The GWOT was open-ended; invading and democratizing Afghanistan was overly ambitious. To make matters worse, Bush decided that Iraq's leader Saddam Hussein was a bad guy and had to be taken down, which he did.[33] Choosing to invade Iraq and transform it, and doing so before resolving operations in Afghanistan, was a sign of hubris and poor judgment. The barriers to such an ill-considered decision failed.[34] The endless wars were set in motion. Many Americans could reasonably feel that their government lacked competence, that the government misled them about Iraq's possession of weapons of mass destruction, and that US leaders did not treasure US lives enough.

In addition to the international context, the financial crisis of 2008 contributed to Barack Obama's decisive presidential election victory over John McCain, winning both the Electoral College and the popular vote by sizable margins. The Democrats also won control of the Senate and the House of Representatives. Obama gave high priority to passing legislation that would provide medical care to many of the country's uninsured people. In July 2009, Nancy Pelosi, then Speaker of the House, announced Democratic plans to extend the country's healthcare system, with the Affordable Health Care for America Act (ACA). As discussed in chapter 4, the Tea Party movement sprang up to block the legislation. During that summer, House members at local meetings were fiercely attacked if they favored the ACA. In the 2010 midterm elections, the Republicans won control of the House, jeopardizing passage of the bill, but it was passed by Congress on March 23, 2010. Thereafter, Republicans sought court rulings to end or undermine it, and in some Republican-controlled states' provisions of the ACA were manipulated to undermine it. Nevertheless, public approval of the ACA grew. Republican leaders continued to attack Obama and oppose his actions. Nevertheless, in the 2012 presidential election, Obama clearly defeated Mitt Romney, taking 51.1 percent of the popular vote and 332 of the 538 electoral votes. Democrats also made a net gain of two Senate seats, retaining the majority. In the House of Representatives, however, despite picking up eight seats, Democrats failed to gain a majority. Despite plentiful funding from wealthy donors, particularly right-wing donors, the Republicans were widely defeated.[35]

Clearly, the class elite, insofar as there is one, is not always entirely the ruling power, although Obama and the Democrats did not threaten any primary concerns of the wealthiest members of the class elite. Obama continued to face consistent resistance from the Republican Party, led by Senator Mitch McConnell who had a simple post-election message for President Obama: "Move toward the GOP or get no help from its lawmakers."[36] Subsequently, in the 2014 midterm elections, Republicans continued to control the House of Representatives and they won control of the Senate. Obama found it very difficult to pass legislation and resorted to executive orders. Overall, Obama did take measures that modestly reduced class, status, and power inequalities. Yet those modest changes did not overcome the forces driving inequalities and even spurred some backlashes.

In 2012, Thomas Mann and Norman Ornstein, careful political analysts, wrote that unlike the Democratic Party, the Republican Party had "become ideologically extremist, . . . scornful of compromise; unpersuaded by conventional understanding of facts, evidence and science; and dismissive of the legitimacy of its political opposition, all but declaring war on the government."[37]

In terms of the approach of this book, the Republican Party focused on striving for distributive power against the Democrats. It had taken on aspects of being a collective identity, which was more salient for them, relative to the US collectivity as a whole. This was associated with viewing the opposition as an enemy and promoting a sense of loyalty to one's own threatened community. This can justify breaking norms of good conduct and believing false allegations about the antagonists. These became signs of solidarity and loyalty, particularly for the right-wing radicals.

Insofar as that process entailed rejecting the primacy of traditional US aspirations to forge a more perfect union with liberty and justice for all, Republican Party conflicts against Democrats were often waged destructively, for nonconstructive goals. According to the approach taken here, such conduct in America tends to be self-destructive, or at least less constructive than a more inclusive approach would be, were it taken. Fighting very aggressively often provokes strong resistance from expanding coalitions and a backlash.

The right wing in the Republican Party had become dominant, which enabled Trump to win the party's nomination for president. The Republican Party had opposed the New Deal, the welfare state, and taxes fairly consistently. Its right wing added a twist, regarding power inequalities of African Americans, as well as of class inequalities. The twist was to appeal to White supremacist sentiments, suggesting that social welfare benefits went to poor Blacks at the expense of Whites. This is consistent with a mistaken zero-sum perspective. Dog whistles indicating sympathy to White supremacist advocates followed. In reality, extensive social benefits, as are common in most other well-to-do countries, would benefit Whites as well as African Americans and other people of color.[38]

In the run-up to the 2016 presidential race, unsurprisingly, Hillary Clinton won the Democratic nomination for president. She had the appropriate credentials to carry on the progress the Democrats had made with Obama. However, Hillary Clinton had some weaknesses. She had been personally attacked for very many years by right-wingers and Republican Party leaders. As secretary of state, she had antagonized Vladimir Putin, president of Russia. Consequently, Russian agents covertly used various media and other channels to disparage and undermine Hilary Clinton's election campaign.

Surprisingly, the Republican Party nominated someone lacking the usual credentials for the presidency, Donald J. Trump. He campaigned by attacking the political establishment, both Republican and Democratic. Despite Trump's dismissive treatment of the Bush family and his harsh treatment of rivals, the Republican Party largely fell into line and backed Trump's venomous campaign against Hillary Clinton. After all, Republican Party leaders had made extreme allegations against the Clintons when Bill Clinton was president and against

Hillary Clinton when she served as secretary of state. Fox News carried endless stories against the Clintons. Trump's charges against Hillary Clinton were in line with what many Republicans had come to believe: that she was somehow corrupt.

In the 2016 election campaign, this TV personality, who portrayed himself to be a hugely wealthy real estate builder, used the rhetoric of a crude and tough leftist populist.[39] He expressed the sense of grievance felt by people fearful of declining status. There were many allegations of sexual misconduct, which were brushed aside by devoted supporters or even admired by some as demonstrations of his manliness.

Republican opposition to the ACA, or Obamacare, had been extreme, with Republican governors taking actions to undermine its provisions. Trump pledged that he would repeal and replace Obamacare with something better, but never did come up with describing what his alternative would be. He adhered to the general right-wing attacks against big government interference and high taxes, promising to free the country from terrible regulations. He spoke out against non-White people from south of the border entering and remaining in America. He adopted extreme positions on cultural issues, particularly relating to abortion and homosexuality, which won him support from Christian evangelical leaders and their followers.

At the time of the election, both Trump and Clinton were viewed unfavorably. On a 10-point favorability scale, Trump's image was worse than Clinton's, with 61 percent viewing him negatively on the scale compared with 52 percent for Clinton.[40]

To almost everyone's surprise, Trump won the election. He won the Electoral College vote, while losing the popular vote, 62,984,828 for Trump and 65,853,514 for Clinton.[41] He polled better than Clinton among Whites, men, older (over 44) people, voters with less education (less than college graduate), and evangelical Christians.[42] Republicans retained control of the Senate and House of Representatives, but in the 2018 midterm election, they lost control of the House. As president, Trump kept to the issues about which he campaigned. He did not alter his positions to order to expand his base. Significantly, Republicans generally yielded and accepted his policies, even when they differed from traditional liberal Republican positions. As will be explored later, some of the appeals to his base resonated with the right-wing members of the Republican Party.

An opening for Trump's leftist language at his rallies is partially attributable to a shift in the Democratic Party. The party had moved away from being the party for the working class and had shifted significantly into a party that fosters meritocracy.[43] Its leaders were drawn from the talented pool of women and minority people. The Democratic Party, which had spoken of workers and unions constantly, shifted to speaking of and for the middle class.

Trump had not planned nor prepared for the presidency, so he moved slowly to make major appointments.[44] Many of those he chose were wholly unprepared, but they were seen as loyal to him. He moved quickly to take personal control of the national government. He claimed to put "America First," by which he meant cutting or reducing engagement with international organizations, alliances, and traditional allies. He acted unilaterally to get better trade deals, with questionable effectiveness. Throughout his presidency he continued to hold large rallies to enthusiastic crowds. This was a new phenomenon. He had a base of supporters who adored him and believed whatever he said.

In addition to assuming policy positions in line with right-wing Republicans, Trump emphasized positions that he thought would win support from particular segments of the population. Many of those positions were consistent with his long-held prejudices, which he expressed in leftist populist terms.[45] He decried America's "trade deficits," globalization, foreign countries, immigrants, and condescending elites. He blamed those factors for the plight of White male workers and promised to correct the way those factors hurt people in the United States. Finally, he flaunted his brazen vulgarity, which also won him huge mass media and social media attention.

Once in power, Trump hastily set out to keep his promises and indulge his prejudices and his personal grievances and desires. He did succeed in getting a budget bill passed that included major tax reductions for corporations and for persons earning high incomes. He gave a high priority to vindictively overturning every policy action Obama had executed. He succeeded in appointing many ultraconservative, young judges to fill vacancies, including three Supreme Court appointments. Although for his first two years as president both Houses of Congress were in Republican control, he had few legislative victories, and his wishes were often unfulfilled. Resistance was strong and sometimes effective; for example, in regard to immigration, the courts ruled that some of his presidential orders and actions were unconstitutional. Sometimes, insiders resisted and at times did not execute what he ordered.

Certainly, during his presidency, many people liked some things that Trump did, even if some of them disliked his manners, but his approval rating remained below 50 percent.[46] When respondents were asked, "Do you approve or disapprove of the way Donald Trump is handling his job as president?," the proportion who approved of his job handling fluctuated between 35 percent and 49 percent.

The appearance of a new virus pandemic, in 2020, necessarily was a huge challenge to the US government. But Trump missed the opportunity it provided to win widespread approval. On January 28, 2020, his national security adviser, Robert O'Brien, began informing him of the grave events that were coming. Trump, however, chose to play it down.[47] Even when leading public health figures began to explain what practices should be followed, Trump often undercut the guidance. Trump's strategy seemed to be to dampen the reaction to the pandemic, in order not to disrupt the improving economy. His conduct, however, politicized the handling of the COVID-19 pandemic, which of course hampered an effective, life-saving response. Divisive language was adopted by Trump followers that extolled freedom from government controls. Republican Party leaders, adhering to Trump, pursued a distributive power policy and attacked Democratic Party leaders. The Democrats retaliated and political polarization deepened. Public health experts used the mass media to explain the pandemic and public health developments.

In the November 2020 presidential election, former vice president Joe Biden and Senator Kamala Harris defeated President Donald Trump and Vice President Mike Pence. The Democratic ticket won 306 electoral votes and 81,268,924 popular votes, and the Republican ticket won 232 electoral votes and 74, 216,154 popular votes. Trump outlandishly insisted he had won the election, and his lawyers challenged the results in many states, but judges dismissed the challenges due to a lack of evidence. Despite all evidence and certifications from Republican as well as Democratic state election officials, Trump persisted in asserting the election was stolen, and most Republican leaders and voters agreed.

The official process to accept the results moved on to the final certification by Congress on January 6, 2021, when the process was violently interrupted by an insurrection attempt. After the disrupters were driven from the Capitol building, the members of Congress certified the election results. The inauguration of Biden and Harris as president and vice-president, respectively, occurred on schedule on January 20, 2021.

Trump's claims of a stolen election were without precedent and difficult to explain. A possible explanation is that he was delusional and believed a vast conspiracy had actually "stolen" the election. However, he never spelled out how that could have happened or how the congressional votes were not also stolen, although being on the same ballots. Other explanations assume he knew the vote was correct, but claiming otherwise might lead to an overturning of the election results. When that failed, the claim was a test of loyalty and/or an expression of grievance that would appeal to his base. A plausible explanation is difficult to find. Explanations of how so many Republican Party leaders and voters accepted Trump's claims are elusive. Short-term fears and feelings of loyalty to a cultish collective identity probably played important roles.

President Joe Biden, on entering office, acted as if a normal transition of power had occurred. He operated in accord with the US Constitution and the established norms of conduct between presidents and Congresses. On the whole, most Republican members of Congress interacted with the president and vice president as if they recognized the Democrats' legitimacy in office. Moreover, Biden claimed to strive for collective power, enhancing America's power, and avoiding efforts at gaining distributional power. He emphasized advancing benefits for all Americans and avoiding setting them against each other. Biden's approach to conflict engagement has seemed congruent with elements of the constructive conflict approach. That will be examined in the next two chapters, when the Biden Administration's domestic policy will be discussed in the context of what is occurring and what is needed in the near future.

Specific Conflicts

After the end of World War II, major conflicts related to political power continued to be waged in ways that extended beyond official institutionalized procedures. Several of the conflicts were related to policies about America's power inequalities. They often included direct actions by citizens engaged in protest demonstrations or other actions outside of official procedures. Often, bitter and destructive conflicts related to distributive power fights emerged between the Republican and Democratic Parties. In addition, distributive power fights also occurred among various collective identity groups, each fighting for power against other collective identity groups. They might be seeking government assistance that would empower them against those who indulge in unwanted domination. Many of these conflicts were discussed in the earlier chapters about class inequality, African American rights, and women's rights. They may be framed as distributive or collective power conflicts.

The mobilization of voters for elections for national, state, and city offices are crucial in US politics. Political party leaders usually oversee turnout campaigns for their candidates, which entails advertising on all mass media platforms. In addition, donors and interested nongovernmental organizations support or carry out organizing efforts, including going door to door, telephoning, and mailing literature. Messaging on social media is also now important. Active voter mobilization efforts can be highly effective, particularly in local elections, given the generally low voter turnouts in these contests. This proved to be the case in the election of relatively progressive Democrats in the elections after Trump's presidency.

There is considerable experience and analyses of ways to organize at the grassroots level in support of desired outcomes. This often entails work by and/or with community organizations such as churches, unions, and other voluntary organizations. Considerable research has been done in academic institutions, think tanks, and peace and conflict resolution organizations about how to wage conflicts effectively and constructively. The judgments made in this book are based on the evidence from such research. That evidence and familiarity with what constitutes best practices are taught in centers across the country. In 2019, over 1,300 persons received degrees in peace and conflict resolution studies, and the number who have had some courses and trainings is a multitude of times greater.[48] Ideas from the conflict resolution field are widespread, such as win-win outcomes, de-escalating conflicts, and attentive listening.

Despite this reality, as frequently noted in this book, bad practices in waging conflicts abound. Some bad practices are ingrained, practiced, and allowed, without much penalty. Consider dirty tricks in politics. They are often deployed by some political figures and their supporters in election campaigns, although they are properly considered unethical.[49] They include actions that interfere with the electoral process, engage in unfair competition, and conduct false negative whisper campaigns. The 2016 election was marked by dirty tricks.[50] Trump notoriously insisted on casting doubt about Obama's US citizenship, offering no evidence and never retracting his falsehoods. Major conflicts, early in the period covered in this book, exemplify the dangers of such conduct.

Red Scare and McCarthyism

A major instance of US conflicts about the extent and sources of subversive actions was the Red Scare and McCarthyism, which erupted soon after the Cold War began. The start of the Cold War, the Communist gaining control of China, the quick acquisition of nuclear bombs by the Soviet Union, and the global spread of Communist ideology prompted fears that Communist agents in the United States bore much responsibility for such developments. Of course, the US government took actions to contain and repel covert Communist actions, but some people greatly exaggerated the threat and even charged that many Democratic officials were Communist agents or aided them in the past or currently.

On March 21, 1947, President Truman issued an executive order, establishing a program to check the loyalty of federal employees.[51] Loyalty boards were instituted in every department and agency of the federal government. The boards undertook to review all employees. If "reasonable grounds" to doubt an employee's loyalty were found, the employee would be dismissed. In addition, the House Un-American Activities Committee (HUAC) became a permanent committee in 1945. Chaired by Democratic Representative Edward J. Hart, it was charged to investigate suspected threats of subversion or propaganda that attacked the US form of government. In 1947, the committee undertook hearings into alleged Communist propaganda and influence in the government, universities, and, notably, the Hollywood motion picture industry. Some industry persons cooperated with the committee, but many did not. Ultimately, the studios boycotted more than three hundred persons in the industry, including directors, actors, and particularly screenwriters. Studio executives explained that wartime films could be considered pro-Soviet propaganda, but noted that the films were valued in the context of the Allied war effort, and were made at the request of government officials.

In a speech on February 7, 1950, Senator Joseph McCarthy (R-WI) claimed that he had a list of 205 names of spy ring members and Communist Party members who were in the State Department. He scapegoated Secretary of State Dean Acheson for the "loss of China."[52] That speech won him national fame, and he went on giving speeches alleging different numbers of Communists. His bullying, extravagant, agitated manner got attention. Moreover, his message was appealing to some people due to his attacks against the elites, and he also attracted resentful Whites who feared that their status was declining.[53] He also resorted to coercion to get power, threatening and then effectively attacking political opponents with abuse while campaigning for candidates who supported him. In addition, he mobilized followers to attack those who challenged him.[54] The power he sought was distributive, and largely for his own self, which also made him vulnerable.

In 1953, McCarthy gained a new platform as chairman of the Senate Permanent Subcommittee on Investigations. He quickly took control, holding numerous hearings, in public and in closed sessions. His lack of consultation with other subcommittee members resulted in the resignation of the three Democratic members, and Republican senators also stopped attending. McCarthy and the chief counsel, Roy Cohn, largely by themselves, ruthlessly interrogated witnesses.

On March 6, 1954, Edward R. Murrow, then one of the nation's most respected journalists, devoted a widely watched episode of his CBS television program to Joe McCarthy's misdeeds. Using McCarthy's own statements, Murrow documented McCarthy's recklessness with the truth and viscous attacks on his critics, which posed a great threat to US democracy. Granted a chance to respond, McCarthy replied ineffectively.

In the spring of 1954, McCarthy charged inadequate security at a secret US Army facility. The Army responded that the senator had sought special treatment for a subcommittee aide who had recently been drafted. McCarthy temporarily removed himself as chairman and special hearings about this controversy were televised. The Army hired the lawyer Joseph Welch to present its case. At the session on June 9, 1954, McCarthy claimed that one of Welch's attorneys had ties to a Communist organization.[55] Welch interrupted, but McCarthy continued his attack. Welch angrily interrupted again, "You have done enough. Have you no sense of decency?" That exchange ended McCarthy's career.

On December 2, 1954, the Senate censured McCarthy by a vote of 67–22. He continued to speak against Communism and socialism, but he was greatly diminished. Shortly afterward, on May 2, 1957, at the age of forty-eight, he died.

Officially the cause of death was hepatitis. Some biographers say this was caused or exacerbated by alcoholism.[56] In the wake of the downfall of McCarthy, the prestige of HUAC began a gradual decline in the late 1950s. By 1959, the committee was being widely denounced. The norms of decent political conduct had held and been reinforced.

McCarthy's exploits and McCarthyism's growth had profound immediate effects and some enduring consequences. Many people in the movie industry, higher education, and government service had their careers cut short, and the results of what their work might have contributed to America were lost. The example of the notoriety and impact that a person could achieve through brazen bullying conduct might inspire others, but it should be a warning that such conduct can be counterproductive, and even self-destructive.

Nongovernmental Organizations and Contentions

Power in America is exercised directly and indirectly by private citizens. Acting as individuals and small groups, they may mobilize large numbers of people to take direct policy actions and also to influence government officials to implement new policies, on the left or on the right. Such relations are obviously an arena of conflicts. In addition, there are many high-ranking persons in nongovernmental organizations who themselves also exercise power over subordinate members, as is the case in large corporations. That was discussed early in the book in relation to analyses of trade unions. People in many occupations strive to characterize themselves as being professionals in order to control their work.[57] People in other occupations or otherwise in lower power positions, such as students, may strive to increase their freedom and relative power.

Student Movements and Protests

In the 1960s, many social movements were underway in the United States, interacting with each other, as examined in earlier chapters. College and university students were significantly engaged in two kinds of social movements contending against different adversaries.[58] One kind of social movement strove against political forces sustaining wars and societal injustices. The other kind of movement was against the authoritarian rule of specific university and college administrations. Consequently, many conflicts overlapped with each other, which hampered transforming each of them alone. The contagion of the conflicts became evident, as an uprising in one university would demonstrate the possibilities of protest and embolden similar actions elsewhere. Hasty confrontations, however, could entail unplanned destructive escalations.

Many college and graduate students supported the civil rights campaign and the antiwar movement. In 1960, at the University of Michigan, Ann Arbor, campus, the Student League for Industrial Democracy reconstituted itself as the Students for a Democratic Society (SDS) and adopted the Port Huron Statement.[59] The statement decried the injustices prevalent in the country and called for corrective action by students and universities. The SDS quickly expanded and functioned with considerable local autonomy. Many members collaborated with people in external left-wing organizations against war and racism.

An early example of rebellions at specific academic institutions happened in the 1964–65 academic year. Students on the Berkeley campus of the University of California insisted on the right to make literature available and raise money supporting various causes. A Free Speech Movement was underway.[60] The dean declared the existing university regulations banned such conduct, and that this ban would be enforced. Civil disobedience actions followed, including sit-ins to undertake negotiations. However, hundreds of students were arrested and taken to jail for a few hours. More protests followed. A new acting chancellor replaced the chancellor who resigned. New rules, allowing for information tables and free discussion, were established.

In the mid-1960s, SDS became more focused on anti–Vietnam War activities. The FBI and other law enforcement agencies infiltrated the movement and sought to disrupt and neutralize it.[61] At the end of the 1960s, SDS was divided about how radical it should be and it splintered at the group's 1969 national convention. One faction evolved into the Weathermen, discussed in the next chapter.

Meanwhile, students at many colleges and universities sought to play greater roles in the way academic departments and schools functioned. They organized locally and undertook to negotiate rule changes. Often, faculty and administrators readily acceded to the negotiations and sometimes sit-ins or other nonviolent actions. Sometimes, however, the occupation of academic buildings escalated to police forcefully removing students.

The wave of student protests in the US during the 1960s culminated in 1968, when large-scale student and youth protests also erupted in many other countries around the world.[62] Indeed, great protest marches were held, notably on May Day in Paris, which was joined by many workers. Many of the protests in western and eastern European countries, and elsewhere, were focused on opposition to war, and particularly on US military engagement in Vietnam. However, local issues were also grounds of protest, including antiracism, anticapitalism, rejection of authoritarianism, and for civil rights. Protests in some countries emboldened young people in other countries to rise up and seek to rectify the denial of desired rights.[63]

In the US, the protests in 1968 at Columbia University are notable. Two issues were paramount: (1) unrevealed institutional involvement in supporting the war in Vietnam, and (2) university plans to build a gymnasium in nearby Morningside Park. SDS students were outraged by the first matter and activists from the Student Afro Society (SAS) about the second. Administration failure to address these matters resulted in student occupation of Hamilton, a university building. The supporters of each issue separated in strategy, with the SAS group drawing in supporters from neighboring Harlem and other local communities. White SDS students moved to occupy another building. There was general support for the goals of the protesters by the various segments of Columbia University, if not of the building occupations. Abruptly, early in the morning of April 30, 1968, New York police stormed the two buildings which; were occupied by White students; it was done violently. Many protesters and some police were injured, and over seven hundred protesters were arrested. Protests and fighting erupted again on May 17–22, 1968.

The protests and their violent suppression were costly for the Columbia University administration and for the university as a whole; but they also resulted in much improved practices. The two original issues were resolved as the protesters wished. Plans for a gymnasium were changed to build it underground on campus grounds. Work with federal agencies would not include classified work. The standing of the university was damaged, resulting in losses in donations and student applications, for many years. Probably, if either side had undertaken some de-escalating moves, greatly injurious damages resulting from the mutually disastrous police intervention would have been avoided.

It is hard to assess consequences of the 1960s or of 1968, a year with many shocking events in the United States. I venture a few observations.[64] The consequences contributed to a loosening of civil behavior and a somewhat more egalitarian manner in American universities. Many of the activists became more committed to progressive engagement and helped move US academia somewhat to the left. In US colleges and universities, men began to become less likely to wear ties. The shifts also generated some backlash and contributed to societal division. Perhaps if the fights were waged less intensely, and more readily negotiated, the results would have been less divisive.

Nongovernmental Actors Directly Engaged in Contentions with Government Officials

Two important examples of individuals arousing public concerns and dramatic change arose in the 1960s. Rachel Carson's influential book *Silent Spring* was published in 1962.[65] She documented the terrible environmental damage that pesticides were doing and the failure of the government to regulate corporate practices that were causing great environmental damages. Many people were energized by the idea of bringing about policy changes to salvage the environment. In 1963 the Clean Air Act was initially enacted to reduce and control air pollution nationwide. Amended many times, it continues as the United States' primary federal air quality law. It is one of the country's first and most influential modern environmental laws. Carson inspired a broad environmental movement, which is marked annually across the country, with Earth Day, since April 22, 1970.

Second, Ralph Nader also published a book that generated great public attention and significant social change against resistance. In 1965, *Unsafe at Any Speed* was published.[66] With the book and his wide-ranging actions, Nader challenged corporations and government agencies to be more accountable to the public. Undoubtedly, as a result, cars have more safety features. He founded Public Citizen, the Center for Study of Responsive Law, and the U.S. Public Interest Research Group, an organization that oversees a federation of state progressive advocacy groups known as "PIRGs." These organizations and others he inspired engage in investigations, lobbying, and other activities to protect consumers and the environment from threats to them.

Other individuals have been associated with forming organizations to influence politics. In 1958, an enduring radical conservative organization was founded by Robert W. Welch: the John Birch Society (JBS).[67] Welch and his associates quickly developed local chapters across the country, stressing anti-Communism and limited government. They tried to influence the Republican Party and ultimately the general public to adopt far-right ideas and policies. They largely used persuasive methods in pursuing distributive power for their ideological beliefs, but modes of influence included use of exaggeration, falsehoods, and reference to conspiracy theories.

The JBS had and continues to have many features that other far-right US organizations tend to have.[68] It has had close connections with extremist racist organizations. This included the White Citizens Council (WCC), a violent White supremacist organization.[69] The WCC was founded in response to the 1954 Supreme Court ruling against segregated public schools. In 1963, a WCC member was found responsible for the murder of civil rights leader Medgar Evers. The JBS supported US military strength, but it consistently opposed overseas military interventions. It called for the United States to withdraw from the United Nations. Members indulged in setting forth bizarre conspiracy lies, such as that Dwight D. Eisenhower was a Communist agent.

In the 1950s and 1960s, leading conservative persons and periodicals, such as William F. Buckley Jr. and the *National Review*, regarded the JBS as a fringe organization in the conservative movement, and feared it would radicalize the

US Right. Many mainstream Republican Party leaders attacked the JBS and successfully held it at bay, but the right wing was winning some internal Republican Party fights, such as when Barry Goldwater won the Republican nomination for president. The organization's influence was high in the 1970s. Its membership and influence declined after the Vietnam War ended. This continued through the 1980s and 1990s following Welch's death in 1985 and the Cold War's end in 1991.

The JBS began a resurgence in the mid-2010s and claimed its ideas shaped the modern conservative movement and especially the rise and rule of Donald Trump.[70] Certainly, members published many works attacking a wide range of political figures, including Franklin D. Roosevelt, George W. Bush, and Bill Clinton. It came closer to the mainstream Republican Party, which itself had shifted to the right. It was a cosponsor of the 2010 Conservative Political Action Conference (CPAC) and cooperated in many undertakings with prominent members of the Trump administration.

It might seem that JBS has been extremely successful in gaining more distributive power, and that the means used may not be congruent with the constructive conflict approach, but it is effective for its narrow purposes. I believe that in many ways the JBS's actions and their consequences have not been broadly constructive. First, it is important to consider consequences for US society as a whole. Fierce attacks that deny the rights and concerns of opponents obviously degrade the relationships and reduce the mutual gains that more constructive interactions could produce. They intensify the divisions among people with different collective identifies, which increases everyone's fears and anxieties. This contributes to mistrust and polarization, hampering collaborative relations in multiple arenas. The power gained by leaders and members of JBS at the expense of others actually tends to reduce the collective power that the United States could likely gain if more collective, and shared, power had been sought. The US aspirations of freedom and justice for all are reduced by increasing inequality in class, status, and power societal dimensions. Finally, positing wild accusations and conspiracy theories degrades civil discourse and useful collaboration, which is costly to collective US well-being. They hamper producing a larger pie with widespread gains.

Major conflicts have been waged in relationship to US foreign policies. One set of these conflicts has pertained to protests against US government military engagements overseas. Protests of involvement in the civil war in Vietnam occurred during the Johnson and Nixon administrations. President Johnson had drawn the US into an ever-expanding military intervention. He saw it undermining his domestic plans, and was troubled by the escalating war, but he could not forgo believing that military success in South Vietnam was necessary to avoid surrendering the Pacific to Communism.[71] He saw no way out that would not make him vulnerable to Republican charges that he had "lost" Vietnam.[72]

The public, which had initially approved of the US military intervention in Vietnam's civil war, generally disapproved by 1968.[73] For example, this question was posed regularly by Gallup: "In view of developments since we entered the fighting in Vietnam, do you think the U.S. made a mistake sending troops to fight in Vietnam?" In August 1965, 61 percent said No, by July 1967 it had fallen to 48 percent, and by October 1968 it was 37 percent. In May 1971, only 28 percent said it was not a mistake. Policy preferences were more complex. Before 1968, preferences to withdraw rose, but even more respondents favored escalation; after 1968 the percent of respondents favoring withdrawal rose sharply, reaching 62 percent in August 1972.

The contentions related to these changing beliefs and preferences were largely waged within the political system, but also in protests on the streets. As Johnson pursued the war, his anticipated campaign to run for re-election in 1968 was challenged by Senator Eugene McCarthy, who made a strong showing in the New Hampshire primary election. McCarthy's campaign was focused on ending the engagement in the Vietnam War by withdrawing US forces. Former attorney general Robert Kennedy soon also entered the race for the Democratic nomination for the presidency. He had a plan that involved the withdrawal of US military forces from Vietnam.[74] On March 31, Johnson announced that he would not seek re-election, and soon thereafter Vice President Hubert Humphrey entered the race. With little time to campaign, Humphrey sought to win delegates and endorsements from powerful leaders. During the spring of 1968, Kennedy was campaigning effectively, but he was assassinated on June 6, 1968.[75] Ultimately, at the Democratic Convention, Hubert Humphrey's party connections were too strong for McCarthy to win, and McCarthy's support was split by Senator George McGovern's entry in the race just before the Convention.

The institutional political processes were not suited to manage the differences of the deeply divided Democratic Party leaders and the numerous nongovernmental organizations opposing the Democratic leaders in the context of a divisive war and a traumatic assassination. This led to days of street protests and riots, August 23–28.[76] Many countercultural and anti–Vietnam War protest groups had planned demonstrations in response to the convention in Chicago. Chicago's powerful Democratic mayor, Richard J. Daley, promised to maintain law and order. Early on, there were some peaceful demonstrations. But as the demonstrations continued and intensified, the protesters were met by the Chicago Police Department in the streets and parks of Chicago. This included indiscriminate police violence against protesters, reporters, photographers, and bystanders, which was soon called a "police riot." During the evening of August 28, with the police riot underway on Michigan Avenue in front of the Democratic Party's convention headquarters, television cameras broadcast the antiwar protesters chanting, "The whole world is watching."

The events were a disaster for the Democratic Party. They contributed to mistrust and rancor among different segments of the party. The Democratic Party suffered a loss in regard among the public at large. The events also weakened its position for the general election in November, and they damaged the US society, embittering social relations and modeling poor civic conduct.

Greater coordination and mutual understanding among and between the protest organizations and Democratic Party factions might have mitigated the destructive qualities of the multiple, interlocked conflicts associated with the convention. In preparing for protests at the convention, the activists in the many organizations might have studied and planned on acting nonviolently and maintaining discipline to avoid hostile escalation and violence. The city government might have prepared better arrangements for camping and holding demonstrations by activists. Finally, the city police might have better prepared to manage large numbers of protesters without recourse to any kinds of violence.

During Richard Nixon's presidency, many significant conflicts were waged. I will discuss two major conflicts, conducted differently. First, I consider the widespread 1970 protests related to an escalation of the Vietnam War. Second, I consider the institutionalized process of handling the Watergate scandal that culminated in Nixon's resignation in 1974.

Once elected, Nixon had continued the war in Vietnam in various ways, including a major escalation announced on April 30, 1970. He sent troops into Cambodia to cut off supply routes to North Vietnamese forces fighting in South Vietnam. Students at colleges and universities across the country launched large protests, including at Kent State University, in Ohio. On May 2, 1970, the Republican governor of Ohio, James A. Rhodes, said of the students, "They're worse than the Brown shirts and the Communist element and also the night riders and vigilantes. . . . We're going to eradicate the problem, we're not going to treat the symptoms."[77] Following such extremist language, on May 4, 1970, National Guard troops opened fire on the protesting students at Kent State; four students were killed and nine were wounded.[78] Students throughout the country responded by occupying campus buildings and halting regular classes.

The shock of National Guardsmen shooting down unarmed students might have generated an outcry that mutual regard for civil life needed to be restored. Instead, some political leaders chose to intensify the conflict, approving of the violence. Ronald Reagan, then governor of California, said, "If it takes a blood bath, then let's get it over with."[79] Nixon ordered intensive surveillance by the FBI and other intelligence agencies to investigate the assumed connections between antiwar demonstrations and Communism.

The way students protested, and administrators, faculty, and police responded, varied greatly from campus to campus.[80] Teaching at Syracuse University then, I can attest to what happened there when students closed down the university and occupied several buildings day and night. The students not only protested the war, but also sought a greater role in university governance and to close the university, ending the semester, with the students receiving course credit. Some faculty spoke at the rallies, supporting the national protest movement. Many faculty members wore arm bands, indicating their readiness to mediate or ameliorate any disputes that might arise. Large open faculty meetings were held about student demands. Ultimately, the university administration decided to end the semester. Throughout the student occupation, the city police chief, Tom Sardino, took a careful and respectful approach. He kept uniformed police off the campus and talked and listened to all sides; he spent a couple of nights with the students occupying the administration building.

One student demand had consequences pertinent to the theme of this book. Many students firmly said: "Teach us nonviolence." That summer, I and other faculty members held noncredit workshops related to nonviolent action. After a year of ad hoc social change classes, Neil Katz was hired to teach courses on nonviolence. That soon led to the creation of a new undergraduate option, the Program on Nonviolent Conflict and Change. That was a forerunner of the Program on the Analysis and Resolution of Conflicts (PARC) established in 1986.

The consequences of student and other antiwar protests were not so benign across the country. The division within the country in some ways was intensified. The Democratic Party shifted more to the left, nominating George McGovern to run against Nixon in the 1972 presidential election. Nixon, running as a law-and-order president, won overwhelmingly. The country seemed to have shifted toward the right. As progress toward greater justice and equality and away from war seemed blocked, some marginal groups did become more radical in goals and means. In reaction to the perceived leftward move and disorderly protests, some people were offended and more hostile to the protesters. Opposition to the war was decried as unpatriotic and hippie conduct was viewed as outrageous. Some liberals were also disenchanted with the New Left, the counterculture, and the anti–Vietnam War protests, and even of elements of Johnson's domestic policies. Some became neoconservatives, increasingly hawkish in foreign policy.

Another Nixon-era conflict was managed and resolved largely by institutional procedures: the Watergate scandal.[81] It arose from the Nixon administration's covert attempts to cover up its involvement in a June 17, 1972, break-in at the Democratic National Committee headquarters at the Watergate Office Building in Washington, DC. Five perpetrators, seeking material that might be used against Democrats, were arrested. They were connected to Nixon's Committee to Re-elect the President. The House of Representatives granted its Judiciary Committee investigative authority and the US Senate then created a special investigative committee. The consequent hearings were broadcast nationwide, arousing great attention. Witnesses, including co-conspirators, testified that the president had approved cover-up plans for the administration's involvement in the break-in. Surprisingly, the existence of a voice-activated taping system in the Oval Office was revealed. As the investigations went on, month after month, the evidence of criminal conduct grew. Slowly, but certainly, Republican members of Congress began to indicate that impeachment was appropriate. On July 24, 1974, the US Supreme Court ruled that Nixon must release the Oval Office tapes to investigators.[82] On July 27, 1974, the House Judiciary Committee approved articles of impeachment against Nixon for obstruction of justice, abuse of power, and contempt of Congress.[83] Public opinion rose sharply to 57 percent supporting impeachment.[84]

With Nixon's complicity in the cover-up made evident and his political support completely eroded, Nixon resigned from office on August 9, 1974. He resigned before being impeached by the House and removed from office by a Senate trial. Afterward, the new president, former vice president Gerald Ford, pardoned him on September 8, 1974, contrary to public opinion. Over decades, the public gradually moved to approve of the pardon.[85] Indicating the magnitude of the crimes committed, sixty-nine people were indicted and forty-eight people—many of whom were high Nixon administration officials—were convicted.

It might be said that the constitutional procedures worked, and illegal conduct was punished, but the response to the scandal was not celebratory. Rather, the sordid conduct of a Republican President and many members of his administration was salient for many people. The Republican Party was diminished, and in the 1974 and 1976 elections, the Republican Party suffered notable defeats.[86] The Watergate scandal also had other consequence, varying among different people. Thus, persons who had voted for Nixon for president were markedly disappointed in him and initially sharply lowered their trust in the executive branch of government between 1973 and 1975, yet they recovered their trust by 1976.[87] McGovern voters initially had less trust, but it steadily increased between 1973 and 1976, although remaining somewhat lower than among Nixon voters. Confidence in the press media initially was higher among McGovern than among Nixon voters, and tended to rise in the same time period. Overall, given other presidential scandals, this one may not have contributed greatly to public distrust of the government in the long run.

In any case, trust in government and with each other has declined to a low level in the United States. This has had tragic consequences. A large-scale research project examined the per capita deaths from COVID-19 in 177 countries from January 1, 2020, to September 30, 2021.[88] There was great variation among the countries. The US had a relatively high death rate, along with other economically advanced countries with high proportions of elderly people. However, among the very many other possible factors examined, the one variable that was most related to per capita deaths was, surprisingly, trust. The US, with its low level of trust, had very high per capita deaths, while the diverse other countries had varying death rates, related to their trust. How to deal with an airborne virus epidemic is widely known in government and public health circles: avoid closed spaces, crowded spaces, and close-contact settings. When policies to foster such conduct were urged by trusted others, and followed, infections would be low. In the US, trust in government and others was low and some people fostered distrust. The result of this distrust in the highly divided American society in those months was deadly for many people.

Conclusions

Having political power that serves the US people has been generally sought in America. Having political institutions reflect the public will is desired. Having political officials who are chosen by the public is the essence of the democratic form of government in the United States. However, in reality, members of the wealthy elite have tended to have a disproportionate large role in electing people to office. In some significant ways, political power had become more equal between 1945 and the end of the 1970s. This is attributable to the progress made by people in previously lower-ranked status categories in gaining more political influence. Nevertheless, the staggering increase in class inequality since the early 1980s tends to magnify the influence of the very wealthiest. This has been masked by the increase in status of previously low-ranking strata in the society, notably women and non-White ethnics.

Moreover, the increase in status for some communities has generated resentments among some of the middle-status persons who feel they are losing power and their relatively higher status. Some members of the class elite may see benefits in that phenomenon and support the groups feeling resentful. The changes in the political parties and their relations discussed in this chapter are associated with these resentments. Some collaboration grew between right-wing White supremacists and some class elitists in attacking the progress of non-Whites and women, as well as steps toward greater class equality. Struggles about political power in America are generally channeled through relations among political parties. We have seen that many badly conducted conflicts arise in the exercise of political power, even in the context of institutionalized procedures to conduct conflicts. Too often the conflicts are about distributive fights, which are framed as win-lose contentions. Collective power fights, framed as having possible win-win outcomes, are naturally more likely to be constructively conducted. The resort to violence is associated with distributive power conflicts. The role of violence in US conflicts pertaining to political power is examined in the next chapter.

9

Issues in Recent Conflicts about Changing Power Inequalities

Changes in the distribution of power are difficult to measure. People are likely to disagree about who is gaining power and who is losing it. Actually, this is a source of fighting. The disagreements about what is an underlying difference complicates matters. It makes constructive ways of fighting difficult. In this chapter I discuss some unfortunate aspects of waging conflicts that arise from such matters, and which in turn make matters worse. I go to examine the conditions that allowed Donald Trump to become president and conduct himself in highly authoritarian ways. I begin by discussing a serious aspect of fighting badly—violence.

Violence in American Conflicts

In earlier chapters, I discussed many overt conflicts related to class inequalities and political power, and also to status inequalities and political power, particularly White and African American inequalities in political power. At this point, I turn to discuss organizations and conflicts that are prone to incorporate violence or the threat of violence. Violence has been part of many conflicts in American history, notably since European colonialization of territories where Indigenous peoples lived, and later in association with enslaved people from Africa. Numerous local vigilante actions and terror attacks were carried out against one or more non-White ethnic groups. Terror attacks destroyed the reconstruction that was achieved in the South shortly after end of the Civil War, leading to the establishment of the Jim Crow system in the South.

Even in the period analyzed in this book, since 1945, numerous organizations have come and gone that were characterized by using or threatening the use of violent coercion to intimidate, dominate, or expel other Americans of different ethnicities, religions, or ideologies. Such groups may celebrate their own religion, ethnicity, or ideology, while denigrating others who differ as un-American. The strength of those sentiments may justify extreme conduct against them. Some American cultural strains may even support such views. The attractiveness of guns for many Americans is an additional feature for many members of such organizations. Possessing guns make some people feel safer and stronger.

Fighting Better. Louis Kriesberg, Oxford University Press. © Oxford University Press 2023.
DOI: 10.1093/oso/9780197674796.003.0009

The Second Amendment to the US Constitution is interpreted by some people as validating such feelings. The celebration of an individualistic conception of liberty and freedom is another strain of American culture that supports a rejection of governmental or other external authority. Of course, such ideas may be held strongly, faintly, or not at all by different Americans.

Obviously, in most conflicts within the country, few people, and then only rarely, actually try to use deadly violence to advance their views. A little more frequently, people may use nondeadly violent coercion. Most likely, people use persuasion in trying to advance their goals. Perhaps surprisingly, people frequently simply assert their views with no expectation of changing anything except to feel better by simply expressing how they feel and what they think. These distinctions should be kept in mind in analyzing people in a conflict.

In American history, some members of left-wing as well as right-wing groups have resorted to violent acts, including attacks resulting in deaths, sometimes in countering armed actions taken by police or other officials. This was notoriously the case in class conflicts between some labor unions and companies in the 1897–1911 period.[1] Since then, labor relations have become more regulated and nonviolently contested, as discussed in chapters 2–4. On the other hand, in recent decades, resistance by large categories of people who resent and oppose other kinds of people improving their status has become manifest in coercive and even violent actions.

Another form of violence that has scarred American history is that of mass or crowd riots. In chapter 5, I discussed the many riots in the 1960s in African American neighborhoods, which were essentially African American protests of conditions they suffered. There have also been riots by White crowds that harm, intimidate, and expel African Americans. Grievously, a massacre of the people in a prosperous African American community in Tulsa, Oklahoma, occurred from May 31 through June 1, 1921, a tragedy that was excluded from White history for decades.

Detroit, Michigan, has been the site of numerous riots, beginning in the June 1943 riot. Detroit was the locus of great industrial expansion notable for automobiles and then military defense. Workers flocked to Detroit from many countries and from the southern United States. African Americans, however, faced severe housing segregation, which forced them to live in poorly maintained yet expensive housing in crowded African American neighborhoods. On June 20, 1943, a fight broke out between some African Americans and Whites spending Sunday on Belle Isle, Detroit's large park in the middle of the Detroit River. Fighting soon spread to the mainland, and rumors about attacks swept across the city, raising tensions that had been high and threatening to boil over into violence for months.[2] The rioters were largely representative of the employed Detroit male population.[3] The riot was suppressed only after six thousand federal troops entered the city. Thirty-four persons were killed, twenty-five of them Black, and most by the White police force.[4]

Following recurrent riots in Detroit, early on Sunday morning, July 23, 1967, a large-scale rebellion erupted in Detroit, composed mainly of confrontations between Black residents and the Detroit Police Department. The precipitating event was a police raid of an unlicensed, after-hours bar. It went on for five days. Governor George W. Romney ordered the Michigan Army National Guard into Detroit to help end the riot. President Lyndon B. Johnson sent in US troops. The result was 43 dead, 1,189 injured, over 7,200 arrests, and more than 400 buildings destroyed.

In 1949, a more ideologically driven riot happened, the Peekskill Riot, in Westchester County, New York, a blue-collar town.[5] The riot comprised a series of violent attacks by mobs of Whites directed against African Americans and Jews attending a civil rights benefit concert where Paul Robeson was appearing. More than 150 people were injured in the riot. Robeson had transformed himself from being primarily a singer into a political figure, opposing the Ku Klux Klan and other right-wing groups and supporting positions and groups that would be viewed as Communist in the McCarthyism era. Robeson had appeared before the House Un-American Activities Committee in opposition to a bill requiring Communists to register as foreign agents, and months before the riots he attended the Soviet-sponsored World Peace Conference in Paris.

Crowds and mobs resorting to violence have been common in some circumstances in the United States. Sometimes they have escalated from peaceful demonstrations; sometimes in response to harsh suppressive conduct by police or other government agents. Sometimes mobs and crowds are incited to resort to intimidating conduct and to violence by leaders. Crowds can become emotionally charged, and passionate persons may engage in behavior that they otherwise would not contemplate. Some members of a crowd or mob may join intentionally to act wildly and attack or injure opponents or destroy property.

Violence in riots and mob actions are unlikely to be a constructive method of waging a conflict. Specific objectives tend to be absent, and negotiations are hampered, sometimes seeming impossible. New grievances are likely to be generated. Third parties are likely to oppose the side seen as initiating the violence. In short, violent conduct tends to be counterproductive, even if making some trouble is often important in bringing about change.[6]

African American involvement in violent riots has been frequent in much of American history, often in the form of Whites attacking them. In some cases, there has been some violence by both sides. Police or other official intervention usually has been by White officials, and African Americans have been disproportionally injured or killed. Large-scale violence aimed at property by African Americans was relatively frequent in the 1960s and 1970s, as noted in chapter 5.

Since then, they have become infrequent, exceptions being riots in Miami in 1980, Los Angeles in 1992, and Cincinnati in 2001. They have been replaced by largely peaceful protest demonstrations and marches, as discussed in the section about Black Lives Matter, in chapter 5.

Attention must also be given to violent overreactions by official authorities to citizen actions deemed to be wrong. In such cases great regret is expressed by the responsible authorities. Unfortunately, the overreactions have lingering effects, both upon the injured community and for people inclined to oppose government controls in general. I mention three such tragic cases: the MOVE bombing in 1985, the Ruby Ridge shooting in 1992, and the Waco siege in 1993.

The 1985 MOVE bombing was the culmination of contentions in Philadelphia, Pennsylvania, between the Philadelphia Police Department and MOVE, which was a Black liberation group founded in 1972 by John Africa.[7] MOVE members sought to live in a communal arrangement, in a preindustrial lifestyle, and they were deeply religious, believing in the ethical treatment of all living creatures. The religious quality of the community is noteworthy; it meant that they had a different worldview than their neighbors or the police. Differences in worldviews between communities can be a basis for conflict and for difficulties in understanding each other in negotiating their disagreements.[8]

The troubles began in 1977, with complaints from neighbors about MOVE's lifestyle and bullhorn-amplified protests in which its members participated. This led the police to obtain a court order requiring the group to vacate their house. When informed, MOVE members agreed to turn in their firearms and leave peacefully if members arrested during prior demonstrations were released from jail. The police complied with the demand but MOVE refused to abandon their house or surrender their weapons. The standoff resulted in 1978 in the death of one police officer and injuries to sixteen other officers and firefighters. Nine MOVE members were convicted of killing the officer and were sentenced to life imprisonment.

The police, seeming to lack any alternative options, decided to greatly escalate the violence. They ordered a police helicopter to drop two bombs onto the roof of the MOVE townhouse. The resulting fire killed six MOVE members and five of their children, and destroyed sixty-five houses in the neighborhood, while the Philadelphia Fire Department let the fires burn out of control.

Of course, alternative options are conceivable. For example, the city authorities might have been more engaged and, as go-betweens, listened to the MOVE leaders, giving their religious beliefs and perspectives some respect. Furthermore, in negotiations and conversations it might have become possible to construct a path so MOVE would transfer its residence to a location more suitable for their lifestyle.[9] Similarly, MOVE leaders might have been more attentive to police and neighbors' concerns and arranged aid to relocate to a more congenial setting for their residence.

Ruby Ridge was the site of an eleven-day siege in Boundary County, Idaho. It began on August 21, 1992, when deputies of the United States Marshals Service (USMS) tried to arrest Randy Weaver after he failed to appear on firearms charges.[10] During the 1980s Weaver, a former Iowa factory worker and US Army Green Beret, with his wife Vicki and their young son and three daughters, moved to northern Idaho. They wanted to homeschool the children and escape what they saw as a corrupted world. In 1978, Vicki began to dream of living on a mountaintop and believed that the apocalypse was coming soon. They had a personal worldview quite different from the worldview of the US government security forces. The Weavers bought twenty acres of land on Ruby Ridge in 1983 and built a cabin, without running water or electricity.

An Aryan Nations settlement was close enough to the cabin for the Weavers to sometimes socialize with the people there. The Aryan Nations is an anti-Semitic, neo-Nazi, White supremacist organization, identified as a terrorist organization by the FBI. Weaver dismissed their views and efforts to recruit him. A government undercover informant in the organization tried to convince Weaver to also work for the government as an informant. Weaver declined and the recruiter tried to get leverage on him by arranging Weaver's illegal sale of a sawed-off shotgun. Weaver refused to surrender for that matter, and was joined by members of his family and a friend, Kevin Harris. The Hostage Rescue Team (HRT) of the FBI became involved as the siege developed, somehow believing that they were dealing with a terrorist foe.

When the Marshals Service reconnoitered the Weaver property, six US Marshals encountered Harris and Sammy, Weaver's fourteen-year-old son, in woods near the family cabin. They were armed, as members of the Weaver family usually were, for hunting, and accompanied by the Weavers' dog. Accounts of what happened at the encounter vary, but the results were clear. Many shots were fired and the dog, Sammy Weaver, and Deputy US Marshal William Francis Degan were killed. Harris returned to the cabin and with Vicki retrieved Sammy's body and placed it in the shed. The next morning, Weaver and Harris went to check the shed, but Weaver was shot; as they dived back to the cabin, Vicki opened the door, was shot and, unknown to the sniper, killed. The shot also critically wounded Harris. The FBI had four hundred agents based below, where news media and local sympathizers of the Weavers had gathered. After days of no communication with the Weavers, the FBI realized they needed an intermediary to negotiate a way out. A White nationalist and former Green Beret known by Weaver, James "Bro" Gritz, undertook to talk with Randy. Bro learned that Vicki was dead, and he told the FBI that they "had screwed up." Shocked, the FBI tried to be conciliatory. On August 30, Bro helped arrange for Harris to come out and get medical treatment and to take Vicki's body from the cabin. The next day, Bro convinced Weaver to leave the cabin with his three daughters.

In 1993, Weaver and Harris were charged with first-degree murder for the death of Degan, but they were acquitted by the jury. In 1995, Weaver and Harris sued the US government and won a $3.1 million settlement. The siege at Ruby Ridge lent support to persons who viewed the US federal government as an enemy that had to be resisted. It prompted the constitutional militia movement and left many Americans with a deep distrust of the government's leadership.[11] Clearly, the FBI response to the Weaver case was greatly disproportionate. In a large national bureaucracy, rules are necessary to maintain order and uniformity in decisions, but that may interfere in giving enough attention to the unique peculiarities of any single case. Obviously, the view that Weaver posed a great danger was misleading. In any case, the rapid escalation in bringing deadly force to bear was self-defeating. The FBI recognized its many mistakes in this case and teaches about them in its training.

The 1993 Waco siege is the third instance of law enforcement overreach that I cite in this work. It refers to the loss of lives at a compound in Waco, Texas, that belonged to a religious community, the "Branch Davidians." They had their own worldview and believed that the End Time was near.[12] The siege was conducted by the federal government, Texas state law officers, and the US military, against the Branch Davidians, who were led by David Koresh and were headquartered at a ranch in Axtell, Texas, thirteen miles northeast of Waco. The Bureau of Alcohol, Tobacco, and Firearms (ATF) suspected that the group had a stockpile of illegal weapons, and obtained a search warrant for the compound and arrest warrants for Koresh, as well as a few of the group's members.[13] President Bill Clinton later characterized the circumstances in these terms: "The sect's messianic leader . . . believed he was Christ reincarnate, the only person who knew the secret of the seven seals of the book of Revelation. Koresh had almost hypnotic mind control over the men, women, and children who followed him; a large arsenal of weapons, which he was prepared to use; and enough food to hold out for a long time."[14]

The siege began early Sunday, February 28, 1993, when armed ATF agents tried to arrest Koresh and others in the compound.[15] A TV cameraman unwittingly tipped off a Davidian of the coming raid shortly before it occurred. The agents were met by gunfire, and in the ensuing shootout, four agents were killed and sixteen were wounded; six Davidians were killed, and Koresh and others were wounded. The FBI took over control of the siege for the fifty-one days it lasted. Negotiations ensued, conducted by FBI negotiators, while other FBI agents, notably the Hostage Rescue Team (HRT), managed the siege. Initially, the government chose a waiting strategy and sought a negotiated solution, but that was not well coordinated with the siege operations. For example, on Monday, ten children were sent out of the compound, but the FBI's armored vehicles moved in closer around the compound and the phone line was cut except for Koresh to talk with negotiators. Koresh was highly agitated by those actions. The next day, Koresh agreed to surrender if a sermon he prepared were sent out nationally; it was sent out on the Christian Broadcasting Network, but Koresh then said God had told him to wait. Day after day, there were long rambling conversations, some conciliatory FBI gestures, cooperative acts by Koresh, like allowing some children to leave, and punitive FBI moves. There was no consistent evolving strategy leading to a shared resolution.

Early in April, the agents on the scene despaired of any surrender by Koresh and concluded that a forceful action to seize the compound was necessary. The costs and burdens of continuing the siege were too high for it to continue. A plan to inject tear gas into the compound was prepared and approval for the plan was sought from Janet Reno, the US attorney general. She consulted with Clinton, seeking his approval. Clinton's account of that discussion is provided in his book, *My Life*.[16] He met with Reno on April 18, who explained the plan to storm the compound the next day using armored vehicles to break holes in the compound and blast tear gas into the compound, which would force everyone inside to surrender within two hours. Clinton recalled a similar situation when he was governor of Arkansas. A right-wing extremist group, with the members' families, had a compound in the mountains of northern Arkansas, living in several cabins with trapdoors to a dugout from which they could fire. The FBI, state police, and other law enforcement personnel wanted to storm the compound to arrest two persons accused of murder. Clinton determined he would not proceed without more information and ordered a helicopter overflight by an experienced veteran. The veteran concluded from his inspection that fifty men would be killed in an assault on the compound. Clinton decided to block any supplies getting to the families in the compound, and eventually they surrendered. Regarding Waco, his instinct was to wait, but Reno explained the costs of the standoff and the threats of child sexual abuse and mass suicide, since Koresh was crazy. Despite his instincts, he told Reno to do what she thought was right.

On April 19, 1993, at 5:59 a.m., the Davidians were informed over a loudspeaker that they were under arrest and should come out.[17] Three minutes later, two FBI combat engineering vehicles moved forward and began spraying tear gas from a nozzle attached to a boom. Two minutes later, the Davidians began shooting. Other FBI vehicles moved forward, firing "Ferret Rounds," to penetrate barriers, such as windows and doors, and spray more gas. By 6:30 the whole building was being gassed and walls opened by attacking vehicles. At about noon, fires erupted in three or more places, probably set by the Davidians. Nine Davidians fled and were arrested. The fires engulfed the buildings and, ultimately, about eighty persons died, some, like Koresh, by bullets. The event was widely regarded as a tragic fiasco.

Interestingly, MOVE, Ruby Ridge, and Waco share some similar mistakes. In all three cases, great government escalations toward reliance on militarized operations occurred. Having fearsome capacities seemed a way to escape being stuck in an intractable fight. Furthermore, in all three cases, the government forces overestimated the threat of the people they sought to arrest. In addition, the resistant groups were not viewed with regard or respect, but rather as outlandish and even crazy. The differences were not seen as matters of differing worldviews, but of differences between truth and righteousness on the government side and bizarre error and criminality on the other side.

More constructive strategies by the police and other officials might have been taken if these mistakes were avoided. Thus, attention to nonviolent alternatives, even of persuasion, and their possible benefits might have proved better for everyone. Some regard for the perspective and concerns of the antagonists might have made other nonviolent options credible and effective. Showing respect for shared humanity might have increased some mutual trust. If rapid escalation of violence were avoided, the cost of waiting would have been kept lower. Some of the same mistakes were made by members of the nonofficial and weaker sides, on a smaller scale. Reducing some of those mistakes might have resulted in less terrible outcomes for them. It is noteworthy that in each of the three cases, one or more police or other security officials were killed at the beginning of the encounter. For the authorities, the perpetrators would have to be held accountable.

Right-Wing Militant Organizations

America has long been known as a country distinguished by a highly active civic life, with numerous voluntary associations.[18] Public participation in voluntary associations in all manner of spheres is a hallmark of American democracy. Yet some of these many organizations have been antidemocratic in their goals and means. Historically, right-wing extremist organizations have often been antagonistic to one or more ethnicity or religion, other than White Anglo-Saxon Protestants (WASPs). Left-wing extremist organizations have usually fought against class domination during limited periods in American history, and they have been suppressed by officials, as happened with the Palmer Raids after World War I.[19]

Right-wing militia groups that have been hostile to African Americans have existed since their actions to overthrow Reconstruction in the South after the Civil War ended. Some versions of the Ku Klux Klan have persisted in different degrees since then. Many other right-wing militia groups have appeared with animosity against various targeted people. They include Christian Identity and the neo-Nazi Aryan Nations, and other anti-Semitic groups. In addition, some groups oppose heavy-handed government interference in their lives and have armed themselves to defend against it. Some radical persons or groups fight Communists. Some believe in conspiracy theories and attribute evil power to entities that do not exist but are attributed to some group or category of people.

An influential figure in several such endeavors was Louis Beam. Beam served in Vietnam as a helicopter machine gunner and was a fierce anti-Communist. He returned to the United States, joined the Klan, and engaged in many militia actions.[20] The FBI tracked him and other prominent figures in the radical racist right, and they believed that they had considerable evidence that would result in the conviction of several extremists. In 1987, Beam and thirteen other principal leaders prominent on the radical racist right were indicted for seditious conspiracy to overthrow the government. Despite the evidence, the all-White jury was not convinced and Beam and his co-defendants, including Aryan Nations leader Richard Butler, were acquitted in the federal trial held in Fort Smith, Arkansas. The FBI was discouraged by the acquittal and held off in trying other cases.[21]

Beam's writing became very influential in the White supremacist community. In 1992, he published a revised earlier essay about "leaderless resistance." The article proposed that White supremacists not plan actions in large groups, but instead take action in small cells of one to six men, minimizing the risks and damages of infiltration by law enforcement officials. These "lone wolves" would "act when the time was ripe." He went on to publish material on his personal website. Beam remained affiliated with the Aryan Nations as an "ambassador at large" for many years. The group was greatly weakened by a 2000 civil judgment against it in a case brought by the Southern Poverty Law Center.

Resistance to extremist groups is conducted by federal, state, and local government agencies, and also by nongovernmental organizations. Government agencies pursue laws against discrimination, hate crimes, and other violations of protected human rights. Nongovernmental organizations also take measures to counter injurious actions taken by extremist, hate groups. The work of the Southern Poverty Law Center was discussed in chapter 5. In addition, the Anti-Defamation League (ADL) focuses on countering anti-Semitism, violent acts by extremist groups, and threats to American security.[22]

On the whole, for decades before Trump's presidency, extremist, right-wing persons and groups were generally viewed as illegitimate and disreputable. National leaders of both major parties opposed or at least kept their distance from such people and their rhetoric. Trump's conduct in his campaign for the 2016 Republican presidential nomination, and then in his campaign as the party nominee, was different. He indulged in vulgarity, racist language, and denigration of opponents. This was attractive to some followers and emboldened White supremacist, anti-Semitic, and antigovernment militia members and isolated individuals. Many people were appalled at such conduct by such a candidate and then by a president.

Immediately after becoming president, Trump issued poorly written bans on persons coming from many predominately Muslim states. American international airports became scenes of large outraged crowds. Courts intervened and the bans were revised and revised again, until the courts accepted a version. Other radical acts and gestures ensued, and resistance grew.

The violent event in Charlottesville, Virginia, on August 11–12, 2017, was a telling case. It can be traced back to covert right-wing militant groups with antigovernment, White supremacist, anti-Semitic, and other radical sentiments. Some earlier violent acts had been committed by "lone wolves," one or two persons only loosely connected with or influenced by such groups and ideas. Most notably, this was true of the truck bombing of the Federal Building in Oklahoma City on April 19, 1995. That bombing was perpetrated by two antigovernment extremists, Timothy McVeigh and Terry Nichols, killing 168 people. April 19 was the day in 1993 when the religious compound in Waco, Texas, burned. McVeigh had visited the siege and after his Oklahoma bombing said events in Waco had contributed to his motivation to commit his attack.

Some of the many mass shootings in America can be attributed to persons with feelings and beliefs of hatred against particular categories of people. This includes the killings at the Emanuel African Methodist Episcopal Church in Charleston, South Carolina, on June 17, 2015, and at the Tree of Life synagogue in Pittsburgh on October 27, 2018. The radicalization of individuals who commit terrorist acts often occurs through membership in extremist organizations, and also through extremist talk and accounts on social media. Radicalization can suffice for individuals to carry out horrible acts. Such individuals and groups may be part of the Alternative Right, known as the "alt-right," a set of far-right ideologies, groups, and individuals whose core belief is that "White identity" is under attack by multicultural forces using "political correctness" and "social justice" to undermine White people and "their" civilization.

After Trump's election and initial actions, right-wing militia members and supporters were thrilled. Leaders Jason Kessler and Richard Spender decided to organize a Unite the Right rally in Charlottesville, Virginia. Many right-wing groups came, including members of White nationalist, neo-Nazi, Klan, alt-right, neo-Confederate, and neofascist groups. The goals were to foster their ideologies, unify disparate right-wing militias, and protest the Charlottesville City Council's order to remove Confederate monuments from public spaces.[23]

The first night of the rally, August 11, 2017, many participants marched with torches, chanting, "Jews will not replace us."[24] The next day, the rally participants gathered together, carrying various Confederate flags and a Nazi flag. Many opponents of the rally also had gathered and the police were not able to keep the two sides apart. As fights went on, James Alex Fields Jr. intentionally drove his car into a crowd of people who were nonviolently protesting the rally, killing one person and injuring thirty-five others.[25] He was convicted in a state court for the first-degree murder of 32-year-old Heather Heyer, and other counts and sentenced to life in prison. The rally organizers had overreached.

Trump's handling of the event included observing that there were "fine people on both sides." That may have heartened and attracted extremist right-wingers, but it was an abomination to most people, including many Republicans. The rally was poorly designed to be effective. The perpetrators tried to be intimidating and show how tough they were. However, intimidation is not attractive and displaying Nazi-like images and actions tends to turn off most Americans. Falling into fights and then lacking discipline that resulted in killing one person and injuring others simply made them seem disreputable. It did not serve to unify the diverse groups, so much so that they were not able to repeat any similar action. On Sunday, June 11, 2021, monuments of Robert E. Lee and Thomas "Stonewall" Jackson were removed as crowds cheered.

Political Parties and Trumpism

Conflicts about power in America are largely conducted within established institutionalized procedures. There are laws and norms that guide and contain the conflicts. The formal structure consists of the laws for officeholders in the three primary branches of government at the national, state, and local levels of governance. Political parties are an integral part of the political institutions. As is widely recognized, the Republican and Democratic Parties are highly polarized, and some steps on that road were discussed earlier in this chapter. At this point, I will focus on political party developments during the Trump presidency. To an unparalleled degree, Trump dominated the Republican Party. He demanded agreement and intimidated those who differed; dissenters dropped out of the party. This pushed the Republican Party beyond a simple polarized relationship with the Democratic Party.

For many Republican leaders and supporters, the party had become their primary collective identity. Its triumph would be sought by any and all means. Denials of any evidence challenging their declarations ensued. They expressed grievances, which are related to status losses, that needed to be expressed and satisfied. They are susceptible to symbolic politics, which Trump was able to articulate in ways that appealed to crowds at his rallies.[26]

Despite Trump's appeals to a base of supporters and an obedient Republican Party, as noted earlier, a majority of Americans never approved of the way he handled his presidency. Trump persisted, with no signs of becoming more "presidential." Indeed, he ridiculed that idea and got rid of people around him who might try to guide and control his excesses. As the 2020 presidential election approached, he started to repeat the idea that if he didn't win it was because the election was "rigged." He insisted that, despite all the evidence of polls, if it appeared he lost, the election must have been fraudulent. He would not, and did not, acknowledge his defeat.

On the other hand, the Democratic Party leaders recognized the hyper class inequalities as more significant than the status concerns of some Whites. The attractiveness of progressive policies was already indicated by the Democratic victories in 2018. In the 2020 election campaign, Biden came up with policies to solve problems, emphasizing collective gains attained by united effort, rather than divisive words and deeds. Using a constructive approach, the Democrats solidly won the 2020 election.

Trump's conflict style in political affairs has not been constructive; rather, he has pursued attaining narrow distributive power. To a high degree, he seems to seek power for himself, maybe his close family and friends, and then, as president, his business and the Republican Party. He also seems to rely heavily on coercive inducements and to admire authoritarian leaders in other countries who readily use force. During the multitude of largely nonviolent Black Lives Matter protest marches in 2021, following the murder of George Floyd, Trump escalated the repression of the protests by sending in armed federal forces in some cities. In fact, he wanted to go much further and use US military forces.[27]

Following the November 2020 presidential election, the returns in several states where Trump was defeated were challenged in the courts. The challenges were all dismissed by the courts. Recounts were made where that was appropriate. No evidence of any errors in the counts that would have changed the outcome in any states were ever produced. The counts of the down-ballot results were not disputed. Trump, however, continued to refuse to concede the election and recognize the legitimacy of the election of President Biden and Vice President Harris. Republican Party leaders generally said what Trump said: the election was stolen. Nevertheless, no officials took actions that would have denied the results. The transition moved on, even if slowly.

The final official step in confirming the election results was the congressional certification of the Electoral College vote on January 6, 2021, at the US Capitol. Many Republican members of Congress declared they would vote against certification of Biden-Harris votes from their respective states, but there was no doubt Biden's victory would be certified. Social media buzz began about holding rallies to protest the certification. The certification procedure was to be overseen by

Vice President Pence. Trump pressured Pence to refuse to participate in the certification, but, ultimately, Pence said he would adhere to his constitutional duties and oversee the certification,[28]

The January 6, 2021, insurrection riot was part of a failed effort by Trump and his associates to overturn Joe Biden's election as president. Prior to that, meetings were held in the White House with several Republican House members to discuss how to overturn the election results.[29] Attention was focused on January 6, the constitutionally set day for the certification of the presidential and vice-presidential electoral results. The House and Senate members who could and would object to the certification of the votes from their respective states were discussed. In addition, Trump tried to convince officials overseeing a few state-wide elections to declare there were inaccuracies or even to just find more ballots for him. His effort failed, but he and many followers simply persisted in asserting he had won the election.[30]

Trump determined to have a rally in Washington the morning of January 6, resulting in a disruptive protest at the Capitol. TV channels sent out horrific images of the violent attack, entry, and pursuit of members of Congress by rioters and the Capitol Police's valiant defense against the attack, until, belatedly, reinforcements evicted the rioters. The members of Congress returned from where they had hidden in the Capitol and proceeded to complete the certification, overseen by Vice President Pence. Many Republican members of Congress voted against certification, but not enough to overturn the results.

The rioters included some members of anti-government militias, the Proud Boys and Oath Keepers, who executed disciplined attacks on the Capitol.[31] Others were QAnon believers in bizarre conspiracy theories.[32] Some were there to demonstrate support for Trump, and some for the excitement. An analysis of 377 Americans who joined the insurrectionary riot and were arrested for their behavior at the insurrection is revealing. Robert A. Pape, a professor of political science at the University of Chicago and director of the Chicago Project on Security and Threats, found that these insurrectionists did not come from deep red counties.[33] Markedly, they came disproportionately from counties with relatively greater increases in non-White residents. Such changes frighten some Whites enough to rally against them. Of course, members of the militia groups added a great intensity to the attempt to disrupt and overturn the 2020 presidential election.

Trump had utilized many of the changes in the American political power system since the 1980s in order to capture a high degree of control of the Republican Party, which had become tribal in style. He used the new presidential role he created to act out many of his authoritarian inclinations. He and many Republicans created a new set of conditions, with which he thought he might be able to overthrow the results of an election he lost. He failed, but the changes he and the party are introducing in many states to suppress and manipulate future elections do threaten America's democracy.

Biden and Harris were inaugurated as president and vice president, respectively, without disruption. Despite verbal expressions denying the reality of the Biden and Harris victory, the reality of the outcome has been accepted in practice. The Democrats have control of the executive and legislative branches of government to the degree that is traditional and within prior norms for the size of their victory. The significance of an attempt to overthrow a duly elected US president will depend on how the insurrection comes to be understood. Options are discussed in the next chapter.

Conclusions

The changes in power inequality in America since 1945 are complex. They are influenced by, and in turn influence, rising class inequality and significant declining status inequality. The changing disequilibrium among these three ranking systems underlay many conflicts. That makes consideration of how to contend constructively very important. The progress toward greater equality for previously relatively low-status people should be seen as progress toward attaining a more perfect American union. Nevertheless, some people are aggrieved by such changes. Conflicts are inevitable and conducting conflicts well is important. Unfortunately, American history is replete with badly managed conflicts and considerable violence related to that progress toward greater equality.

The increased ranking of people in previously subordinated members of several collective identities, as discussed in chapters 5–7, marks progress in fulfilling American values and aspirations. That progress was made, significantly, by applications of constructive conflict methods. Nevertheless, some people have experienced those gains as losses to them. They resent that. The political power system could provide instruments to mitigate resentments and destructive ways to express the resentment.

Many political power developments discussed in this chapter have not ameliorated the problems identified here. In fact, they have exacerbated them, as illustrated by political party tribalism, the recourse to fostering falsehoods, and the reliance on coercion and even violence. The American democratic system was challenged, even diminished, as it rarely has been previously. Furthermore, the exponentially increasing income and wealth inequality in America tends to distort the workings of the political system, directing the government to serve the interests and concerns of the wealthiest, rather than of the people who were held back or harmed by technological, environmental, and social changes. In the next and concluding chapter, I will discuss possible better ways of fighting that might alleviate the problems all Americans face together.

10

Recovering and Advancing Equality in the Future

The accounts in the preceding chapters bear evidence that despite notable progress in the United States since 1945, there also have been setbacks. Class inequality has risen more and more since the early 1980s. On the other hand, greater
equality in status of many categories of people with shared collective identities
has occurred, although in varying degrees. Progress toward greater equality in
political power has also occurred in some ways, but in recent years, democracy
has deteriorated significantly. Changes in each of these dimensions of equality
interact with each other, sometimes for the better and sometimes for the worse.
In this final chapter, I consider ways in which conflicts can be waged better in
order to overcome recent setbacks, and how they can be waged to make further
progress in the future. I focus on US advancement along each dimension, noting
each dimension's interaction with the other two.

The analyses made in this book are based on a constructive conflict approach, laid out in seven core ideas in the first chapter. They are identified again
here; hopefully, they are more meaningful after being applied in assessing past
conflicts:

1. Many conflicts are conducted constructively relying largely on legitimate,
   institutionalized procedures or supplementing them.
2. Importantly, constructive conflicts are generally conducted using blends
   of persuasion and positive sanctions, and some coercion, minimizing
   violence.
3. Opposing parties in constructive social conflicts recognize that they are not
   homogeneous, unitary actors; rather, each consists of shifting components.
4. Members of each side in a conflict social construct their conflict, which can
   contribute to being constructive by viewing the conflict as an aspect of a
   broader relationship.
5. A constructive approach generally entails opponents noticing and considering each other's concerns, which can result in some mutual, but unequal,
   benefits.

Fighting Better. Louis Kriesberg, Oxford University Press. © Oxford University Press 2023.
DOI: 10.1093/oso/9780197674796.003.0010

6. Constructive conflicts are usually importantly interconnected, including recurring over time and with members of each side also engaged in many other conflicts, checking overzealous focus on one conflict.
7. Recognizing that conflicts are not static, the approach fosters constructive conduct by utilizing fresh changes within any side or in the conflict's context.

In making suggestions about fighting conflicts in better ways, I do so regarding different positions—national, state, and local governmental offices, as well as nongovernmental actors. The suggestions usually pertain to proposed goals of some proponents in particular conflicts. A reader is invited to support or oppose any suggested goal or action to further the goal and learn more about it. The reader may extend support by donating money to proponents, by joining some collective action, or by starting some initiative. I do not propose any comprehensive action plan for any reader. What any one of us should do depends on her or his resources, circumstances, and understandings.

By 2021, the United States was beset by four severe challenges, and many destructive conflicts about how to meet the challenges. The four great challenges are: (1) intensifying global warming, (2) recovering from a pandemic, (3) adapting to rapidly changing global economic and political contexts, and (4) overcoming threats to the US democratic system and the failures to constructively manage the inevitable conflicts about the other great challenges—those failures are also challenging.

This book has examined how conflicts in the United States have been conducted destructively but sometimes constructively about issues related to three primary dimensions of societies: class, status, and power inequality. The findings of that examination should help fashion better ways of fighting to reduce inequities in class, status, and power.

The four challenging developments identified above are likely to significantly impact future changes in class, status, and power inequality in the United States. Two developments are contextual. First, global warming is having increasingly devastating effects across the world, in the form of rising sea levels, droughts and fires, more frequent hurricanes and floods, and decreasing biodiversity. Second, the COVID-19 pandemic has sickened and killed millions of people globally and hugely disrupted the economies of the world's countries. The other two developments relate more specifically to the United States. Third, in recent years the major role the US has played in foreign conflicts, which had drawn US attention away from domestic concerns, is changing and needs fresh thinking. Global interconnections continue to grow, major actors increase in relevance, and illiberal ideologies gain converts. Fourth, the great intensification in the antagonistic divisions within US society has hampered dealing effectively with global warming, the new pandemic, and the changing foreign engagements. Those other developments might be viewed as threats to the country as a whole and be a cause for unity, not division and tribalism.

Progress in class, status, and power equity will be impacted by these developments. The effects, however, are not immutable; they depend on the way inevitable conflicts are waged. The challenges can be good opportunities if conflicts are waged constructively.

Reducing Class Inequality

The COVID-19 pandemic was a development that had great immediate and perhaps lasting impacts upon class inequality in the United States. Related events, prior to Biden's administration, were discussed at the end of chapter 4. What the pandemic revealed and what it changed in class inequality are key to understanding its effects. The great vulnerability of low-income people to COVID-19 became apparent as they suffered higher rates of illness and death than did the middle and upper classes. Also, the closing down of much of the country's economy tended to harm low-income people much more than people who had good incomes prior to the locked-down economy. Indeed, the stock market rose to new heights and some corporations and individuals profited hugely. This enhanced the ill effects of the pandemic on class equality, as the poorer members of society were more vulnerable to becoming ill and dying than were higher-income persons.

By 2020, global warming had advanced far enough for the country to suffer frequent raging fires and droughts in the country's Northwest. Frequent terrible hurricanes and floods struck the people in the southern and eastern states. Lives were lost and the costs of recovery rose. Again, these effects were more damaging to low-income persons who lacked resources to recover their losses. The political antagonism also hampered taking actions to mitigate the damages resulting from global warming. The increasing costs of recovery and of building preventive infrastructure could generate programs that improve the well-being of low-income people. Republicans, who had held out in denying that human activity was fueling the rising temperatures of the earth, are beginning to acknowledge the reality they had denied. Sentiments have increased favoring actions to improve the country's resilience and to slow, even halt, global warming before civilized human life on earth is untenable.

The high level of societal polarization and political dysfunctionality hampered and may continue to hamper programs and actions to raise class equality. There are, however, ways to move toward greater equality, which would also reduce the society's polarization and political malfunctioning. That is a hoped-for opportunity. As legislation is passed in a bipartisan manner to protect the well-being of persons who suffer losses due to the pandemic-weakened economy, the political systems gain more capacity to function. This was the case in the bipartisan passage in March 2020 of the $2 trillion CARES Act, signed by President Trump, as discussed in chapter 4.

Finally, reducing military international engagement, insofar as that occurs, could free up resources and attention for domestic affairs, and therefore foster actions that increase class equality by reducing income inequality. Preparing for war and engaging in foreign wars have been sources of high profits for well-connected corporations and their leaders and investors.[1] The highly active military role that the US has played in the world has contributed to diverting attention and resources, which otherwise might be used to reduce the hyper US class inequality. Insofar as that actually declines, attention might be brought to that possibility.

Admittedly, international concerns and domestic issues need be addressed. US international engagement has long been overly militarized.[2] In foreign policy, Biden is inclined to endorse the mainstream approach and has chosen advisers who also do. Moreover, members of Congress have generally favored spending money on new military equipment. In December 2021, Biden proposed a Pentagon budget larger than the preceding one. Nevertheless, Republican and Democratic members of Congress insisted on spending even more.[3]

The horrible tragedy of the November 2022 Russian invasion of Ukraine had many unattended antecedents and will have many grave consequences. A more constructive approach by many of the major and minor actors in the events that produced the tragedy might have averted it. The constructive approach could help mitigate some of the dire consequences that will follow. In any case, Vladimir V. Putin's dreadful conduct toward Ukraine demonstrates the dangers of autocracy.

Some overt fights against increasing class inequality were noted in chapters 2–4, but that increase occurred largely within the institutionalized political system. The constructive conflict approach stresses that conflicts are interconnected and that as one rises in salience, others tend to decline in significance for the same partisans. The expanding war in Vietnam quite directly stifled the previously major struggle for reducing poverty and inequality in the United States. The Cold War against Soviet Communism varied in intensity, rising during President Reagan's first term, when class inequality began its steep rise. The end of the Cold War failed to result in a peace dividend, and the Global War on Terrorism was launched after September 11, 2001. That mistakenly conceived and implemented adventure, and related endless wars, contributed to political failures domestically.[4]

Significantly, by the twentieth anniversary of the September 11 attacks, the US had ended its war in Iraq and withdrawn in defeat from Afghanistan. Grave misgivings were widely expressed, even by leading Republicans, about the mistaken invasion of Iraq and the harms the US suffered because of the way the war against terrorism had been waged. Preferences to attend to domestic concerns rose.[5]

These four profound developments are converging. That may exacerbate the ill effects of each development. However, aspects of these developments can be used to support actions that reduce the extreme class inequality in the United States. The ideas of the constructive conflict approach can suggest strategies to reduce the extreme class inequality and high proportion of Americans in poverty. A central idea is that conflicts can be waged by using persuasive inducements to win over allies and reduce the number of adherents on the opposing side. Of course, persuasive efforts are not always constructive in a conflict. They may be taken to rally supporters to the detriment of the adversary, and they may be based on fraudulent information. Indeed, a great antagonistic division in US society emerged gradually between adherents of radical right-wing thinking, as used by Trump, and a range of opponents. There are many positions and policies that are in dispute, yet also some shared concerns.

Many people in the United States had come to believe that great class inequality is the natural and inevitable feature of human societies and of US prosperity.[6] They believed the great differences in income and wealth were for the most part attributable to differences in hard work and skill. Some people seemed to believe that rich people needed to have the incentive of getting richer to work hard while poor people needed to be driven to work hard by punishment if they do not. It became a hallmark of Republican economic orthodoxy that, contrary to evidence, cutting taxes for the rich would lead to greater investment and good jobs for many people.

As discussed in chapter 3, economists like John Maynard Keynes and Milton Friedman differed about the role of the government and the free market in maintaining good economic progress. Friedman's confidence in minimizing government regulation of a free market was happily accepted by Republicans. The evidence over the years has clearly shown that a "trickle-down" approach is defective, and some government regulation is needed to maintain prosperity.[7] However, this debate is part of the ideological war that has been salient for many years. More novel persuasive efforts are needed to be effective; perhaps speaking of government protection or consumer protection or freeing the market would be more attractive than government regulation.

Those persons favoring little government intervention in the market and low taxes for corporations and high-income earners have been relatively successful in gaining agreement from many Republican political leaders. This was not only a matter of persuasion. As noted in chapter 8, financial contributions tended to be given generously to politicians who would vote against taxes and regulations. Furthermore, elements of a unitary elite do exist. High-positioned persons in class, status, or power societal dimensions often move from leadership in one dimension to high positions in another dimension. Thus, wealthy persons or highly paid corporate figures may enter high positions in government, and vice versa. In addition to the same person moving from a high position in one dimension to a high position in another dimension, being in one high position may help family members to get a high position in the same or another dimension.[8] Actions of a person in a high position in one dimension may be taken to benefit themselves in future moves or some of their relatives in their careers. For example, for many years, tax lawyers at large accounting firms have taken senior jobs at the Treasury Department for two or three years and then returned to their former firms, taking on lucrative senior positions.[9] While at the Treasury Department, some write tax provisions that could yield tax breaks for the clients of their former firm. There are only a few laws to prevent such self-aggrandizing moves from a government position to a corporate one. Official and nonofficial oversight are needed to prevent such patterns of conduct and to sustain the integrity of tax laws.

Broadly conveyed factual evidence should be part of campaigns to influence the public at large. Each party has argued that its strategy would be economically beneficial for the economy as a whole. However, analyses of the performance of the economy during the periods of Republican compared to Democratic presidents clearly finds much faster economic growth occurs during Democratic than Republican presidencies. Inflation-adjusted GDP grew about one and a half times faster under Democrats than under Republicans.[10] Even greater differences were found for private-sector job growth: businesses added jobs at an almost two and a half times faster rate under Democrats than under Republicans, on average.

Government policies to salvage the economy that had so largely closed down in response to COVID-19 demonstrated the value of government spending and tax credits for low-income people and the economy as a whole. Many ways are available that can reduce the extreme class inequality that exists. I will note some of these ways and discuss how conflicts might be waged constructively to establish effective reforms. I start with noting possible federal government tax policies that reduce extreme income inequality.

One obvious way to increase equality is to tax income and wealth progressively; that is, taxing higher incomes and greater wealth at higher rates and redistributing the government expenditures to benefit lower-income persons.[11] In response to the widespread economic shutdown, the federal government undertook to provide funding for expanding needed medical care, medical care loans, and payments for food and housing to persons who had little or no incomes.

A broad safety net became available to everyone that also served to sustain a necessary level of economic activity. Consequently, many people who had lived in dire poverty received increased income. By September 2021, the poverty rate had dropped to the lowest it had ever been: to 9.1 percent from the 11.8 percent it had been in 2019.[12] Provisions for children improved as well.

This demonstrated the possibility of quickly interrupting the ongoing cycle of poverty in which many people were trapped. Too often, people trying to escape poverty fail when becoming sick, losing a working car, or needing care for an ill relative. A safety net can provide a necessary way to persevere and escape poverty or a low income. In the first year of the Biden administration, major bills were proposed to deal with immediate and long-term challenges. The bills included programs to deal with COVID-19, global warming, long-neglected infrastructure needs, inadequate child-care provisions, and extreme income and wealth inequality.[13]

Trump's populist style and anti-elitism, if not his policies, had been attractive to non-college educated persons. It would be useful for the Democratic Party to return to its appeals to the working class, including the celebration of hands-on labor. Restoring pride in manual skills can help mitigate status resentments. Biden finds giving such recognition natural. Trump had aroused some concerns that contributed to the identity matters that many Democrats had come to stress, turning away from neoliberalism. All this enhanced the strength and attractiveness of "progressive" ideas and policies.

In November 2020, the Democratic Party won control of all three branches of government, but narrowly. Even before being inaugurated, Biden announced plans for a $1.9 trillion stimulus bill, the American Rescue Plan Act of 2021. Ten Republican senators proposed a $600 billion COVID-19 relief bill counterproposal. The Democratic bill was passed by the House on February 27, 2021, and it advanced swiftly through the appropriate House and Senate steps and was signed by Biden on March 11, 2021. Clearly, this swift action was intended to demonstrate that the federal government could take action and make a difference. The broad provisions of the bill also highlighted the benefits of exercising collective power and not mere distributive power serving narrow party interests. Those shared benefits should be given attention and contrasted to narrow self-serving gains of opponents. Those who forgo the great gains that could be made by a few because of the idiosyncrasies of the COVID-impacted markets should be recognized.

Primary components of the bill included expanded unemployment benefits, phasing out direct payments to high-income taxpayers, and extending emergency paid leave and food stamps.[14] Tax provisions included expanding child and dependent care tax credits and earned income tax credits. The bill also provided grants to small businesses; funds to help state, local, and tribal governments meet budget shortfalls; and funding for education and housing. Finally, it provided funds to help cover some public health expenditures.

The impact of the bill in reducing poverty was remarkable.[15] The expanded Child Tax Credit could itself reduce child poverty by up to 40 percent, and in combination with all COVID-related relief, it could reach a 52 percent reduction in child poverty.[16] The bill was designed to help people in poverty or those with low incomes, and it did so.

An infrastructure bill had long been needed and often urged, even by Trump when he was president. Biden triumphally signed a $1.2 trillion bipartisan infrastructure bill on November 15, 2021. It is formulated to upgrade roads, bridges, water systems, and broadband access.[17] Its bipartisan character was evident by collaboration in writing the bill and by Mitch McConnell's yes vote, along with eighteen other Republican senators and thirteen Republican House members. Infrastructure provisions matter to congressional Republicans as well as Democrats because they often entail improving living conditions for their constituents. Furthermore, corporate leaders generally want good means of transportation.[18] Moreover, anti-Democratic fervor was more focused on culture war issues than basic, mundane government tasks.

Nevertheless, Trump and his stalwart supporters denounced the Republicans who voted for the bill, saying they should be ashamed of helping Democrats.[19] This sign of a splintering Republican Party, of Republican leaders disregarding Trumpian threats and name-calling, was of great value to the Democrats and all those who worried about extremist conduct. Achieving a major bipartisan project makes achieving others more likely. It demonstrates the expression of collective power, rather than contested distributive power.

The passage of the Build Back Better Act will be a bigger test. The very narrow political majority the Democrats held in the House and especially in the Senate, following the 2020 election, enabled less progressive senators to set aside some provisions Biden and the Progressive caucus wanted to achieve.

The bills are costly and will require increases in government revenues in the future. Of course, there is the to-be-expected expansion in the national economy that will yield increased tax revenues. Some increases in tax rates for the very rich were made and they should be targeted further to increase class equality. This should include increases in the tax rate for the highest income tax bracket, in accord with earlier years. Similarly, the tax rate for the largest estates should be significantly raised, also in accord with earlier years. Certainly, the estate tax rate paid upon the death of the estate owner can be avoided by various arrangements. Nevertheless, the rate should be increased. A new annual wealth tax is a useful contribution to greater class equality and perhaps a symbolic aid to greater intergenerational mobility, which would also be aided by a larger safety net and higher worker wages. Measuring wealth that might be taxed is undoubtedly a complicated and difficult matter, but there are possible ways of measuring some kinds of wealth.[20] Focusing on physical property simplifies the matter. Senator Elizabeth Warren of Massachusetts, in her campaign for the presidential nomination in the 2020 election, proposed a wealth tax and sustained her support for it.[21] An enhanced estate tax was part of the Build Back Better Act, which failed to be passed in the Senate.

Finally, a simple and readily justifiable way for the federal government to collect more tax money is to reduce illegal tax avoidance.[22] This is an important task for the Internal Revenue Service (IRS). However, notoriously, due to a lack of resources, the IRS has failed to audit many hundreds of thousands of wealthy persons who failed to file tax returns, and has failed to force the collection of owed taxes. Republican Party leaders have reviled the IRS and cut its resources. In the future, the IRS should be given the resources to impose the laws and its successes should be celebrated. This was a provision in the Inflation Reduction Act, signed by President Biden on August 23, 2022.

More public attention to the ways the tax structure sustains inequality, and specific ways that might be corrected, is needed. The hidden conflict between the very rich and the not-so-rich should be made more overt. A broad coalition of diverse organizations might undertake a campaign for more equitable tax policies. More governmental and media attention to local, state, and national coalitions, such as the Poor People's Campaign, are needed.

Taxation laws are handled by governments and are contested in the context of political party institutions; therefore, persuasive efforts are prominent. I discuss some persuasive points that oppose, and others that support, progressive taxes. One somewhat conventional idea about earnings that serves to support right-wing anti-tax advocates is that "owners" of enterprises who "make profits" should possess the profits. However, other views are possible, including that employees should share in the profits. Some corporations do have profit-sharing provisions for the workers making a product or delivering a service. Sometimes this is done in regard to increases in productivity. Other organizational structures also exist, such as cooperatives, which are owned by the customers or by the producers, discussed later in this chapter.

Another consideration about who owns the profits is that economic transactions generally have "external costs," also known as spillovers and third-party costs. They occur from production and consumption processes, but the costs are generally not paid for only by the persons in the transaction. For example, purchasing consumer goods commonly creates waste in terms of packaging, along with other environmental costs, including carbon emissions resulting from associated production and transportation processes. These costs are generally paid for by the public at large, subsidizing those who get the profits.

More generally, profit-making enterprises need publicly supported material infrastructure and appropriately capable workers to function. More recognition of that reality would support policies that would lessen the poverty and hyper class inequality in the United States. For example, a carbon tax, a proposal to counter global warming, might raise funds for improving the living conditions of lower-income Americans. Another avenue to reducing class inequality is to reduce tax credits to favored industries and increase taxes on industries that produce relatively high external costs.

In general, a persuasive way to win support for reducing poverty and income inequality is to stress the burdens and costs of poverty and inequality borne by the society as a whole. The trauma of COVID-19 revealed the extra dangers of illnesses and deaths resulting from poor living conditions and inadequate medical support. Attention to improving living conditions of the poor and low-income members of the society also promises benefits to all. People who are a burden due to their failures to be productive members of society would become able to help expand material and nonmaterial benefits to other society members.

The argument should be made by persons in lower- and higher-class positions, recognizing the self-harms that hyper-inequality inflicts on everyone. Pity and empathy for people with very low incomes need not be the paramount reasons for reducing the extreme income and wealth inequality in America. Particularly in the face of the challenging developments identified earlier in this chapter, collective gains are desperately needed.

Another constructive conflict idea is that conflicts are generally interlocked, which often contributes to any given conflict being intractable. To wage a conflict constructively, with the goal of reducing class inequality, it may be useful to frame the conflict in more attractive ways. For example, persons lacking a college education were aided when, in 2017, IBM began an extensive technical apprenticeship program in software engineering.[23] Or it may entail expanding successful programs for training and retraining potential workers in other new and growing spheres of employment and enterprise. In responding to damage from global warming, programs should develop better and more secure housing in the often vulnerable areas in which low-income people live.

The campaign for public support for child care exemplifies many ideas of fighting better for reducing poverty and increasing class equality. It appeals to a wide array of supporters, including low-income people, women, persons concerned about the well-being of young children, and would-be child-care providers. It should also appeal to employers who could select employees from a larger pool. In short, its benefits are evident for a wide spectrum of people, including employers, educators, and healthcare providers, as well as children. The effective resistance to government-supported child care rests on the primacy Republican Party leaders give to preventing any government expansions, even popular ones. The long fight about the Affordable Care Act even after its enactment is illustrative.[24]

Finally, and importantly, the rights of employees to organize, to form unions, and to bargain collectively contributes to higher wages and better working conditions. Federal laws in the years of the New Deal affirmed those rights. However, as early as 1947, with the Taft-Hartley Act, those rights began to be circumscribed. As discussed in chapter 2, the proportion of workers in trade unions began to decline. Changes in industrial technologies and international trade contributed to the decline in union membership, but do not wholly account for it. National laws should be passed to restore worker rights to organize and fight to win better wages. The widespread benefits of protecting workers' rights and good working conditions should be noted, not only by unions, but by government agencies and nongovernmental organizations. Again, COVID-19 revealed the health risks of unsafe working conditions, notably in the meat-packing industry.[25]

Changes in markets and businesses offer new opportunities to earn good incomes, but they may also present jobs that are exploited by others. This is the case with the emergence of the "gig economy," which first referred to white-collar workers offering consultancies and part-time work within a digital marketplace.[26] It also enabled employers to reduce labor costs by using independent contractors for tasks previously done by internal employees, avoiding healthcare or other benefits that would otherwise be provided. This has occurred to some extent, and there are many examples of gig workers being used in the corporate world. The real growth in the gig economy has been in the unskilled market, which occurred during the pandemic. Reforms might include extending law to include gig workers, regulating work contracts, and removing barriers preventing gig workers from forming unions.

These suggestions are highly relevant for actions at the national level by the federal government, but some are amenable to action by state and city governments as well. They pertain to state and city budget allocations as well as taxes. As discussed in chapter 6, states vary greatly in the degree to which they allocate funds to local school districts to increase support to less affluent districts. States and cities vary greatly in tax policies, welfare programs, and long-term economic planning. More equitable budgetary provisions should be implemented in many state and local budgets.

Nongovernmental efforts to reduce class inequality are also important and could be increased in many localities. Protest demonstrations and other actions at the state and city levels for increasing the minimum hourly wage have been effective. Furthermore, some people who had worked for low wages and poor working conditions at the outset of the COVID-19 pandemic took direct action. They acted out their objections, seeking better jobs or leaving the workforce.

Labor shortages appeared and so did somewhat higher wages.[27] In the fall of 2021, major strikes erupted across the country. Workers complained of harsh working conditions, exceedingly long hours of work, inadequate benefits, and stagnant wages, while the corporate employers profited.[28] A new period of union activism may be arriving, aided by more favorable public views and by federal actions, such as those evident in the report released in October 2022 by the White House Task Force on Worker Organizing and Empowerment.[29] The US Department of Labor is working with the Task Force to achieve many goals, including increasing workers' awareness of their organizing and bargaining rights, establishing a resource center on unions and bargaining, protecting workers who are organizing from illegal retaliation, and providing information about the role of unions in the United States. There are numerous laws that hamper workers' rights to organize and be protected, which warrant correction.[30] For example, right-to-work laws exist in twenty-seven states. These laws make it illegal for unions to collect fees from nonmembers in unionized companies, although those workers receive the benefits of collective bargaining agreements.

Furthermore, private corporate structures and policies could be reformed in ways to enhance class equality. As discussed in chapter 3, wage-salary differences increased dramatically beginning in 1980. The great differential in employee benefits and payments are not necessary, but may be difficult to decrease once in place. A shareholders' movement to stop and reverse this practice might be influential, but difficult to mount. Broadening the composition of the boards of directors could be a path to moving corporate policies to foster more class equality. Modest increases in the diversity of corporate directors in 2019 raised that possibility.[31] Most remarkably, the percentage of women joining boards steadily increased from 11 percent in 2011 to 45 percent of the stock market in 2019 (based on the Russell 3000 Index, a benchmark of the entire US stock market.) Ethnic diversity also increased, but more slowly and to a lesser degree. It would be useful to increase the diversity in terms of previous experience with the corporate economy. This might include employees and consumers as well as stockholders being on boards.

A novel and successful form of such corporate structures is to be found in Germany, called codetermination (*mitbestimmung*).[32] It has existed since 1976 in corporations with more than two thousand employees; shareholders and workers are represented equally on the supervisory board, but the chairperson, always a shareholder representative, has the decisive vote in the event of a deadlock. The arrangement is somewhat different in the coal, iron, and steel industries, where the chair is an added shareholder, so that the shareholders can always outvote the employees. In addition, in companies with at least five members of staff, the employees may elect a works council, which would represent the interests of all employees to the employer. A works council can conclude "works agreements" with the employer, which are legally valid agreements that regulate working conditions in the company. The great success of German industry is significantly attributable to codetermination.[33] Furthermore, trade unions in Germany are strong significant actors.

In the United States, there are various worker-owned and worker-managed productive organizations, many in the form of cooperatives.[34] Cooperatives have long been important in agriculture: there are over three thousand agricultural co-ops in the United States presently.[35] Any business can be so constituted. Worker cooperatives are also found in the service and retail sectors, including accommodation, food service, healthcare, manufacturing, and engineering fields. These organizations generally operate democratically, many according to principles based on the 1844 Rochdale Society concepts, which were updated in 1995.[36] More attention to the benefits of such egalitarian ways to organize work would be useful.

Given the high rate of technological innovation and marked changes in production processes around the world, the engagement of all levels of corporation members in shaping policies would be advantageous. Information from different perspectives is likely to be sounder than if only one perspective determines policy. Morale of all employees may be expected to benefit from such recognition. Some aspects of such practices may exist to some degree in some US-based corporations. A good degree of worker engagement in corporate decision-making would have important benefits and should be a goal for unions and unorganized workers.

Another manifestation of creating structures with broad engagement and forms of self-management may be seen in community development that is collaborative and builds community wealth.[37] For example, in September 2021, Lori Lightfoot, the mayor of Chicago, proposed using $15 million of Chicago's allotment from the Biden administration's 2021 American Rescue Plan (noted in chapter 4) to invest in cooperatives or land trusts with shared equity. Housing ownership would remain in the community. It would not involve tax breaks to large corporations that would profit from large-scale development. The idea of community wealth building began being developed in 2005 by the Democracy Collaborative and implemented in Cleveland in 2008.

The COVID-19 experience has shaken up many people's concepts of work and jobs. If that results in greater autonomy and dignity for many employed people, civil relations in American society will be greatly enhanced. Specific objectives need to be fashioned and proposed as goals to be pursued. As noted in chapter 3, there are numerous think tanks that are increasingly engaged in analyzing economic problems and ways to overcome them.

Advancing class equality furthers US ideals of fairness for all people in the country. Severe poverty is a denial of basic human rights. The country as a whole will be a healthier and happier place when class equality is considerably greater than it is now. There are signs of organized efforts by the previously diminished unions and many unorganized collective actions spurred by COVID-19 events that resulted in new efforts to increase class equity. In the fall of 2021, workers whose wages had stagnated, and were seeing corporate profits soaring, threatened and conducted a great wave of strikes. Workers who found their working conditions were harsh, and even injurious, chose to walk away. In many businesses, wages and work conditions began to improve some.

Fighting to Reduce Status Inequalities

As discussed in chapters 5, 6, and 7, the status of many collectivities previously suffering low regard from other society members has risen in significant degrees. However, such changes have also resulted in resentments and backlashes. Nevertheless, going forward, further progress in improving low status identities is likely. In this section, I discuss how more advances may be achieved constructively and so minimize resentments and backlashes. The advancements in women's and African American's status have meant substantial advances for the United States as a whole and can be celebrated by men and Whites as well as non-Whites. This is apparent in US successes in the Olympics and in US leadership in popular music and culture. Advances in status have also contributed to stronger roles in the political party system by women, African Americans, and other collectivities. Nevertheless, the progress has spurred resistance and resentment from some people, and many conflicts persist.

Advancing Women's Status

Since 1945, women have greatly increased their capacity to tell their stories and express their grievances, and they have gained power to make desired changes. It is a challenge, however, to make such changes without arousing resentments and fears from some men. Actually, there are many strategic approaches that can and have been effective to fight against backward steps. One strategy is to make claims for more respect and better social and material conditions in solidarity with men, demonstrating the absence of anti-male sentiments. Given the centrality of families bonding, men and women, and the shared interests resulting from that, conflicts can minimize zero-sum, win-lose formulations.

A primary arena for raising the status of women is to free them to equally enter the market of paid employment. Great progress in this area has been made and further progress is likely, because as young early entrants move up in diverse fields, increasing numbers will attain leading positions. That may result in resorting to strikes, protest marches, litigation, and taking direct actions to live as if the circumstances were as desired.

Providing support for the care of children is such a matter. Early feminists recognized this, and some argued for modifying the work prescriptions that required minimally being at the job, five days a week, eight hours a day. Over time, there was some relaxation of the strict gender roles, which caused men to be largely absent from the care of young children. Increasingly, men were liberated to enjoy the pleasures as well as the obligations of spending time with their young children. But this did not suffice, in most families, to free the wife-mother from being the primary child-care provider. In most economically developed countries, childcare and preschool arrangements were publicly available. This went with broad safety-net provisions in those countries.

The Republican Party has generally been ideologically opposed to such government policies, consistent with wanting a small federal government that would not provide attractive benefits and not require increased tax revenues. Clearly, this is associated with and contributes to class inequality. For years, many Democrats have advocated government supported child-care and preschool provisions. The 2020 election brought an increase in progressive Democrats into office, particularly in the House of Representatives.[38] COVID-19 and school lockdowns created crises in many homes, which often resulted in working mothers abandoning their jobs. Government rescue operations helped, but the need for establishing comprehensive child-care and preschool programs became more apparent. Furthermore, in some industries, offices were closed, and work was done at home using zoomed communications. The concept of a radically changed job was realized. Anticipating some ongoing working at home in the future, fathers as well as mothers saw child-care provisions as desirable. This was a salient issue for many Democrats, but not for many Republicans.

Conflicts at some level about who chooses to deny or permit abortions are likely to persist. In chapter 7, I analyzed the episodes in the long contention about legally allowing abortions and setting rules about it. I reported a broad public consensus supporting the 1973 Supreme Court decision, in *Roe v. Wade*, that the US Constitution protects a pregnant woman's right to have an abortion, within the first two trimesters. Nevertheless, some Republican leaders and some religious communities were fiercely opposed to the decision and that became a Republican principle. As a candidate, and as the president, Trump assured Christian evangelicals that his judicial appointments would overturn the decision. In states governed by Republican governors and state legislatures, a race went on to enact laws that would make it extremely difficult to have a lawful abortion.

A fierce escalation of the conflict about abortion arose early in 2022, when the radically conservative Supreme Court majority took on a case that might result in overturning the *Roe v. Wade* decision. A draft of a coming radical decision overturning the 1973 decision, written by Justice Samuel Alito, was leaked. This set off a storm of protests, opposing and supporting what came to be 5-4 decision, which the Supreme Court handed down on June 24, 2022. There are possible options that may ameliorate the intensity and destructiveness of the conflict. Efforts of some people on opposing sides to converse with each other and work together to promote policies that would minimize abortions could emerge. This would include reducing unwanted pregnancies with family planning. It might also include greater opportunities to facilitate adoptions of newborn infants whose mothers believed that they could not care for them. In addition, a good safety net might enable a mother to care for her infant. If the intense conflict were transformed, pains and tragedies could be reduced. Cooling the heat of the conflict would be a step forward.

There is another way forward, which is advancing. In 2000, the Food and Drug Administration approved an abortion medication, consisting of mifepristone and misoprostol.[39] A doctor must write the prescription for a patient. In some states this was prohibited, but usage is increasing as information and accessibility increases. The prescription can be filled abroad, and it may be provided to be kept at home. It is advisable that the physician is informed as to when the medication is taken, even if only by telephone. Some people may forever be dismayed that such abortions occur and seek to make it impossible, but the rancor and extreme attention to it may decline when those efforts are widely rejected.

The salience of the fight about legislation regarding abortion may well become widely recognized as destructive to civil life. The burdens of the destructive fighting may bring about recourse to more constructive collaborative options.

Finally, conflicts about protecting women from harassment, abuse, and violence are ongoing. Protection against harassment in places of employment had been limited because of the widespread provision that any dispute about alleged harassment must be settled by arbitration, and recourse to court action was not allowed. In February 2022, an act disallowing such forced arbitration provisions was passed in the House of Representatives, 335–97, and it was passed in the Senate by a voice vote.[40] This bipartisan bill was achieved by bipartisan leadership, general recognition of its correctness, and the revelations from the #MeToo movement. The broadcast journalist Gretchen Carlson, following her struggle regarding forced arbitration, effectively lobbied Congress for five years.

In March 2022, President Biden signed into law the Violence Against Women Act Reauthorization Act of 2022. This legislation was finally passed by Congress as part of an Omnibus appropriations package.[41]

Increasing Equality for African Americans.

During the years of Trump's administration, as discussed in chapter 6, there were some challenges and setbacks to the progress African Americans had achieved in previous years. African Americans disproportionally fell ill and died from COVID-19, due to relatively more of them living in poor housing, in difficult employment circumstances, and other living conditions. The economic downturn resulting from responding to COVID-19 also disproportionally set back African Americans.

Trump bears some responsibility in augmenting the impediments to African Americans progress toward equality. The overall ravages of the pandemic probably would have been less if Trump had not tried to downplay its gravity. In addition, Trump's encouragement to White supremacists contributed to their efforts to check the growing trend of many Whites to reconcile with African Americans. It also distorted and limited the gains that the Black Lives Matter protests sought and accomplished.

The effects of these White supremacists' actions to reverse African American gains, however, were limited and even counterproductive. The appearance of White supremacist militia groups aroused strong resistance. Their overt organized appearances culminated in violence at the January 6 insurrection. Arrests and trials of perpetrators of the attack on the Capitol followed, and insurrection instigators and plotters were identified. FBI and police surveillance of overt and covert White supremacist and antigovernment militias expanded.

More dangerously, Trump dominates the Republican Party leadership, which would focus on his return to the presidency. Consistent with that, at Trump's instigation in 2020, extremist Republicans raised the specter of critical race theory (CRT).[42] This academic approach, which emerged in the 1970s, emphasizes the intersection of race and US law, taking a constructivist perspective. It is critical of the mainstream liberal approach to racism and social justice. It is to be understood in the context of graduate school study, not in elementary curriculums. But right-wing agitators used the words as a weapon in another culture war battle and employed them to attack schooling or even public attention to learning about unjust treatment of African Americans and other minorities, as well as efforts to overcome them in the country's past or present.

The outbreak of attacks on CRT began after Christopher Rufo learned of training sessions in Seattle's municipal system intended to reduce unaware White racism. He wrote articles about ideas such as "White privilege" that were spreading rapidly throughout the federal government. Such outrageous affairs were linked to CRT. Tucker Carlson invited Rufo to his program on Fox News. Then President Trump, seeing the program, quickly signed an executive order banning the use of CRT by federal departments and contractors in diversity training. Joe Biden rescinded the order on his first day as president. But the war was on. In some Republican-controlled states and in many school districts, laws and regulations banned training and course material deemed to be exemplars of CRT or accounts of unjust treatment of African Americans. Meetings of school boards across the country have been scenes of screaming fights about school curriculums.

It will be useful to constructively transform such wildly destructive local meetings. There is a wide range of material about ways to reduce, avoid, or transform disruptive meetings.[43] They include having meetings chaired or facilitated by skilled outsiders, consulting with a local community dispute resolution center, ensuring that attenders agree upon rules of discussion, setting time limits, requiring that no one speak more than once until all who wish to speak have done so (this avoids speakers arguing with each other rather than addressing the whole meeting), leaders modeling being calm and listening well, and, finally, closing off disorderly meetings.

The tactic of disrupting local governing meetings will be discussed in the next section of this chapter. At this point, I consider the content of the message being sent. It is an attack against those people who would champion recognizing systemic discrimination against African Americans in the past and present and the enduring deprivations that impact the US society. Undoubtedly, that message is effective in some localities for some periods. But the integration that has occurred and the shared sense of the nation's promise is too strong to be turned back. The effort to reject the reality of and desire for diversity is self-limiting.

An important contribution to explaining why a powerful fascist movement did not arise during the Great Depression in the US should be noted. The few would-be fascist groups remained small because they attacked people of color, Jews, and Catholics. They represented too few people to attract a large following. The contemporary right-wing White supremacists also have self-limiting messages. I think the gains that African Americans have made, along with several other developments, make further gains likely. The likelihood could increase depending on the actions taken by those who do not support the messages of White supremacists.

First, it must be recognized that efforts of African Americans themselves have been, and probably will be, in the forefront in making the changes they desire. In terms of simple status, the great cultural achievements (in both popular and high cultural areas) have been important and are increasingly so. There are more and more shoulders that provide platforms upon which new generations can stand in museums, concert halls, film studios, and social media. For example, consider that the first opera of the 2021 Metropolitan Opera season was *Fire Shut Up in My Bones*, with a score by the jazz trumpeter and composer Terence Blanchard, based on a book by Charles Blow. It was the first opera by a Black musician presented at the Metropolitan in its 138 years.

Of course, achievements are not always recognized without contentions pursued by the achievers and their allies. For example, the Academy of Motion Picture Arts and Sciences annually honors the best films and best persons in various film production roles. The distribution of the Oscars was generally awarded to Whites by the White Academy reviewers. This lack of diversity was increasingly criticized by African Americans and others associated with the movie industry. The lack of diversity in the list of awards announced in February 2015, which bypassed important work by African Americans, evoked a storm of ridicule, complaint, and anger. The Academy expanded and broadened the diversity of reviewers and set out to avoid bias.

African Americans have fought for more equality in a multitude of ways. The analyses of this should help inform future struggles. Large-scale peaceful protest can be effective at the national, state, and local levels, particularly when government entities are friendly. Effectiveness is enhanced when specific goals are set forward. The Black Lives Matter movement might be insignificant as only an expression of wounded feelings. But the origins and specific demands focused on desired changes in the way police relate to African Americans. Police policies and conduct are determined at the national, state, and community levels. Recognizing that, many BLM leaders at each level are and will be negotiating about police conduct in dealing with African Americans.

Many areas of life are below decent American standards for many African Americans, such as residential housing, medical care, childcare, and education. At the national level, the Biden administration is trying to create a safety net that is adequate for every American and is trying to rectify some particularly egregious past discriminatory government policies. African American organizations should seek to participate in fashioning those policies, should mobilize support for those they like, and should be engaged in the implementation of the policies. Building good relations between government officials and nongovernmental leaders is crucial in making progress. This must be done at the state and local levels as well at the national level. The increasing public awareness of past discriminatory policies is resulting in some official efforts to make amends, even in the form of reparations, as discussed in chapter 6. The massive infrastructure work that is beginning provides an opportunity to remedy some of the damages caused by earlier highway building and by housing restrictions that contributed to environmentally unhealthy neighborhoods. Work to improve the infrastructure needed to resist the effects of global warming should be done in ways that improve poor neighborhoods. Poor Whites and African Americans often reside in areas that are vulnerable to increasingly frequent flooding and pollution.

Important work can and should be done at the local level and even in small neighborhoods. In several cities across the country, small, abandoned neighborhoods with junk and dilapidated houses have been cleaned up and transformed into well-groomed green areas with trees. Surprisingly, considerable research has found that crime levels declined in the areas that had been made over and had trees.[44] Crime levels did not rise elsewhere, they simply declined in the newly orderly and attractive space.

Much work by private citizens has always gone on to foster more equality between African Americans and Whites. That work is cumulative but may vary considering contrary factors. I have noted the increased activity of White supremacist militias and even some approval from extremist officials and political leaders. Furthermore, social media serves as a way for people in such organizations to reassure each other of their valor and correctness. In addition, some right-wing Republican Party leaders, commentators, Fox television figures, and social media platforms are fighting against people and messages they assert are devaluing Whites and wrongly advancing equity for African Americans. This is exemplified by attacks on critical race theory discussed earlier.

Another example is the 1619 Project, a journalistic project to foster a shared history, by placing slavery and its consequences as well as African American contributions at the center of the country's national narrative. It was developed by Nikole Hannah-Jones and other New York Times writers, and was first published in the New York Times Magazine in August 2019, marking the 400th anniversary of the arrival of enslaved Africans in the English colony of Virginia. On May 4, 2020, Nikole Hannah-Jones was awarded the 2020 Pulitzer Prize for Commentary for her introductory essay to the 1619 Project. The work generated a great clash between people celebrating it and others decrying it, including major political actions at the highest level.[45] Certainly, debates about interpretations of historical events can be avenues for more comprehensive and better-founded understandings. Politicized disputes, with vituperative language, and imposition of censorship, however, hamper attaining truthful understanding. In this instance, in November 2020, President Trump established the 1776 Commission by executive order, requesting eighteen conservative leaders to prepare an opposing response to the 1619 Project. The 1776 Report, released in January 2021, was widely criticized for factual errors.[46] The commission was terminated by Biden on his first day as president.

Constructive discussions generally entail mutual respect or at least civility, honesty, and the absence of coercion. Such modalities were not always observed in heated exchanges often intending to arouse and mobilize supporters, not convincing opposing proponents.

The surge in right-wing militia activity during the Trump presidency presents many challenges to increasing equality and preserving democracy. Of course, the

FBI can and does keep watch and brings charges against perpetrators of crimes. Also, nongovernmental organizations conduct investigations and bring civil charges against militia groups for misconduct, notably the Southern Poverty Law Center and the Legal Defense Fund.

A novel way of countering highly coercive actions of militia groups is for harmed persons to bring civil charges against perpetrators. In October 2021, in Charlottesville, Virginia, twenty-four organizers of the August 2017 rally were subjected to a civil trial. The plaintiffs accused the organizers of plotting to foment the violence that injured them. The trial had a devastating effect on the plotters and the militia organizations they led.[47] This is a newly available threat to militia groups that engage in injurious conduct.

News media should conduct investigations as well. An account of KKK members and Daryl Davis, an African American jazz pianist, warrants retelling and updates. For many years, Davis has been meeting with KKK members of various ranks and engaging in long conversations with them. In numerous cases, these dialogues resulted in a Klansman changing his beliefs and quitting the KKK.[48]

Thought should be given about reassuring Whites that their status is good, that they are respected. It should be recognized that many Whites are sharing in advancing a greater and better United States and advancing its ideals. It is important and true that a more equal country will be a more productive and accomplished one, about which all Americans can be proud. Americans collectively lose if bias excludes any category of people from equal opportunities. The country would be better if we had more and better highly gifted contributors. Improving the status of non-Whites is not necessarily wholly a zero-sum matter. Thinking in terms of one's American identity or God-believer identity, pride can be taken by American Whites. Indeed, many Americans do celebrate the country's diversity and the global glory it receives for it.

Increasing status equality in the ranking of people who differ in ethnicity, gender, and members of other collective identities furthers the US ideals of liberty and justice for all. It advances freedom for people to be and act as they wish, so long as they do not injure or interfere with such freedom for others. The resulting diversity of ways of being and acting is generally productive.

Fighting for More Equal Political Power

I have observed a few times in this work that the trend lines of class and status inequalities have crossed to some degree since 1945, with contrary effects upon power equality. Political power changes have their own trajectory and they also impact class and status developments, while class and status changes have important implications about the distribution of political power in the United States. The power of the majority of the population to determine political policies has waxed and waned since 1945. The two major political parties have been the primary agents in that complex interaction. Often, they have competed to gain collective power and exercise it for most of the nation, but to varying degrees they have also competed for distributive power for themselves and their major supporters. This section of the chapter is divided into three contemporary matters: the Republican and Democratic Parties, major political structural features, and nongovernmental developments and organizations.

The Republican and Democratic Parties

For several decades, the Republican Party has consisted of members with varying positions along the moderate conservative to extreme right-wing continuum. Over time, however, right-wing advocates increasingly came to dominate the party, and it became increasingly hostile to the Democratic Party. It lessened interest in seeking collective power and shifted to seeking primarily distributive power for itself. A path was created for Trump to win the nomination for the presidency, and as president he dominated the Party. Even after his defeat, he maintained great control over most Republicans. In all his conduct he largely relied on approaches contrary to the constructive conflict approach. An assessment of the effectiveness of Trump's conduct, therefore, provides a test of the validity of the constructive conflict approach.

Trump's defeat in 2020, by a large majority of the popular vote, demonstrated a failure of his distributive and divisive policies and conduct. Most remarkably, Trump and most of the political party he leads denied the 2020 presidential election results. Of course, in practice, the reality of the election results was accepted by just about everyone. The congressional results were accepted, even while some Republican representatives asserted the belief that a few seats were stolen. In some states where legislatures were controlled by the Republican Party, new rules for registering and voting in the 2022 election seem designed to suppress votes that are likely to be Democratic. This has been attempted in small and large ways, and sometimes these efforts have been blocked by court decisions.

Furthermore, after the 2020 elections, experienced election officials and volunteers were harassed and even threatened in many localities so they would resign and be replaced by less knowledgeable overseers or ones more likely to overturn undesired results. The Trump-inspired threats were particularly widespread and grave in Georgia.[49] Law enforcement must be enhanced to suppress such coercive conduct.[50] These actions were designed to reduce future Democratic votes, which undermines US democracy and imposes rules by authoritarian leaders.[51] Several writers have noted the declines in democracy in countries across the world, for example in Hungary and Poland. Analysts have attributed the shifts away from liberal democracies to more authoritarian, right-wing governments to the failures of neoliberal economics. A great reliance on the free market and neglect of government regulation had become very influential doctrines decades earlier. For example, Johnathan Hopkin argues that with the liberal market ideology, beginning in 1970, self-aggrandizing capitalist rules expanded.[52] The analysis in this book on the rise of hyper class inequality is consistent with such views.

It is noteworthy that Thomas Piketty, a leading analyst of the increasing wealth inequality since the 1980s, thinks the United States should and will undergo considerable wealth redistribution, which is needed to overcome the burdens and lack of overall growth resulting from great wealth disparities.[53] This would be in accordance with the great declines in inequality that have been happening in the world for hundreds of years, and notably since the French and American revolutions.

I will not try to predict what the future of the Republican Party will be. It may remain, led by Trump or another would-be Trump, a strong antagonist to the Democratic Party. Or it may splinter into an authoritarian, right-wing populist party and into a restored elitist conservative moralistic party. Or it may dwindle into a marginal cultish Republican Party and a new party that emerges as a modern conservative party. What I will do is discuss actions that may be taken by Republicans and outsiders to make the party a contributor to increasing equality in political power; that is, to enhance American ideals of liberty and justice for all.

Two major Republican leaders initially spoke out clearly denouncing Trump's complicity in the January 6 insurrection and his insistence that he had won the election against Biden, but then quickly fell in line and acted as if that was not true. Mitch McConnell, the US Senate minority leader, and Kevin McCarthy, the minority leader in the US House of Representatives, each did that.

One action that numerous Republican Party figures did was to express their lack of support for Trump at his 2016 and 2020 elections. For example, former president George W. Bush voted in 2016 for neither Donald Trump nor Hillary Clinton, and in 2020 he voted for Condoleezza Rice for president. Many Republican officials have resigned from their positions or announced they would not seek re-election. Many of them did so without criticizing Trump. For example, Speaker of the House Paul Ryan announced his decision not to run for re-election, explaining that he disliked the divisive identity politics that was so prominent and that it was not for him.[54] He left the House after the triumph of getting a budget bill passed, which he viewed as a great accomplishment.

Very many other Republican officials left office and decried Trump's misconduct as president. They and former Republican-aligned commentators appeared on cable news programs and wrote books revealing Trump's mistakes, inadequacies, and narcissistic qualities. The charges against Trump grew even greater after the failed insurrection attempt. Some turned away from Republican conservatism entirely.[55] Most dissenters, however, did not undertake collective political action.

Notably, however, some dissenters did organize and joined together to fight against Trump and his control of the Republican Party, or against the Republican Party as it had become. The Never Trump movement emerged to stop the Republican Party's nomination of Trump for president. The founding leaders included Colin Powell, a former secretary of state, but otherwise they were mostly commentators, namely Joe Scarborough, George Will, Steve Schmidt, Jennifer Rubin, and William Kristol.[56] They, and several other Republican-orientated individuals and organizations, strove to block Trump from becoming the Republican Party's nominee for president. Having failed, they and other normally Republican voters went on to vote for a variety of other candidates in the 2016 and 2020 elections. Despite the widespread opposition, there was no consensus about an alternative, and the party organization succumbed to Trump's leadership.

Another major Republican anti-Trump organization is the Lincoln Project (https://lincolnproject.us/). Some former Never Trumpers, including Steve Schmidt, joined with other Republican opponents to found a project that would attack Trump in a Trumpian manner, but also with humor. It produced ads and material for social media but had limited effects.[57]

Finally, there might be efforts to start a new Republican Party, without Trumpism, despite Trump's approval rating among self-identified Republicans having remained extremely high.[58] One such effort was announced on May 13, 2021, in A Call for American Renewal by 150 members and former members of the Republican Party. "It calls for strengthening the rule of law and increasing government ethics."[59] Leaders include Miles Taylor, former member of Congress Barbara Comstock, former governor Christine Todd Whitman, and William F. B. O'Reilly. In October 2021, leaders of this movement decided that the most effective way to defeat Trumpism was to vote for moderate Democrats when they oppose Trump and Republican candidates "who embrace Trump's lies," in the 2022 and 2024 elections.[60]

This is probably the most effective and constructive way to change the Republican Party and protect the US electoral system. An interesting public opinion survey, however, suggests another option: a distinctive third party. In early 2001, a national Gallup poll asked respondents whether the two-party system is doing an adequate job, or if a third party is needed.[61] Since the question was first posed in October 2003, a record-breaking 62 percent of respondents said a third party was needed. Among independents, 70 percent; among Democrats, 46 percent; and among Republicans, 63 percent so responded. Clearly, there is sentiment for a third party, particularly among Republicans. Perhaps a moderate-conservative party orientation that was not defined primarily as anti-Trump would be attractive.

As revelations about the complicity of many Republican Party leaders in trying to overturn the defeat of Trump in 2020 goes on and Trump makes ever more clearly that he cares about himself above party or country, the Republican Party will be splintered and weakened. Only elections that are subjected to extraordinarily politicized procedures could yield Trumpian national election victories, as discussed further in this chapter.

As the other side in an ongoing political conflict, the Democratic Party is more in accord with traditional institutional ways than the Republican Party. Indeed, the orientation of the Democratic Party, even as it sought to defeat the Republicans, was to advance collective power for the United States. This was the case even if the party was inclined to draw on neoliberal economic and social thinking, as was notably so during the Clinton administration. A shift in the party toward a more progressive orientation was evident in the competition for the Democratic Party presidential nomination in 2016 between Hillary Clinton and Bernie Sanders.

To a surprising degree, a social democratic, older man from Vermont garnered a wide youthful crowd of supporters. To some extent, Trump's faux populist rhetoric and thuggish manner may have helped Bernie Sanders relative to Hillary Clinton. Sanders was fiercer and more dogmatic and emphasized workers' needs. Hillary Clinton's defeat in 2016 tended to reinforce Sanders' more progressive arguments. Although he did not win the nomination in 2020, and the presumed moderate Joe Biden was nominated, a substantial segment of the Democratic Party wanted to take government action to increase class, status, and power equality.

The Democratic Party in Congress is ideologically divided into two major caucuses, the Congressional Progressive Caucus (CPC) and the New Democratic Coalition (NDC), founded in 1991 and 1997, respectively. Although they have about equal membership, the CPC has been more influential in the formation of the Biden administration's legislation. Pramila Jayapal, as chair of the CPC, was highly engaged in the final negotiations regarding the Build Back Better bills. Given the very narrow majority the Democrats had in the House and Senate, individual senators could play very powerful roles. In the Senate, Joe Manchin, from West Virginia, did that.

Internal dissension and conflicts are often linked to external contentions. In this case, it enhanced the inclination of the Republican leader in the Senate to stay united in refusing any allowance for cooperation. The party division and antagonism that was thereby increased was contrary to the best interest of the US as a whole.

The Republicans in the Senate, led by Mitch McConnell, refused to cooperate on any significant Democratic bills, so they would not be able to claim any legislative accomplishments. Clearly, this was a distributive power strategy. Despite that obstacle, Biden did get major bills passed. This was presented as demonstrating that the US government can act. The transparency of the process should build greater trust in the government.

US political parties generally contend within rules and norms of the institutionalized political system. Sometimes they are altered in an open, recognized manner. However, too often, in recent years, they are violated in covert as well as open manners. This has been the case at schoolboard meetings relating to the ways schools manage responding to COVID-19 and to what is taught in school courses.[62] As noted earlier in this chapter, fierce intimidation, with threats of violence, have been directed at election officials who oversaw the 2020 election results that marked Trump's defeat. Such conduct obviously undermines the rights of US citizens to choose their government officeholders. Such interference with US elections must be ended. Perpetrators need to be held accountable.

Party independents could well play greater roles in political affairs. They are numerous, and even if diverse, multiple groups could draw attention to issues and forms of conduct that would contribute to more constructive fighting. For example, many people feel disgust, fear, and anxiety with widespread uncivil conduct, including demonizing adversaries, threatening violence, and violating long-standing norms of mutual respect. Independents could play roles in condemning such conduct though good government organizations or simply join with some other people to express the condemnation in local settings.

The striking rise in homicides and lethal attacks on police could spur gun control legislation at all levels of government. Local actions, such as gun buybacks, are helpful. Other actions are possible, including mediating or facilitating adversaries who are interacting destructively. Speaking out as independents might help cool down some outlandish partisan conduct in fear of losing votes.

Structural Political Features

There are several features of the US political system that are enshrined in the Constitution, in Supreme Court decisions, and in enduring norms. I discuss a few of them. Federalism is a major feature of the US political system, with many benefits. It allows many political issues to be handled by people closer to those issues than if all political issues were centrally determined. Federalism allows for local decisions, considering local environmental and societal qualities. However, the Constitution was written so that for some national purposes, all states are equal. Thus, each state, regardless of population size, has two senators, while the number of House representatives in each state is proportional to the population of the state. Each state has as many electors as the number of senators and representatives it has. In addition, since 1961, the District of Columbia has three electors. The vote of the Electoral College electors determines the election of the president and vice president.

According to the Constitution, the president and vice president are chosen by the Electoral College after a national election. Since the states with the smallest populations have a somewhat greater voting weight, if they tend to vote for one party, it is possible that candidates for president and vice president may win a majority of the electoral votes (270), but fewer of the national votes. Most recently, that happened in 2000, when Al Gore lost to George W. Bush, and in 2016, when Hillary Clinton lost to Donald J. Trump.

That is clearly unfair to voters from the larger populated states. Moreover, it biases the presidential campaigns, which focus on swing states.[63] Campaign themes and events focus on a few states such as Ohio, Florida, and Virginia. Correcting this matter by eliminating the Electoral College would require a constitutional amendment. Another solution is in progress: a compact among states to elect the president by national popular vote. The states would direct their electors to vote for the candidates who had won the popular vote. So far, a national popular vote bill has been enacted in sixteen jurisdictions possessing 195 electoral votes (see www.NationalPopularVote.com). This provision certainly deserves more attention and supportive action.

Another voting inequity that warrants correction is the denial of full voting rights to the citizens living in the District of Columbia. That could best be rectified by statehood for the district, and it should be enacted by Congress. A small federal district could be created to fulfill constitutional provisions. The fight for statehood has been waged by DC citizens since its early history.[64] The lack of statehood hampered an appropriate response to COVID-19 efforts, so much of which was done through state operations. Moreover, the difficulties in quickly mounting a strong defense against the January 6 attack on the Capitol provides new arguments for statehood. In April 2021, the House voted along party lines to grant statehood to Washington, DC.[65]

Another structural feature of US governance is the great power of the Supreme Court. A majority vote of the nine justices can override legislation that was widely supported by Congress, the president, and the public. This has happened, generally in a conservative direction.[66] A few ways for the Court to make decisions more reflective of contemporary thinking have been raised. One way is that the justices themselves take more cognizance of public sensibility. This may have happened in the past, during the years of the New Deal legislation. Increasing the number of judges is another way. President Biden established a commission to consider possible changes to the Supreme Court.[67] The members of the commission indicated some interest in possibly establishing terms of service for the judges, in one or another manner. That would be a modest improvement and may occur in several years.

Finally, dissatisfaction and misgivings have risen about the growing primacy of the presidency relative to the legislative branch of government. This is undoubtedly a problem in foreign policy. Even in domestic policy, a more effective legislative operation could enhance its role beneficially.[68]

Gerrymandering is the last structural political matter I note. While voters would like to choose their elective officeholders, politicians would like to choose their voters. The practice of gerrymandering is an important way for a party that controls the state legislature and the governor's office to be able to adjust the legislative district boundaries to maximize safe electoral districts for its candidates. Depending on a state's constitution, laws, and judicial oversight, many ways of gerrymandering may be employed.[69] Every ten years, when a national census is made, the Constitution requires the reallocation of congressional seats and districts. The Republican Party has been especially aggressive in maximizing state legislatures and executing gerrymandering that would favor its electoral strength. The Democratic Party seeks to enact national legislation to constrain extreme gerrymandering and make voting more equitable.

Nongovernmental Developments

A strong civil society is a bulwark against authoritarianism. The civil society interacts with institutionalized political power arrangements, collaboratively and contentiously. The four major developments noted at the outset of this chapter also challenge the civic society. I focus here, however, on recent developments that may hamper or may help nongovernmental movements and organizations to enhance class, status, and power equality in the future, in the larger context.

First, I discuss the electoral process at the national, state, and city levels, as impacted by the persistence of Trump and many Republican leaders to deny the legitimacy of the election of President Biden and Vice President Harris in the 2020 election. In one matter, they sought to change voting procedures to heighten the chances of Republican victories in later elections, a threat to law and order and to democracy. This was confounded with other issues that aroused public attention and protests. One was federal sanctioning of vaccinations and masking to counter the spread of COVID-19. Another set of issues relates to schools: demands that schools not set rules about ways to manage responding to

COVID-19, and demands to control curriculum so as to exclude discussions of material deemed hurtful to Whites, as noted earlier.

These are matters of contentious campaigns in culture wars, which involve profound feelings of disrespected identities for some people and associated with resentments associated with not being well respected. Such feelings are difficult to be assuaged. Yet they may be tempered if people having such feelings are not derided or belittled. Their concerns can be recognized and alleviated, even if policy preferences differ.

For example, the pride in the US spirit of independence can be celebrated when it results in brushing aside allegiance to authoritarian directives and instead entails freely joining in accord with freely agreed-upon policies. Differences of opinion are useful to reach good collective decisions. Another approach may be useful in moderating White resentments about loss of respect when attention and even privileges are accorded to non-Whites. The great contributions many Whites have made to overcoming all kinds of racist discrimination is to be celebrated. Whites fought and died to save the Union and end slavery. Whites elected representatives who passed the Thirteenth, Fourteenth, and Fifteenth Amendments to the Constitution, ensuring equal rights for all. Many Whites joined and supported the civil rights movement in the South. Even now, many Whites are striving to reduce the inequities that remain. This is done out of feelings of fairness and out of a sense of simple justice. It is also done, as stressed in this book, out of self-interest. Living in the United States will be better for everyone as class, status, and power equality increases.

There are many events, organizations, and movements that do and will contribute to enhancing equality. Many have been cited and discussed throughout this book. I only mention a few more here, and some possible new ones. Some good government organizations have been active for many decades. The League of Women Voters (https://www.lwv.org/) was founded in 1920, after women had won the fight for women's right to vote. It has a large and active membership in over seven hundred state and local leagues. It educates and mobilizes voters at elections and campaigns to protect voting rights. Common Cause (https://www.commoncause.org) was founded in 1970 as a nonpartisan organization that advocates for government reforms. Its mission is to advance the core values of US democracy.

Indivisible (https://indivisible.org/) is an important new organization working to advance American democracy. It was established soon after Trump was elected president and remains active following his failure to be re-elected. The Indivisible movement had begun with an online handbook written by congressional staffers with ideas about resisting the anticipated move to the right in the executive branch of the US government. It is a nationally coordinated movement of locally led communities. Local organizations, in congressional districts, meet regularly (by Zoom when appropriate), and attend to local and state matters as well as national and international ones. Knowledge is shared and actions are taken collectively and individually. Lines of communication with officeholders are maintained. Such grass-roots work is an effective way of responding to the too often uncivil disorderly ways of doing politics.

Another increasingly important development that contributes to improving public policies is the increase in the number and diversity of think tanks. The FP Analytics, an independent research division of *Foreign Policy* magazine, conducted an international study of proliferating think tanks responding to increasing authoritarianism and other social problems.[70] The resulting report recognizes increasing challenges in many countries, but also considerable successes in advancing reforms.

A major concern among people in the United States, particularly during the recent years of division, lack of civility, and even violence, is peoples' lack of trust in many institutions and in each other. Thus, in a national survey conducted November 27–December 10, 2018, 75 percent of Americans believed trust in the federal government had been shrinking, and 64 percent believed that trust in each other was shrinking.[71] Moreover, 68 percent of Americans believed it was very important to improve the level of confidence in the government, and 58 percent believed it was very important to improve the level of confidence with each other. Similar percentages believed that the low trust makes it harder to solve problems.

Indeed, trust in government has irregularly, but generally, moved downward since 1958, when the Pew Research Center began such surveys.[72] A relatively large decline began in the 1960s, with the escalation in the Vietnam War and other developments discussed in earlier chapters. Subsequent declines were associated with significant economic downturns. Confidence varies for the major branches of government. It spiked upward for the military with the end of the Cold War and has remained high.[73] Next is the presidency, which spikes with new presidents, and then slopes downward. Confidence for Congress is the lowest and slopes downward. Overall, an October 2021 Gallup survey showed a record decline in confidence in politicians dating back to 1974.[74]

The levels of confidence and changes in confidence levels differ considerably among some US institutions. Confidence in science has been consistently high. Confidence in religion and education were moderately high in the 1970s and have declined only a little. However, confidence in the press and television were modestly high in the 1970s and sharply declined to a low stable level in the 1990s.[75]

People differ in the likelihood that they will have confidence in some US institutions. Of particular interest in the context of this chapter are the differences and the similarities between Republican and Republican-leaning respondents compared to Democratic and Democratic-leaning respondents in a 2021 survey.[76] They are similarly modestly confident in the Supreme Court (39% and 35%), banks (35% and 33%), and the criminal justice system (20% and 19%). They also similarly lack confidence in big business (19% and 17%) and Congress (7% and 17%).

They differ about some institutions. Republican respondents are more confident than Democratic respondents regarding the police (76% and 31%) and the church or organized religion (51% and 26%). On the other hand, Democrats are more confident than Republicans regarding organized labor (39% and 16%), public schools (43% and 20%), and newspapers (35% and 5%).

These findings indicate which institutions are particularly low in trust among only Republicans or Democrats, or among both. Members of those institutions should take these results seriously and take actions accordingly. More good information is needed.

These variations suggest that feelings of confidence in US institutions are influenced by ideological identities and misinformation. The Trumpian rejection of the 2020 presidential election results, with unsubstantiated charges of irregularities, contributes to lowering confidence in US political figures and institutions. Overreliance on social media and Fox News for political and other information certainly is a source of misinformation. Fox News features commentators who spread bizarre stories, and it does so with no pretense of being a conveyor of news.[77] The very numerous right-wing social media platforms include *The Federalist*, Breitbart, Liberty Nation, *American Thinker*, and the *Epoch Times*. All three major cable outlets greatly increased their audiences from 2016 to 2020: Fox rose to about 3.08 million, CNN reached 1.80 million, and MSNBC reached 1.6 million.[78]

Of course, there are innumerable sources for factual, grounded newsworthy information. Sound information can be viewed on television, heard on the radio and podcasts, viewed on the Internet, and read in newspapers, magazines, and books. And most people do make use of such sources. Nevertheless, many people are misled by misinformation, by sources that lie. High schools, civic organizations, and higher education schools should convey ways to assess whether information is truthful or not.

The three primary television news channels generally avoid clearly partisan reporting or commentaries, while providing Democratic and Republican leaders and spokespersons equal time to state their opinions. They have each increased their large audiences for the evening news programs: between 2016 and 2020, ABC's audience rose to 7.6 million viewers, NBC reached 6.5 million, and CBS reached 5 million.[79] In addition, the *NewsHour* program on the Public Broadcasting System (PBS) attracted 1.2 million viewers on average in

2020.[80] It provides detailed nonpartisan reporting and news analyses. Through the *Frontline* series, PBS provides in-depth reports of newsworthy events.

Nonetheless, highly partisan perspectives are allowed to be presented without conveying alternative viewpoints. As discussed in chapter 3, in 1987, President Reagan ended the Fairness Doctrine, which had been introduced in 1949. It had required radio and television broadcasters to devote some airtime to discussions of controversial public matters and provide contrasting views regarding those matters. Undoubtedly, that action has contributed to increasing the level of party polarization in the United States. A new version of the Fairness Doctrine should be established.[81] In 2019, H.R. 4401, the Restore the Fairness Doctrine Act, was introduced in the House.[82] This bill would require a broadcast radio or television licensee to provide an opportunity for discussion of conflicting views on matters of public importance.

Newspapers have long been the way to keep up with important current political and economic news. Of course, some papers appealed to people interested in entertainment and scandals. The Internet, in any case, undermined the market that had sustained local newspapers and even many large urban newspapers. This greatly reduced the opportunity to get information about local matters, reducing the possibility of oversight and engagement by local citizens. Here and there, wealthy persons may purchase an excellent but endangered newspaper and maintain its autonomy.[83]

The COVID-19 pandemic was another blow to sustaining local newspapers. In 2020, the Nieman Foundation for Journalism issued a report titled *4 Ways to Fund—and Save—Local Journalism*.[84] One way is to monetize the crisis is to expand members as supporters and collaborate with local and national civic institutions and philanthropic organizations to provide for needed functions. A second way is to pivot to philanthropy; indeed, support from foundations and philanthropists has greatly increased investigative reporting and other matters since 2019. The third way is to tax other platforms (search engines, social media) to pay for verified information. Finally, nonpartisan government funding could be greatly increased, as is done in many countries and as favored by the US public.

Reliance on the Internet and social media for information has become huge, but it is particularly vulnerable to misuse. YouTube provides information about purchasing, using, or repairing everything. Google provides knowledge about locations everywhere. *Wikipedia* provides up-to-date encyclopedic essays on almost any topic. However, Facebook and Twitter can transmit personal gossip, and stories that are true or false, enlightening or malicious.

Some oversight exists to exclude highly pornographic, dangerously false, and extremely hateful material. Facebook algorithms tend to send circulating items to followers who would like them, as evidenced by people's prior choices. Consequently, people get material they agree with and do not see contrary views.

This contributes to social-political polarization.[85] Congress has held hearings on this phenomenon and pressured Mark Zuckerberg, the founder, chairman, and CEO of Facebook, to take more action to reduce the ill effects of the way it currently functions. Facebook, under pressure, is taking more care. Different strategies may lessen the ill effects and public engagement in choosing and supporting the best strategy would be desirable.

Undoubtedly, there are many avenues for citizens to access comprehensive factual news about national political events. To some degree, this information is seasoned with sensational popular stories and is conveyed from a US-centric perspective. For many citizens, nonpartisan, detailed coverage of news is simply too boring to regularly watch, listen, or read about. High school courses might do more to help enhance skills to interpret and appreciate getting sound information.

Misinformation has been a hallmark of the political tribalization that developed over recent years. As noted earlier, it was intentionally employed, particularly by many Republican political figures and spokespersons, to attack the character of targeted Democrats. This sometimes goes so far as to spread conspiracy theories relating to criminal actions, intending to demonize Democrats. This was notably done in attacking Hillary Clinton.[86] Big money has gone into supporting misinformation, including the lie that Trump won the 2020 presidential election.[87]

False conspiracy theories can be particularly dangerous. This is the case for the "great replacement conspiracy theory," which has wide and deep roots. It holds that plotters cause the decline in the percentage of Christian Whites in the United States and seek their replacement. The plotters are variously identified on social media platforms as Democrats, seeking immigrants to vote for them, Jews, and Muslims. Such allegations are even expressed by commentators such as Tucker Carlson on Fox News. Such allegations have contributed to mass murders, such as those targeting Jews in a Pittsburgh synagogue in October 2018 and targeting African Americans in a Buffalo grocery store in May 2022.[90]

Good information is essential to maintain democracy, civil discourse, and social progress. Furthermore, it is accessible in many sources and outlets, but many people prefer to ignore or disregard much valid information. The misinformation is preferred or embraced by some people for diverse reasons, including that it makes them feel better, they believe in the dispensers of the misinformation, it is a way to sustain their tribal identity, and/or they fear punishment.

Finally, I turn to efforts that are taken and might be undertaken to refute lies, misinformation, and bizarre conspiracy theories. Perhaps most important in dealing with political misinformation is overcoming the persistent denial of Trump's electoral defeat in 2020. Significantly, many important Republican politicians, former officials, and analysts assert that the Republican Party must accept the reality that there is no evidence of electoral malfeasance that would account for Trump's and Pence's election loss. Several Republican members of Congress voted to certify Biden's victory on the night of January 6, 2021, some of them deciding to do so in response to the violent attack on the Capitol.[88] In July 2021, the House of Representatives established the House Select Committee on the January 6 Attack. Two Republican House members joined: Liz Cheney, as vice chair, and Adam Kinzinger. As noted earlier, a scattering of Republican leaders continue to emerge and reject the Big Lie allegation and call for Republicans to turn away from the past election and instead to focus on policies for the next election.

Violence has often been associated with conflicts in US history, and attention to it has risen again since 2016. Right-wing militia groups and promoters of false conspiracy theories undermine democracy and spur violence. It is important that using and threatening violence in political and other civil conflicts not become normalized and not be used by government authorities. That is extraordinarily dangerous, and it must be recognized as unacceptable and in the long run self-destructive.

An important step to counter resorting to violence in social conflicts is to ensure that perpetrators of such violence be legally constrained and punished. To some degree this is happening in many settings. Importantly, by the end of November 2021, more than 700 participants in the attack on the Capitol were arrested and 129 had pleaded guilty.[89] Charges and punishments varied greatly, from misdemeanors to felonies. The trials and related investigations have been very disruptive to right-wing militia groups. Thus, in November 2021, the House committee investigating the Capitol attack issued subpoenas to three militia groups, including the Proud Boys and the Oath Keepers, which are believed to have information about the attack on the Capitol.[90]

In addition, the Department of Justice investigates and charges individuals committing criminal acts relating to political violence. For example, an extremist individual, Brendan Hunt, was found guilty of threatening the lives of several Democratic members of Congress on social media posts, starting in December 2020. He was sentenced to nineteen months in prison in November 2021.[91] Furthermore, within Congress, when threats of violence occur, they are condemned. On November 17, 2021, House Democrats censured Representative Paul Gosar and removed him from two House committees after he posted an anime video of himself killing Representative Alexandria Ocasio-Cortez and attacking President Joe Biden.[92]

One increasing response to false allegations and hurtful actions is for plaintiffs to bring civil charges against persons and groups who harass, threaten, and injure them.[93] One of the most vicious false conspiracy theories has been promoted by Alex Jones.[94] He has claimed for years that the Sandy Hook Elementary

School shooting in December 2012 was a "giant hoax." This resulted in vicious harassment of some parents of children killed in the school shooting, causing the bereaved families to move and hide. He was tried in Texas and Connecticut in mid-2018 and found guilty, but he continues in his false claims and avoids any payments while making legal appeals. Further convictions have followed.

As noted earlier in this chapter, in preparation for forthcoming elections, many Republican Party officials have sought to establish favorable gerrymandering and voting procedures in states that they dominated. In addition, pro-Trump groups harassed experienced election workers to have them drop out and be replaced by more compliant election workers. In order to halt extreme harassment, some targeted persons have filed defamation suits for punitive damages.[95]

A major instance of bringing civil damage charges in a conflict was brought against organizers of the 2017 right-wing rally in Charlottesville, Virginia, and won in November 2021.[96] The twelve plaintiffs had suffered grave injuries when James Alex Fields Jr., one of the defendants, drove his car into a crowd of protesters against the rally, killing Heather Heyer, as discussed in the previous chapter. The suit was heard in the US District Court in Charlottesville. The jurors awarded more than $25 million in damages from the more than twenty-four organizers and organizations that were the defendants. Certainly, many of those defendants have been somewhat diminished in power and influence by the trial. It probably also contributed to further reducing the legitimacy of right-wing militia groups and recourse to violence. Unfortunately, it may have tightened the ties between Trumper Republicans and the right-wing extremists, and deepened the chasm between Trump supporters and many in the rest of the country.

Countering misinformation and false conspiracy theories and prosecuting criminal political falsehoods and violence is necessary, and should be pursued actively with judicial actions, some humor, and good contrary evidence. However, even all of that is not enough. More proactive work is needed. Having good information is not enough. The information must be blended with values and preferences to formulate good social policies, which are pursued. Keystones include gaining and applying collective political power and formulating and striving for mutually beneficial goals.[97] The importance and possibility of framing conflicts so that win-win outcomes are feasible has been a fundamental idea from the beginning years of the conflict resolution movement.[98]

In many ways, the recent antagonistic relations that beset the United States are indeed intractable conflicts. Research into how severe conflicts persist despite efforts to de-escalate, resolve, or transform them has greatly expanded over many years.[99] Intractable conflicts can and do become transformed. An excellent website that provides findings about how intractable conflicts can be transformed using various constructive strategies is Beyond Intractability (https://beyondint ractability.org/). It consists of a wide range of materials by numerous authors in various groupings, including "Constructive Conflict Massive Open Online Seminar" and the "Constructive Conflict Initiative." Of course, much background information is available about the theory and practice of transforming intractable conflicts. The website was initiated and continues to be maintained by Guy Burgess and Heidi Burgess.

Indeed, the ideas and practices of the conflict resolution movement are continuing to spread. There are growing numbers of organizations providing conflict resolution practices at the local, state, and national levels. There are increasing educational institutions offering courses and also MS and PhD degrees in conflict resolution. Applications of the constructive conflict approach and other components of the conflict resolution perspective are becoming part of everyday thinking.

Trump's presidency and overall conduct, of course, have been and remain largely contrary to the constructive conflict approach. I think that he has failed to accomplish many of his stated goals, and the results of what he did accomplish have worsened progress toward more class, status, and political power equality. Biden's presidency and overall conduct, in its beginning, has often been significantly in accord with the constructive conflict approach. This is the case for taking constructive steps that enhance equity of class, status, and political power.

Conclusion

Greater equality in the possession of political power enhances American ideals of democracy. It also enhances class and status equality. More equality in class, status, and power tends to maximize creativity and the production of desired goods and services. More equality will reduce social problems of all kinds. The enhanced status equality that has been achieved is broadly beneficial. Yet some resentments have sadly arisen and have been exploited. That exploitation needs to be countered. At the same time, while fighting for more status equality, ways to minimize resentments should be practiced.

Conflicts usually do not have to be fought destructively. Generally, domestic American conflicts can be conducted constructively and provide pathways to make progress. Certainly, some mutual interests can be found to be shared between adversaries; if they are creatively sought. When wholly zero-sum conflicts are waged by all sides, great losses will result. If only one side rejects the existence of any common interest, long-term self-destructive results are likely.

The analyses of cases in this book should have demonstrated the value of various kinds of inducements, strategies, and tactics. Too often, there is on overreliance on coercion, and even violence. Such reliance arouses resistance and hinders positive conflict transformations. Noncoercive inducements, persuasion, and positive sanctions can often be decisive for one or both sides to advance toward their goals. In any case, when some level of coercion is applied, its effectiveness is increased if it is combined with persuasive actions and the promise of positive benefits.

There are a multitude of tactics that may be combined into strategies whereby people can fight better, more constructively. The persons may be leaders, dissenters, or low-ranking partisans from one side or the other. They may be intermediaries who provide mediating functions or aid in reconciliation.

Since 1945, there have been profound moves toward greater equality in class, status, and power, but also serious setbacks and regressions. At present, the divisions in the United States are such that the future of progress toward a more democratic and egalitarian society is not certain. Too few leaders and groups recognize the benefits of cooperation and the dangers of zero-sum thinking. There is too much reliance on the use of coercion and violence to advance self-serving gains.

The deterioration in American political life has occurred over very many years. Recovering a better civic and political life will also take many years. Despite the somber domestic and global challenges that confront the US, the well-done conflicts of the past promise that more progress is possible in the future. It is possible if we make it so.

Notes

Prologue and Acknowledgments

1. Louis Kriesberg, *Realizing Peace: A Constructive Conflict Approach* (New York: Oxford University Press, 2015).
2. Louis Kriesberg, *Mothers in Poverty* (Chicago: Aldine, 1970).
3. Bruce W. Dayton and Louis Kriesberg, *Constructive Conflicts: From Emergence to Transformation*, 6th ed. (Lanham, MD: Rowman & Littlefield, 2022).
4. Louis Kriesberg, *Louis Kriesberg: Pioneer in Peace and Constructive conflict Resolution Studies*, ed. Hans Guenter Brauch, (Cham, Switzerland: Springer International, 2016).

Chapter 1

1. Catherine Gerard and Louis Kriesberg, eds., *Conflict and Collaboration: For Better or Worse* (Abingdon, UK: Routledge, 2018), https://www.maxwell.syr.edu/about-cas-conflict-and-collaboration/?gclid=EAIaIQobChMI1NaHsOOw8QIV5SmzAB22BwJmEAMYASAAEgI8yvD_BwE.
2. Bruce W. Dayton and Louis Kriesberg, *Constructive Conflicts: From Emergence to Transformation*, 6th ed. (Lanham, MD: Rowman & Littlefield, 2022).
3. For a summary of that evolution, see Louis Kriesberg, "The Evolution of Conflict Resolution," in *Sage Handbook of Conflict Resolution*, ed. Jacob Bercovitch, Victor Kremenyuk, and I. William Zartman (London: SAGE, 2009), 15–32, https://lkriesbe.expressions.syr.edu/wp-content/uploads/Oxford-Research-Encyclopedia-March-2019-pdf.
4. Barbara W. Tuchman, *The March of Folly: From Troy to Vietnam* (New York: Ballantine, 1984).
5. Its original form, in 1892, was: "I pledge allegiance to my Flag and the Republic for which it stands, one nation, indivisible, with liberty and justice for all." In 1923, the words, "the Flag of the United States of America" were inserted, and in 1954, Congress inserted "under God."
6. Louis Kriesberg, "Peace Movements and Government Peace Efforts," in *Research in Social Movements Conflicts and Change*, ed. Louis Kriesberg, Bronislaw Misztal, and Janusz Mucha (Greenwich, CT: JAI Press, 1988), 57–75.
7. Examples during the Cold War are in my article "Noncoercive Inducements in U.S.-Soviet Conflicts: Ending the Occupation of Austria and Nuclear Weapons Tests," *Journal of Political and Military Sociology* 9, no. 1 (1981): 1–16.

8. Roger Fisher, William Ury, and Bruce Patton, *Getting to Yes: Negotiating Agreement without Giving In*, 2nd ed. (New York: Penguin, 1991); Howard Raiffa, *The Art and Science of Negotiation* (Cambridge, MA: Harvard University Press, 1982); Louis Kriesberg and Bruce W. Dayton, *Constructive Conflicts: From Escalation to Resolution*, 5th ed. (Lanham, MD: Rowman & Littlefield, 2017), 261–64; Roy J. Lewicki, David M. Saunders, and John W. Minton, *Essentials of Negotiation* (New York: McGraw-Hill, 2000); Heather McGhee, *The Sum of Us* (New York: One World, 2021).

9. Laura E. Drake and William A. Donohue, "Communicative Framing Theory in Conflict Resolution," *Communication Research* 23, no. 3 (1996): 297–322).

10. Kimberle Crenshaw, "Demarginalizing the Intersection of Race and Sex: A Black Feminist Critique of Antidiscrimination Doctrine, Feminist Theory and Antiracist Politics," *University of Chicago Legal Forum* 1 (1989): 139–67.

11. Lewicki et al., *Essentials of Negotiation*; Raiffa, *The Art and Science of Negotiation*; Anthony Wanis-St. John, *Back Channel Negotiations: Secrecy in the Middle East Peace Process* (Syracuse, NY: Syracuse University Press, 2011); I. William Zartman and Guy Faure, *Escalation and Negotiation in International Conflicts* (Cambridge: Cambridge University Press, 2005).

12. Kumar Rupesinghe, ed., *Conflict Transformation* (New York: St. Martin's, 1995); Louis Kriesberg, Terrell A. Northrup, and Stuart J. Thorson, eds., *Intractable Conflicts and Their Transformation* (Syracuse, NY: Syracuse University Press, 1989); Martina Fischer, Joachim Giessmann, and Beatrix Schmelzle, eds., *Berghof Handbook for Conflict Transformation* (Farmington Hills, MI: Barbara Budrich Publishers, 2011); John Paul Lederach, *Preparing for Peace: Conflict Transformation across Cultures* (Syracuse, NY: Syracuse University Press, 1995).

13. Louis Kriesberg, *International Conflict Resolution: The U.S.–USSR and Middle East Cases* (New Haven, CT: Yale University Press, 1992); *Realizing Peace: A Constructive Conflict Approach* (New York: Oxford University Press, 2015).

14. Max Weber, *The Theory of Social and Economic Organization*, trans. A. M. Henderson and Talcott Parsons (New York: Oxford University Press, 1947).

15. I discussed different theories and relevant research about world and US inequalities in each dimension in Louis Kriesberg, *Social Inequality* (Englewood Cliffs, NJ: Prentice-Hall, 1979).

16. For example, among analysts stressing conflict, see Ralf Dahrendorf, *Class and Class Conflict in Industrial Society* (Stanford, CA: Stanford University Press, 1959); Randall Collins, *Conflict Sociology* (New York: Academic Press, 1975). Among those stressing functionalism, see Talcott Parsons, *The Social System* (Glencoe, IL: The Free Press, 1951); Robert K. Merton, *Social Theory and Social Structure* (New York: Free Press, 1949), rev. and enl. ed. 1957; 3rd ed., enl., 1968.

Chapter 2

1. Mary P. Follett, *Freedom and Co-ordination: Lectures in Business Organization* (New York: Management Publications Trust Limited, 1949).

2. P. J. Roethliberger, William J. Dickson, and Harold A. Wright, *Management and the Worker* (Cambridge, MA: Harvard University Press, 1939).

3. Alex Carey, "The Hawthorne Studies: A Radical Criticism," *American Sociological Review* 32, no. 3 (1967): 403–16.

4. In 1953–56, I taught a sociology course on industrial relations at Columbia University. Unsolicited, I received material from industrial companies, some of which reported how their supervisors were trained to interact with their subordinates. They were told to ask questions and listen to what the workers had to say about their work. The supervisors were assured that this would not entail any abdication of their authority. It seemed to me workers might well view this as manipulative.

5. Jeffrey Haydu, "Managing 'the Labor Problem' in the United States ca. 1897–1911," in *Intractable Conflicts and Their Transformation*, ed. Louis Kriesberg, T. A. Northrup, and S. J. Thorson (Syracuse, NY: Syracuse University Press, 1989), 93–106.

6. Saul Alinsky, *Reveille for Radicals* (New York: Vintage, 1946); Saul Alinsky, *Rules for Radicals* (New York: Random House, 1971).

7. Donald C. Reitzes and Dietrich C. Reitzes, *The Alinsky Legacy: Alive and Kicking* (Greenwich, CT: JAI Press, 1987).

8. Kevin Boyle, *The UAW and the Heyday of American Liberalism, 1945–1968* (Ithaca, NY: Cornell University Press, 1995); John Barnard, *American Vanguard: The United Auto Workers during the Reuther Years, 1935–1970* (Detroit: Wayne State University Press, 2004).

9. https://www.history.com/this-day-inhistory/ford-signs-first-contract-with-autoworkerss-union.

10. Assassination attempts were variously attributed to hostile corrupt union groups or to intransigent corporate origins. Suspicions have been expressed that the plane crash was not accidental. Michael Parenti, *Dirty Truths* (San Francisco: City Lights Books, 1996).

11. Walter Reuther also was president of the CIO from 1952 until 1955, when the AFL and CIO merged. The two labor federations gradually changed and accepted each other's approach to union organization and activities. This was true for their international and domestic policies. Internationally, they both worked with the US government during the Cold War to support non-Communist trade unions around the world. The AFL worked independently and sometimes covertly, while the CIO worked more with the International Confederation of Free Trade Unions (ICFTU), which was established to counter the World Federation of Trade Unions after it had become dominated by the Soviet Union.

   In 1955, I interviewed several union leaders about their unions' engagement in international affairs. On August 31, I interviewed Victor Reuther, the assistant to the president of the CIO (his brother, Walter), who dealt with the ICFTU. On August 25, I interviewed Jay Lovestone, the primary person in the AFL's international anti-Communist work. John P. Windmuller, *American Labor and the International Labor Movement, 1940–1953* (Ithaca, NY: The Institute of International Industrial and Labor Relations, Cornell University, 1954).

12. Randy Shaw, *Beyond the Fields: Cesar Chavez, the UFW, and the Struggle for Justice in the 21st Century* (Berkeley University of California Press, 2008). https://ufw.org.

13. Henry S. Farber, Daniel Herbst, Ilyana Kuziemko, and Suresh Naidu. *Unions and Inequality over the Twentieth Century: New Evidence from Survey Data* (Cambridge, MA: National Bureau of Economic Research, 2018).

14. Gerald Mayer, *Union Membership Trends in the United States* (Washington, DC: Congressional Research Service, 2004).

15. Robert B. Reich, "Foreword," in *The Spirit Level: Why Greater Equality Makes Societies Stronger*, ed. Richard Wilkinson and Karte Pickett (New York and London: Bloomsbury Press, 2010), IX–XII.

16. Robert E. Lane, "The Politics of Consensus in an Age of Affluence," *American Political Science Review* 59, no. 4 (1965): 874–95.

17. Michael Harrington, *The Other America: Poverty in the United States* (New York: Macmillan, 1962); Louis Kriesberg, "Policy Continuity and Change," *Social Problems* 32 (December 1984): 89–102.

18. Guian A. McKee, "Lyndon B. Johnson and the War on Poverty, Vol. 1," Presidential Recordings, Digital Edition (Charlottesville: Miller Center, University of Virginia, 2014), http://prde.upress.virginia.edu/content/WarOnPoverty.

19. Frances Fox and Richard A. Cloward Piven, *Poor People's Movements* (New York: Vintage Books, 1979).

20. Ibid., 332.

21. McKee, "Lyndon B. Johnson and the War on Poverty."

22. Ralph M. Kramer, *Participation of the Poor: Comparative Community Case Studies in the War on Poverty* (Englewood Cliffs, NJ: Prentice-Hall, 1969).

23. Frances Fox Piven and Richard A. Cloward, *Regulating the Poor: The Functions of Public Welfare*, updated edition (New York: Vintage Books, 1993).

24. Irwin Deutscher, Mike Miller, Seymour Bellin, Charles V. Willie, and Helen Icken Safa were particularly important in that research project.

25. The research was funded by grants from the Ford Foundation and the Department of Health, Education, and Welfare. Some findings of the research are in Louis Kriesberg, *Mothers in Poverty: A Study of Fatherless Families* (Chicago: Aldine, 1970).

26. Louis Kriesberg, "The Relationship between Socio-economic Rank and Behavior," *Social Problems* 10, no. 4 (1963): 334–53. Louis Kriesberg and Seymour S. Bellin, "On the Relationship between Attitudes, Circumstances, and Behavior: The Case of Applying for Public Housing," *Sociology and Social Research* 51, no. 4 (1967): 453–67; Irwin Deutscher and Elizabeth J. Thompson, eds., *Among the People: Encounters with the Poor* (New York and London: Basic Books, 1968).

27. Gene Falk and Karen Spar, *Poverty: Major Themes in Past Debates and Current Proposals* (Washington, DC: Congressional Research Service, 2014).

28. Piven and Cloward, *Regulating the Poor*, xiii.

29. Ibid.

30. Daniel P. Moynihan, *The Politics of a Guaranteed Income: The Nixon Administration and the Family Assistance Plan* (New York: Vintage Books, 1973).

31. Jill Quadagno, "Race, Class, and Gender in the U.S. Welfare State: Nixon's Failed Family Assistance Plan," *American Sociological Review* 55, no. 1 (1990): 11–28; *The Color of Welfare: How Racism Undermined the War on Poverty* (New York: Oxford University Press, 1994).

32. Piven, *Poor People's Movements*.

33. Mike Puma, Stephen Bell, Ronna Cook, Camilla Heid, Pam Broene, Frank Jenkins, Mashburn, Andrew, and Downer Jason. *Third Grade Follow-Up to the Head Start Impact Study Final Report, Executive Summary* (Washington, DC: Administration for Children and Families, US Department of Health and Human Services 2012).

34. Martin Luther King Jr., "Martin Luther King Jr. on the Vietnam War," *The Atlantic*, March 2018, https://www.theatlantic.com/magazine/archive/2018/02/martin-luther-king-jr-vietnam/552521/.

35. One organizational manifestation of the movement was the John Birch Society, a radical right-wing organization founded by Robert W. Welch Jr. in 1958. It developed chapters nationwide, reaching its greatest influence in the 1970s. Although regarded as a fringe group even by many Republican conservatives at the time, it left a legacy of belief in unfounded conspiracy theories to explain political developments.

36. Seven persons were tried and convicted for their roles in the protest, a case known as the Chicago Seven. Upon appeal, the convictions were annulled. In 2020 it was the subject of a documentary film, *The Trial of The Chicago Seven*, which captured the spirit of the trial in a dramatized account.

37. William Bundy, *A Tangled Web: The Making of Foreign Policy in the Nixon Presidency* (New York: Hill and Wang, 1998), 35–48.

38. http://www.historycommons.org/context.jsp?item=a050270rhodeseradicate#a050270rhodeseradicate.

39. Charles DeBenedetti and Charles Chatfield, *An American Ordeal: The Antiwar Movement of the Vietnam Era* (Syracuse, NY: Syracuse University Press, 1990).

40. James Eric Eichsteadt, "'Shut It Down': The May 1970 National Student Strike at the University of California at Berkeley, Syracuse University, and the University of Wisconsin-Madison" (PhD diss., History, Syracuse University, 2007, https://surface.syr.edu/hst_etd/7).

Chapter 3

1. Louis Kriesberg, *Social Inequality* (Englewood Cliffs, NJ: Prentice-Hall, 1979).

2. Chad Stone, Danilo Trisi, Arloc Sherman, and Roderick Taylor. "A Guide to Statistics on Historical Trends in Income Inequality," Center on Budget and Policy Priorities, May 15, 2018.

3. Ibid., 11.

4. Drew DeSilver, "Global Inequality: How the U.S. Compares," Pew Research Center, December 19, 2013, http://www.pewresearch.org/fact-tank/2013/12/19/global-inequality-how-the-u-s-compares/.

5. Emmanuel Saez, "Striking It Richer: The Evolution of Top Incomes in the United States," University of California, Berkeley, 2013.

6. Congressional Budget Office, *The Distribution of Household Income 2016*. https://apps.urban.org/features/wealth-inequality-charts/.

7. Congressional Budget Office https://www.cbo.gov/sites/default/files/114th-congress-2015-2016/repor1846-familywealth.pdf.

8. Congressional Budget Office https://www.cbo.gov/publication/51846ttps://inequality.org/facts/wealth-inequality/

9. Ibid.

10. Thomas Piketty, *Capital in the Twenty-First Century*, translated by Arthur Goldhammer (Cambridge, MA: The Belknap Press, 2014), 348–49.

11. Richard Wilkinson and Kate Pickett, *The Spirit Level: Why Greater Equality Makes Societies Stronger* (New York and London: Bloomsbury Press, 2010), 75–85.

12. There is a significant body of research relating to the importance of relative deprivation. See W. G. Runciman, *Relative Deprivation and Social Justice* (Berkeley: University of California Press, 1966); Ted Robert Gurr, *Why Men Rebel* (Princeton, NJ: Princeton University Press, 1970); Louis Kriesberg and Bruce W. Dayton, *Constructive Conflicts: From Escalation to Resolution*, 5th ed. (Lanham, MD: Rowman & Littlefield, 2017), 62–69.

13. Wilkinson and Pickett, *The Spirit Level*, 33–35.

14. https://usafacts.org/data/topics/people-society/health/longevity/life-expectancy/?utm_source=bing&utm_medium=cpc&utm_campaign=ND-StatsData&msclkid=3f0c3cabfb6818e289b05d6e19cad495#explorer.

15. Claudia Goldin and Katz Lawrence, *The Race between Education and Technology* (Cambridge, MA: Harvard University Press, 2008).

16. Josh Bivens and Heidi Shierholz, *What Labor Market Changes Have Generated Inequality and Wage Suppression?* (Washington, DC: Economic Policy Institute, 2018), https://www.epi.org/publication/what-labor-market-changes-have-generated-inequality-and-wage-suppression-employer-power-is-significant-but-largely-constant-whereas-workers-power-has-been-eroded-by-policy-actions/.

17. Lawrence Mishel, Elise Gould, and Josh Bivens, "Wage Stagnation in Nine Charts," Economic Policy Institute, 2015, https://www.epi.org/publication/charting-wage-stagnation/.

18. Robert E. Scott, "The U.S. Trade Deficit: Are We Trading Away Our Future?," Economic Policy Institute, 2002, https://www.epi.org/publication/webfeatures_viewpoints_tradetestimony/.

19. Federal Reserve Bank of St. Louis, "Foreign Automobile Sales in the United States," Federal Reserve Bank of St. Louis, 1970, https://fraser,stlouisfed.org/title/960/item/37804/loc/174839.

20. John Pearley Huffman, "5 Most Notorious Recalls of All Time," *Popular Mechanics*, February 12, 2010, https://www.popularmechanics.com/cars/g261/4345725/.

21. Lawrence Mishel, John Schmitt, and Heidi Shierholz, "Wage Inequality: A Story of Policy Choices," *New Labor Forum* 23 (2014).

22. Rick Peristein, *Reganland: America's Right Turn* (New York: Simon and Schuster, 2020).

23. John N. Palmer and Isabel V. Sawhill, "Perspectives on the Reagan Experiment," in *The Reagan Experiment*, ed. John N. Palmer and Isabel V. Sawhill (Washington, DC: The Urban Institute Press, 1982).

24. The strategy was challenged by many mainstream economists. Empirical evidence does not support it. See https://www.americanprogress.org/issues/economy/news/2012/08/01/11998/the-failure-of-supply-side-economics/.

25. John O'Connor, "US Social Welfare Policy: The Reagan Record and Legacy," *Journal of Social Policy* 27, no. 1 (1998): 37–61.

26. Allen W. Smith, "Ronald Reagan and the Great Social Security Heist," *FedSmith.com*, October 11, 2013, https://www.fedsmith.com/2013/10/11/ronald-reagan-and-the-great-social-security-heist.

27. Robert D. Plotnick, "Changes in Poverty, Income Inequality, and the Standard of Living in the United States during the Reagan Years," *International Journal of Health Services* 23 (1993): 347–58. Frances Fox Piven and Richard A. Cloward, *Regulating the Poor: The Functions of Public Welfare*, updated edition (New York: Vintage Books, 1993), https://doi.org/10.2190/H95U-EX9E-QPM2-XA94

28. Daniel R. Feenberg and James M. Poterba, "Income Inequality and the Incomes of Very High-Income Taxpayers: Evidence from Tax Returns," *Tax Policy and the Economy* 7 (1993), https://www.journals.uchicago.edu/doi/pdfplus/10.1086/tpe.7.20060632.

29. Lawrence Lessig, *Republic, Lost: How Money Corrupts Congress—And a Plan to Stop It*, 2nd ed. (New York: Hachette Book Group, 2016).

30. Public service unions did expand in later years, and were subsequently a new arena for conflict. Thus, Governor Scott Walker of Wisconsin proposed and won legislation in 2011 that radically reduced public employees' collective bargaining power and cut back on their pensions and other benefits.

31. The NLRB settled only about half as many labor complaints of employers' actions as had the board during the previous administration of Democrat Jimmy Carter. Furthermore, settlements upheld employers in three-fourths of the cases, a large increase compared to the preceding administration. See https://truthout.org/articles/ronald-reagan-enemy-of-the-american-worker/.

32. Louis Kriesberg, *Realizing Peace: A Constructive Conflict Approach* (New York: Oxford University Press, 2015).

33. Reagan executed 39 regular vetoes and 39 pocket vetoes, and his vetoes were overridden nine times. See https://www.cop.senate.gov/reference/Legislation/Vetoes/ReaganR.htm.

34. In the academic year 1957–58, I was a Senior Fellow in Law and the Behavioral Sciences at the University of Chicago Law School. Edward Levi was the dean and he explained to me and the other Fellows that he hoped we would demonstrate the contributions each of our disciplines could make to law, as Aaron Director was doing for economics. As a sociologist, I was critical of the presumed possibility and benefits of an autonomous economic market, but I confess I did not try to challenge him or formulate sociological contributions to the practice of law. I regret that.

35. During the first years of his two terms, Reagan intensified his rhetoric and US actions against the Soviet Union and Communism. This included increasing the nuclear arms race and US military intervention in Nicaragua and elsewhere in Central America.

36. James Tobin, "Voodoo Curse: Exorcising the Legacy of Reaganomics," *Harvard International Review* 14 no, 4 (1992): 10–13.

37. Stephen Knott, "George W. Bush: Domestic Affairs," University of Virginia, Miller Center, 2019, https://millercenter.org/president/bush/domestic-affairs;

38. https://thebestschools.org/features/most-influential-think-tanks/.

39. Jason Stahl, *Right Moves: The Conservative Think Tank in American Political Culture since 1945* (Chapel Hill: University of North Carolina Press, 2016);

40. Alexander Hertel-Fernandez, Caroline Tervo, and Theda Skocpol, "How the Koch Brothers Built the Most Powerful Rightwing Group You've Never Heard Of," *Guardian*, September 26, 2018, https://www.theguardian.com/us-news/2018/sep/26/koch-brothers-americans-for-prosperity-rightwing-political-group; Alex Hertel-Fernandex, *State Capture: How Conservative Activists, Big Businesses, and Wealthy Donors Reshaped the American States—and the Nation* (New York: Oxford University Press, 2019).

41. Alan Greenblatt, "Alec Enjoys a New Wave of Influence and Criticism," *Governing*, November 29, 2011, https://www.governing.com/topics/politics/ALEC-enjoys-new-wave-influence-criticism.html,) https://www.governing.com/topics/politics/ALEC-enjoys-new-wave-influence-criticism.html.

42. Ed Pilkington and Suzanne Goldenberg, "State Conservative Groups Plan US-Wide Assault on Education, Health and Tax," *Guardian*, December 5, 2013, https://www.theguardian.com/world/2013/dec/05/state-conservative-groups-assault-education-health-tax.

43. Mary Bottari, "The Two Faces of Janus," *In These Times*, March 2018, 18–27.

44. https://www.nytimes.com/2018/06/27/us/politics/supreme-court-unions-organized-labor.html.

45. Eric Boehm, "After the Supreme Court Said Unions Can't Force Non-members to Pay Dues, Almost All of Them Stopped," *Reason: Free Minds and Free Markets*, April 9, 2019, https://reason.com/2019/04/09/janus-211000-workers-fled-seiu-afscme).

46. Larissa MacFarquhar, "What Money Can Buy: Darren Walker and the Ford Foundation Set Out to Conquer Inequality," *New Yorker*, December 27, 2015, https://www.newyorker.com/magazine/2016/01/04/what-money-can-buy-profiles-larissa-macfarquhar.

47. See Ford website: https://www.fordfoundation.org/work/challenging-inequality/.

48. David Callahan, "Systemic Failure: Four Reasons Philanthropy Keeps Losing the Battle against Inequality," *Inside Philanthropy*, August 22, 2018, https://www.insidephilanthropy.com/home/2018/1/10/systemic-failure-four-reasons-philanthropy-keeps-losing-the-battle-against-inequality.

49. David Kovick, "The Hewlett Foundation's Conflict Resolution Program: Twenty Years of Field-Building, 1984–2004," Hewlett Foundation, 2005, https://www.hewlett.org/wp-content/uploads/2016/08/HewlettConflictResolutionProgram.pdf.

50. Brian D. Polkinghorn, Haleigh La Chance, and Robert La Chance, "Constructing a Baseline Understanding of Developmental Trends in Graduate Conflict Resolution

Programs in the United States," in *Pushing the Boundaries: New Frontiers in Conflict Resolution and Collaboration*, ed. Rachel Fleishman, Catherine Gerard, and Rosemary O'Leary (Bingley, UK: Emerald Group, 2008), 233–65.

51. Joseph Stiglitz received the Nobel Memorial Prize in Economic Sciences in 2001. He is a former senior vice president and chief economist of the World Bank and a former member and chairman of the US President's Council of Economic Advisers.

52. Gideon Lewis-Kraus, "The Change Artists," *New York Times Magazine*, July 24, 2016, 30–35, 47.

53. Kenneth B. Noble, "Reagan Vetoes Measure to Affirm Fairness Policy for Broadcasters," *New York Times*, June 21, 1987, https://nyti.ms/29zHkoV.

54. http://www.imao.us/wp-content/uploads/2008/11/media_project_poll_info.pdfZogby.

55. For example, see https://fair.org/extra/the-way-things-arent/; Al Franken, *Rush Limbaugh Is a Big Fat Idiot and Other Observations* (New York: Delacorte Press, 1996).

56. Brian Stelter, *Hoax: Donald Trump, Fox News, and the Dangerous Distortion of Truth* (New York: Simon and Schuster, 2020).

57. Howard Kurtz, "Doing Something Right; Fox News Sees Ratings Soar, Critics Sore," *Washington Post*, February 5, 2001.

58. Steven Kull, "Misperceptions, the Media and the Iraq War," Center for International Security Studies at Maryland, 2003.

59. https://www.pewresearch.org/fact-tank/2014/09/24/how-social-media-is-reshaping-news/; https://www.journalism.org/2013/11/14/news-use-across-social-media-platforms/.

Chapter 4

1. Paul Krugman, "The Rich, the Right, and the Facts," *American Prospect*, March 12, 2020, https://prospect.org/features/rich-right-facts-deconstructing-income-distribution-debate/

2. Tom Hertz, "Understanding Mobility in America," Center for American Progress, April 26, 2006, https://www.americanprogress.org/issues/economy/news/2006/04/26/1917/understanding-mobility-in-america/.

3. Julia Isaacs, "International Comparisons of Economic Mobility," Brookings Institution, 2008, https://www.brookings.edu/wp-content/uploads/2016/07/02_economic_mobility_sawhill_ch3.pdf.

4. Susan K. Urahn, "Pursuing the American Dream: Economic Mobility across Generations," Pew Charitable Trusts, July 2012, pursuingamericandreampdf.pdf.

5. Hertz, "Understanding Mobility," 6–8.

6. Urahn, "Pursuing the American Dream," 5.

7. Hertz, "Understanding Mobility," 10–12.

8. Other significant variables are health, region of the country, female-headed household, and inheritance.

9. Bruce D. Baker, Danielle Farrie, and David G. Sciarra, "Mind the Gap: 20 Years of Progress and Retrenchment in School Funding and Achievement Gaps," *ETS Research Report Series* 2016, no. 1 (June 2016), https://onlinelibrary.wiley.com/doi/10.1002/ets2.12098.

10. C. Kirabo Jackson, "Does School Spending Matter? The New Literature on an Old Question," National Bureau of Economic Research, April 11, 2022, http://www.nber.org/papers/w25368.

11. Gil Troy, *The Age of Clinton: America in the 1990s* (New York: Thomas Dunne Books, 2015), 82–83.

12. Todd S. Purman, "Clinton Defends Income Tax Credit Against GOP," *The New York Times*, September 19, 1995, p. A1.

13. Alana Semuels, "The End of Welfare as We Know It," *Atlantic*, April 1, 2016, https://www.theatlantic.com/business/archive/2016/04/the-end-of-welfare-as-we-know-it/476322/.

    Donna Shalala, then Secretary of Health and Human Studies, came to Syracuse University to defend this shift. She had studied and worked for many years at Syracuse University's Maxwell School of Citizenship and Public Affairs and faced former colleagues who were critical of the new policy. As I recall, her defense was that it would be easier to get public funds in the future for people who might be in need if they had previously been employed. Many of us were dubious.

14. Clinton came to believe that his refusal to compromise was an error. Bill Clinton, *My Life* (New York: Alfred A. Knopf, 2004), 577.

15. James T. Patterson, *Restless Giant: The United States from Watergate to "Bush v. Gore"* (New York: Oxford University Press, 2005), 350–58.

16. Thomas Frank, *Listen, Liberal: Or What Ever Happened to the Party of the People* (New York: Metropolitan Books, 2016).

17. Bill Adair, "The Peace Dividend Began with a Bush," , January 24, 2008, https://www.politifact.com/truth-o-meter/statements/2008/jan/24/rudy-giuliani/the-peace-dividend-began-with-a-bush/.

18. Leslie H. Gelb, "Foreign Affairs; What Peace Dividend?," *New York Times*, February 21, 1992, A31.

19. Robert Draper, *Dead Certain: The Presidency of George W. Bush* (New York: Free Press, 2007), 119–20, 363–71.

20. Michael Lewis, *The Big Short: Inside the Doomsday Machine* (London: Allen Lane, 2010).

21. John Cassidy, "An Inconvenient Truth: It Was George W. Bush Who Bailed Out the Automakers," *New Yorker*, March 15, 2012, https://www.newyorker.com/news/john-cassidy/an-inconvenient-truth-it-was-george-w-bush-who-bailed-out-the-automakers.

22. Louis Kriesberg, *Realizing Peace: A Constructive Conflict Approach* (New York: Oxford University Press, 2015).

23. Douglas L. Kiner and Francis X. Shen, *The Casualty Gap: The Causes and Consequences of American Wartime Inequalities* (New York: Oxford University Press, 2010).

24. Stephanie Condon, "What's Obama Doing to Your Taxes," CBS News, April 15, 2010, https://www.cbsnews.com/news/whats-obama-doing-to-your-taxes/.

25. Kathy Orton, "Homeowners Get More Time to Take Advantage of HAMP, HARP," *Washington Post*, May 8 2015, http://wapo.st/1IWnFIO?tid=ss_mail&utm_term= .c2333298288b

26. www.zillow.com/research/housing-bust-wealth-gap.

27. Jason Furman, "How Obama Has Narrowed the Income Inequality Gap," *Washington Post*, September 23, 2016, http://wapo.st/2dg6njf?tid=ss_mail&utm_term=.9655f 73451c5.

28. Paul Kirk, a Democrat, was appointed an interim senator to replace Kennedy. Then a Republican, Scott Brown campaigned against the act and surprisingly won the election to fill out Kennedy's term in office, which he began in January 2010. This meant the Democrats would not have the 60 votes needed to block filibusters, and they used budget reconciliation to get the bill passed.

29. Brooks Jackson, "Obama's Final Numbers," FactCheck.org, September 29, 2017, www.factcheck.org/2017/09/obamas-final-numbers/.
    During President George W. Bush's time in office, unemployment had increased by 3.6 percent, while during Clinton's presidency it had fallen by 3.1 percent.

30. Jeffrey D. Sachs, *The End of Poverty: Economic Possibilities for Our Time* (New York: Penguin, 2005).

31. Scott G. McNall, *Cultures of Defiance and Resistance: Social Movements in 21st-Century America* (New York and London: Routledge, 2018), 51–86.

32. Donald C. Reitzes and Dietrich C. Reitzes, *The Alinsky Legacy: Alive and Kicking* (Greenwich, CT: JAI Press, 1987).

33. Theda Skocpol and Vanessa Williamson, *The Tea Party and the Remaking of Republican Conservatism* (New York: Oxford University Press, 2016), 7.

34. Ibid., 9–18.

35. Arlie Russell Hochschild, *Strangers in Their Own Land: Anger and Mourning on the American Right* (New York: New Press, 2016).

36. Jane Mayer, *Dark Money: The Hidden History of the Billionaires* (New York: Anchor Books, 2016).

37. Skocpol and Williamson, *The Tea Party*, 131–33.

38. Louis Kriesberg, "Interactions among Populism, Peace, and Security in Contemporary America," *Sicherheit und Frieden (Security and Peace)* 37, no. 1 (2019): 1–7.

39. McNall, *Cultures of Defiance*, 68–70.

40. hhtp://www.ibtimes.com/occupy-wall-street-anniversary-police-crackdown-movement-cost-new-york.

41. https://www.brookings.edu/research/not-so-demanding-why-occupy-wall-street-need-not-make-demands-yet/; https://www.nytimes.com/2011/10/17/nyregion/occ upy-wall-street-trying-to-settle-on-demands.html.

42. Micah White, *The End of Protest: A Playbook for Revolution* (Toronto: Knopf Canada, 2016).

43. Brian Montopoli, "Poll: 43 Percent Agree with Views of 'Occupy Wall Street,'" CBS News, October 26, 2011, https://www.cbsnews.com/news/poll-43-percent-agree-with-views-of-occupy-wall-street/.

44. *Financial Times*: https://www.ft.com/content/052226f8-f80c-11e0-a419-00144feab49a.

45. http://www.montgomerynews.com/germantowncourier/guest-oped-a-decade-after-the-foreclosure-crisis-northwest-philadelphia/article_d46a8d82-ae3a-11e9-b330-5fabc7d9cc74.html.

46. Joel Cutcher-Gershenfeld, Dan Brook, and Martin Mulloy, "The Decline and Resurgence of the U.S. Auto Industry," Economic Policy Institute Report, May 6, 2015, Https://Www.Epi.Org/Publication/the-Decline-and-Resurgence-of-the-U-S-Auto-Industry/.

47. Ibid.

48. Tadlock Cowan, "Military Base Closures: Socioeconomic Impacts," Congressional Research Service, 2012, https://fas.org/sgp/crs/natsec/RS22147.pdf.

49. Cutcher-Gershenfeld et al., "Decline and Resurgence.

50. There is a large literature about Donald Trump, including a book by his niece: Mary L. Trump, *Too Much and Never Enough* (New York: Simon & Schuster, 2020).

51. https://www.theguardian.com/us-news/2020/may/16/trump-obama-obsession-coronavirus-president.

52. Louis Kriesberg, "Interactions among Populism, Peace, and Security in Contemporary America," *Sicherheit und Frieden (Security and Peace)* 37, no. 1 (2019): 1–7.

53. Louis Kriesberg, "Connecting Theory and Practice in the Peace and Conflict Studies Field," in *Handbook of Peace and Conflict Studies*, ed. Sean Byrne, Thomas Matyok, and Imani Michelle Scott (New York: Routledge, 2020), 35–44.

54. https://apnews.com/article/donald-trump-virus-outbreak-global-trade-trade-policy-mexico-39aadae9a6d18de2b91889f1e552b605.

55. https://www.hinrichfoundation.com/research/wp/sustainable/trade-development-in-crisis-fundamentals/.

56. https://www.americanprogress.org/issues/economy/news/2020/10/28/492473/6-ways-trump-administration-rigging-already-unfair-tax-code/.

57. https://www.nytimes.com/2017/10/29/us/politics/democrats-tax-reform-middle-class.html.

58. https://www.businessinsider.com/warren-buffett-bill-gates-say-trump-tax-cut-plan-wont-help-business-2017-5.

59. https://www.project-syndicate.org/commentary/republican-tax-reform-voodoo-economics-by-joseph-e--stiglitz-2017-10; https://www.nytimes.com/2019/01/01/opinion/the-trump-tax-cut-even-worse-than-youve-heard.html; https://www.bloomberg.com/news/articles/2017-10-10/nobel-economist-takes-a-jab-at-trump-s-confidence-knowledge; and https://www.bloomberg.com/news/articles/2017-10-10/nobel-economist-takes-a-jab-at-trump-s-confidence-knowledge.

60. https://news.gallup.com/opinion/polling-matters/249161/public-opinion-2017-tax-law.aspx.

61. https://theconversation.com/real-pay-data-show-trumps-blue-collar-boom-is-more-of-a-bust-for-us-workers-in-3-charts-131264.

62. Alan S. Blinder and Mark W. Watson, "Presidents and the US Economy: An Econometric Exploration," *American Economic Review* 106, no. 4 (April 2016): 1015–45.

63. Lynn Rhinehart, "Under Trump the NLRB Has Gone Completely Rogue," *The Nation* April 7, 2020. Thenation.com/article/politics/nlrb-workers-rights-trump

64. {https://www.nytimes.com/2019/10/11/business/immigration-cuts-economy.html.

65. https://oxford.universitypressscholarship.com/view/10.1093/oso/9780190872 199.001.0001/oso-9780190872199-chapter-1.

66. https://www.democracyinaction.us/2018/midterms18rxn.html.

67. https://www.pgpf.org/blog/2021/03/heres-everything-congress-has-done-to-resp ond-to-the-coronavirus-so-far.

68. https://www.theguardian.com/us-news/2021/feb/10/us-coronavirus-response-don ald-trump-health-policy.

69. Ezra Klein, "The COVID Policy That Really Mattered Wasn't a Policy," *New York Times*, February 6, 2022.

70. https://www.pewresearch.org/fact-tank/2020/01/09/.

Chapter 5

1. Max Weber, *The Theory of Social and Economic Organization*, trans. A. M. Henderson and Talcott Parsons (New York: Oxford University Press, 1947).

2. John L. Comaroff, "Humanity, Ethnicity, Nationality: Conceptual and Comparative Perspectives on the U.S.S.R.," *Theory and Society* 20 (1991): 661–87; Benedict Anderson, *Imagined Communities: Reflections on the Origin and Spread of Nationalism*, Revised ed. (London: Verso, 1991).

3. Matthew Frye Jacobson, *Whiteness of a Different Color: European Immigrants and the Alchemy of Race* (Cambridge, MA: Harvard University Press, 1999).

4. Charles C. Mann, *1491: New Revelations of the Americas before Columbus*, 2nd ed. (New York: Random House, 2011).

5. Heather McGhee, *The Sum of Us* (New York: One World, 2021).

6. John A. Powell, *Racing to Justice: Transforming Our Conceptions of Self and Other to Build an Inclusive Society* (Bloomington: Indiana University Press, 2012).

7. Kenneth Prewitt, *What Is Your Race? The Census and Our Flawed Efforts to Clarify Americans* (Princeton, NJ: Princeton University Press, 2013), 31.

8. Emma Green, "Are Jews White?," *Atlantic*, December 5, 2016, https://www.theatlan tic.com/politics/archive/2016/12/are-jews-white/509453/2016.

9. Gunnar Myrdal, Richard Sterner, and Arnold Rose, *An American Dilemma: The Negro Problem and American Democracy* (New York and London: Harper & Brothers, 1944).

10. Ibid., xlvii.

11. Eric Foner, *Reconstruction, America's Unfinished Revolution, 1863–1877* (New York: HarperCollins, 1988).

12. C. Vann Woodward, *The Strange Career of Jim Crow*, 2nd ed. (New York: Oxford University Press, 1966); Myrdal, Sterner, and Rose, *An American Dilemma*.

13. Jeffrey D. Gonda, *Unjust Deeds: The Restrictive Covenant Cases and the Making of the Civil Rights Movement* (Chapel Hill: University of North Carolina Press, 2015).

14. As a student in the College of the University of Chicago, in 1947, I joined the many students who rallied to protest such conduct by the university. A vice president met with a mass gathering of students and explained why that conduct was necessary to protect the immense investment of the university by ensuring students would continue to come to study and live in the campus neighborhood. See Richard Rothstein, *The Color of Law: A Forgotten History of How Our Government Segregated America* (New York and London: Liveright, 2017), 105

15. Ibid., 64.

16. Ibid., 65.

17. Ibid.

18. Ibid., 32.

19. Ibid., 71.

20. Ira Katznelson, *When Affirmative Action Was White: An Untold History of Racial Inequality in Twentieth-Century America* (New York: W. W. Norton, 2006); Hilary Herbold, "Never a Level Playing Field: Blacks and the GI Bill," *Journal of Blacks in Higher Education* 6 (Winter 1994): 104–8.

21. Harry S. Truman, *Memoirs by Harry S. Truman*, Vol. 2, *Years of Trial and Hope* (Garden City, NY: Doubleday, 1956), 180–82.

22. Andrew E. Busch, *Truman's Triumphs: The 1948 Election and the Making of Postwar America* (Lawrence: University Press of Kansas, 2012); Harvard Sitkoff, "Harry Truman and the Election of 1948: The Coming of Age of Civil Rights in American Politics," *Journal of Southern History* 37, no. 4 (1971): 597–616.

23. Truman, *Memoirs*, 183.

24. Eric Arnesen, "A. Philip Randolph: Labor and the New Black Politics," in *The Human Tradition in the Civil Rights Movement*, ed. Susan M. Glisson (Lanham, MD: Rowman & Littlefield, 2006), 80–82.

25. Andrew Glass, "Truman Desegregates Armed Forces on Feb. 2, 1948," *Politico*, February 2, 2008, https://www.politico.com/story/2008/02/truman-desegregates-armed-forces-on-feb-2-1948-008258.

26. Harold Cruse, *Plural but Equal: A Critical Study of Blacks and Minorities and America's Plural Society* (New York: William Morrow, 1987).

27. W. E. B. Du Bois, *The Souls of Black Folk* (Chicago: A.C. McClurg, 1903).

28. https://www.historycom/naacp-legal-team/legal-history.

29. Richard Kluger, *Simple Justice: The History of "Brown v. Board of Education" and Black America's Struggle for Equality* (New York: Knopf, 1975).

30. Thomas F. Pettigrew, "Intergroup Contact Theory," *Annual Review of Psychology* 49 (1998): 65–85.

31. August Meier and Elliott M. Rudwick, *CORE: A Study in the Civil Rights Movement, 1942–1968* (Urbana: University of Illinois Press, 1975).

32. Jo Ann Robinson and David J. Garrow, *The Montgomery Bus Boycott and the Women Who Started It* (Knoxville: University of Tennessee Press, 1986).

33. https://www.history.com/topics/Black-history/montgomery-bus-boycott.

34. Theda Skocpol, Ariane Liazos, and Marshall Ganz, *What a Mighty Power We Can Be: African American Fraternal Groups and the Struggle for Racial Equality* (Princeton, NJ: Princeton University Press, 2006).

35. Raymond Arsenault, *Freedom Riders: 1961 and the Struggle for Racial Justice* (New York: Oxford University Press, 2006).

36. Aldon D. Morris, "Birmingham Confrontation Reconsidered: An Analysis of the Dynamics and Tactics of Mobilization," *American Sociological Review* 58, no. 5 (1993): 621–36.

37. Arthur M. Schlesinger Jr., *A Thousand Days* (Boston: Houghton Mifflin, 1965), 958–59

38. Ibid., 972–73.

39. Robert Caro, *The Power Broker* (New York: Vintage, 1974), 1014.

40. https://www.democracynow.org/2014/11/18/the_fbi_vs_martin_luther_king; also see https://kinginstitute.stanford.edu/encyclopedia/federal-bureau-investigation-fbi.

41. Ted Robert Gurr, *Why Men Rebel* (Princeton, NJ: Princeton University Press, 1970).

42. Jules J. Wanderer, "An Index of Riot Severity and Some Correlates," *American Journal of Sociology* 74 (March 1969): 500–505.

43. https://www.theguardian.com/commentisfree/2008/apr/04/thelegacyofthe1968riots.

44. National Advisory Commission on Civil Disorders (Kerner Commission), *Report of the National Commission on Civil Disorders* (New York: Bantam, 1968), 1–2.

45. Steven M. Gillon, *Separate and Unequal: The Kerner Commission and the Unraveling of American Liberalism* (New York: Basic Books, 2018), 247–48.

46. Stephanie M. H. Camp, "Black Is Beautiful: An American History," *Journal of Southern History* 81, no. 3 (2015): 675–90.

47. Malcom X and Martin Luther King Jr. were moving closer in many ways. See https://www.cnn.com/2010/LIVING/05/19/Malcolmx.king.

48. https://www.pbs.org/independentlens/films/the-Black-panthers-vanguard-of-the-revolution/.

49. https://www.zinnedproject.org/news/tdih/cointelpro-exposed/.

50. https://en.wikipedia.org/wiki/Fred_Hampton`#cite_note-75.

51. Nikole Hannah-Jones, "It Was Never about Busing," *New York Times*, July 14, 2019.

52. U.S. Commission on Civil Rights, "Process of Change: The Story of School Desegregation in Syracuse, New York," Washington, DC: U.S. Commission on Civil Rights, 1968; Zoe Cornwall, *Human Rights in Syracuse: Two Memorable Decades* (Syracuse, NY: Human Rights Commission of Syracuse and Onondaga County, 1989), 82–93.

53. Cornwall, *Human Rights in Syracuse*, 82–107.

54. Hannah-Jones, "It Was Never about Busing"; Sean F. Reardon and Ann Owens, "60 Years after *Brown*: Trends and Consequences of School Segregation," *Annual Review of Sociology* 40 (2014): 199–218.

55. Reardon and Owens, "60 Years after *Brown*."

56. Gillon, *Separate and Unequal*, 286–87.

57. Gregory D. Squires, *The Fight for Fair Housing: Causes, Consequences, and Future Implications of the 1968 Federal Fair Housing Act* (New York: Routledge, 2017).

58. Rothstein, *The Color of Law*, 127–31.

59. Sheryll Cashin, *The Failures of Integration: How Race and Class Are Undermining the American Dream* (New York: Public Affairs, 2004), 115–17

60. Rothstein, *The Color of Law*, 127.

61. Meier and Rudwick, *CORE: A Study in the Civil Rights Movement*, 306–7.

62. Joseph V. Ganley, "City Whipping Racial Housing Woes," *Herald American*, September 22, 1963, 1, 13.

63. Meier and Rudwick, *CORE: A Study in the Civil Rights Movement*, 308. Wiley went on to higher offices in CORE but left to help found the National Welfare Rights Organization, discussed in chapter 2.

64. Ibid., 127, 128.

65. Some of the faculty, including me, were associated with the Community Action work associated with the Great Society and War on Poverty activities identified in chapter 2. The actions and background are given in: Cornwall, *Human Rights in Syracuse*, 10–13.

Chapter 6

1. The National Opinion Research Center (NORC), at the University of Chicago, has conducted the General Social Survey for many years, tracking the responses to many identical questions. The Associated Press-NORC Center for Public Affairs Research and the General Social Survey (GSS) Staff, *Changing Attitudes about Racial Inequality* (Chicago: AP-NORC Center, 2019), https://apnorc.org/projects/changing-attitudes-about-racial-inequality/.

2. Heather Long and Andrew Van Dam, "Financial Chasm Separates Blacks from Whites," *New York Times*, June 7, 2020, B2 .

3. Sean F. Reardon and Ann Owens, "60 Years after *Brown*: Trends and Consequences of School Segregation," *Annual Review of Sociology* 40 (2014): 199–218.

4. Rucker Johnson, *Children of the Dream: Why School Integration Works* (New York: Basic Books, 2019).

5. Erica Frankenberg, Jongyeon Ee, Jennifer B. Ayscue, and Gary Orfield, *Harming Our Common Future: America's Segregated Schools 65 Years after* Brown (Los Angeles: The Civil Rights Project, University of California), 7, www.civilrightsproject.ucla.edu

6. Center for American Progress, K-12 Education Policy report on funding https://www.americanprogress.org/issues/education-k-12/reports/2018/11/13/460397/quality-approach-school-funding/. For example, the Kirwan Commission on Innovation and Excellence in Education is an initiative to develop major funding and policy reforms to improve the quality of Maryland's public education system.

7. https://www.pewresearch.org/fact-tank/2015/09/30/how-u-s-immigration-laws-and-rules-have-changed-through-history/.

8. Frankenberg et al., *Harming Our Common Future*, Figure 1; Richard Rothstein, *The Color of Law: A Forgotten History of How Our Government Segregated America* (New York and London: Liveright, 2017), xii.

9. Justin Driver, *The Schoolhouse Gate: Public Education, the Supreme Court, and the Battle for the American Mind* (New York: Pantheon, 2018).

10. The majority consisted of Warren E. Burger (Chief Justice), Potter Stewart, Harry Blackman, Lewis F. Powell Jr, and William Rehnquist. The dissenters were William O. Douglas, William J. Brennan Jr., Byron White, and Thurgood Marshall.

11. Rothstein, *The Color of Law*, xii.

12. Nikole Hannah-Jones, "It Was Never about Busing," *New York Times*, July 14, 2019.

13. Prior to this initiative, Boston, like many other large cities in the 1960s and 1970s, had significant declines in population. Whites were moving out into expanding suburbs, while African Americans and other minority groups were moving into the city. This was making reducing school segregation increasingly difficult.

14. School desegregation in Boston; a staff report prepared for the hearing of the U.S. Commission on Civil Rights in Boston, Massachusetts, June 1975.
United States Commission on Civil Rights.
Oversized –Paperback. ii, 155p. Cat.No: 94109 United States Commission on Civil Rights. [Washington]: The Commission, 1975. Oversized Paperback. ii.

15. For current information, see website of METCO, Inc. (https://metcoinc.org) and About Metco (http://www.doe.mass.edu/metco/).

16. In 1948, Willie graduated Morehouse College as a classmate of Martin Luther King, and in 1957 he received his PhD in sociology from Syracuse University, where he had a distinguished career until he left for Harvard University in 1974. See https://www.syracuse.com/news/2022/01/todays-obituary-charles-willie-94-syracuses-university-first-black-tenured-professor.html.

17. https://www.bostonglobe.com/2022/01/16/metro/charles-v-willie-an-architect-bostons-controlled-choice-school-desegregation-plan-dies-94/.

18. Jonathan Kozol, *The Shame of the Nation: The Restoration of Apartheid Schooling in America* (New York: Three Rivers Press, 2006).

19. Urban Institute, "How Has Education Funding Changed Over Time?," 2017, http://apps.urban.org/features/education-funding-trends/. Cited in Carmel Martin et al., "A Quality Approach to School Funding; Lessons Learned from School Finance Litigation," Center for American Progress, 2018, 8, https://www.americanprogress.org/article/quality-approach-school-funding/.

20. Bruce Baker, Danielle Farrie, Theresa Luhm, and David G. Sciarra, "Is School Funding Fair? A National Report Card," in *National Report Card* (Newark, NJ: Education Law Center, and Rutgers University Graduate School of Education, 2016), https://edlawcenter.org/assets/files/pdfs/publications/National_Report_Card_2016.pdf).

21. Martin et al., "A Quality Approach to School Funding."

22. Margaret A. Jorgensen and Jenny Hoffmann, "History of the No Child Left Behind Act of 2001 (NCLB)," in *Assessment Report* (San Antonia, TX: Pearson Education, 2003).

23. Ways of calculating adequate yearly progress under NCLB were made readily available for school teachers and administrators.

24. Valerie Strauss, "34 Problems with Standardized Tests," *Washington Post*, April 19, 2017, https://www.washingtonpost.com/news/answer-sheet/wp/2017/04/19/

34-problems-with-standardized-tests/; Marion Brady, *What's Worth Learning?* (Charlotte, NC: Information Age Publishing, 2010).

25. Quinn Mulholland, "The Case against Standardized Testing," *Harvard Political Review*, May 14, 2015, http://harvardpolitics.com/united-states/case-standardized-testing/.

26. Pew Research Center, "Demographic Trends and Economic Well-Being," Pew Research Center, 2016, https://www.pewsocialtrends.org/2016/06/27/1-demograp hic-trends-and-economic-well-being/.

27. Andrew McGill, "The Missing Black Students at Elite American Universities," *Atlantic*, November 24, 2015, https://www.theatlantic.com/notes/2015/11/the-miss ing-black-students-at-elite-american-universities-contd/417585/). Also see Kevin Carey, "A Detailed Look at the Downsides of California's Ban on Affirmative Action," *New York Times*, August 21, 2020, https://www.nytimes.com/2020/08/21/upshot/ 00up-affirmative-action-california-study.html.

28. David Kirp, *The College Dropout Scandal* (New York: Oxford University Press, 2020).

29. James V. Koch, *The Impoverishment of the American College Student* (Washington, DC: Brookings Institution, 2020).

30. /https://www.bing.com/search?q=HBCU+colleges++black+students+proporti ons&FORM=AFSCVO&PC=AFSC.

31. Dick Startz, "When It Comes to Student Success, HBCUs Do More with Less," Brookings Institution, January 18, 2021, https://www.brookings.edu/blog/brown-center-chalkboard/2021/01/18/when-it-comes-to-student-success-hbcus-do-more-with-less/#:~:text=HBCUs%20receive%20much%20less%20revenue,to%20only%20 %244%2C900%20at%20HBCUs.

32. Michael Hansen, Elizabeth Mann Levesque, Jon Valant, and Diana Quintero, *2018 Brown Center Report on American Education: Trends in NAEP Math, Reading, and Civics Scores* (Washington, DC: Brookings Institution, 2018), https://www.brookings. edu/research/2018-brown-center-report-on-american-education-trends-in-naep-math-reading-and-civics-scores/.

33. Karen Bogenschneider and Carol Johnson, "Family Involvement in Education," Policy Institute for Family Impact Seminars, 2015, https://www.purdue.edu/hhs/ hdfs/fii/wp-content/uploads/2015/07/s_wifis20c02.pdf.

34. Christopher Jencks and Meredith Phillips, "The Black-White Test Score Gap: Why It Persists and What Can Be Done," Brookings Institution, March 1, 1998, https://www. brookings.edu/articles/the-black-white-test-score-gap-why-it-persists-and-what-can-be-done/.

35. Ibid.

36. Elizabeth A. Canning, Katherine Muenks, Dorainne J. Green, and Mary C. Murphy. "Faculty Who Believe Ability Is Fixed Have Larger Racial Achievement Gaps and Inspire Less Student Motivation in Their Classes," *Science Advances* 5, no. 2 (2019), https://advances.sciencemag.org/content/5/2/eaau4734.

37. D. Deming, "Early Childhood Intervention and Life-Cycle Skill Development: Evidence from Head Start," *American Economic Journal: Applied Economics*. 1, no. 3 (2009): 111–34..

38. Sneha Elngo, Jorge Luis García, James J. Heckman, and Andrés Hojman, "Early Childhood Education," NBER Working Paper 21766, National Bureau of Economic Research, 2015, https://www.nber.org/papers/w21766.pdf.

39. David Madland and Alex Rowell, *Attacks on Public-Sector Unions Harm States: How Act 10 Has Affected Education in Wisconsin* (Washington, DC: Center for American Progress Action Fund, 2017), https://www.americanprogressaction.org/issues/econ omy/reports/2017/11/15/169146/attacks-public-sector-unions-harm-states-act-10-affected-education-wisconsin/.

40. Madland and Rowell, "Attacks on Public-Sector Unions."

41. Dave Umhoefer, "For Unions in Wisconsin, a Fast and Hard Fall since Act 10," *Milwaukee Journal Sentinel*, October 9, 2016, 9–9.

42. Robert Samuels, "Walker's Anti-union Law Has Labor Reeling in Wisconsin," *Washington Post*, February 22, 2015, https://www.washingtonpost.com/politics/in-wisconsin-walkers-anti-union-law-has-crippled-labor-movement/2015/02/22/1eb3ef82-b6f1-11e4-aa05-1ce812b3fdd2_story.html.

43. F. Howard Nelson and Michael Rosen, *Are Teachers' Unions Hurting American Education?* (Milwaukee, WI: Institute for Wisconsin's Future, 1996), http://68.77.48.18/RandD/Other/Collective%20Bargaining%20Study.pdf; Abigail Wydra, "Teachers' Unions Improve Student Achievement: Insights from California Charter Schools," *Chicago Policy Review*, January 20, 2018, https://chicagopolicyreview.org/2018/01/20/teachers-unions-improve-student-achievement-insights-from-califor nia-charter-schools/.

44. Dave Kamper, "Facing Wisconsin-Style Attack, Iowans Stick to the Union," *Labornotes*, November 16, 2017, https://labornotes.org/2017/11/facing-wisconsin-style-attack-iowans-stick-union.

45. Jonathan Sader and Shannon Rieger, "Patterns and Trends of Residential Integration in the United States since 2000," Research Brief, Harvard Joint Center for Housing Studies, 2017, https://www.jchs.harvard.edu/sites/default/files/09192017_spader-rieger_integration_brief.pdf); Douglas S. Massey and Andrew B. Gross, "Explaining Trends in Racial Segregation, 1970–1980," *Urban Affairs Quarterly* 27, no. 1 (1991): 13–35.

46. Massey and Andrew B. Gross, "Explaining Trends in Racial Segregation, 1970-1980," *Urban Affairs Quarterly* 27 (1) (1991).

47. Louis Kriesberg, Bronislaw Miszta, and Janusz Mucha, eds., *Social Movements as a Factor of Change in the Contemporary World* (Greenwich, CT: JAI Press, 1988). This was evident to me from my meetings and observations in Warsaw and Krakow in October 1985.

48. William Julius Wilson, *The Truly Disadvantaged: The Inner City, the Underclass, and Public Policy* (Chicago: University of Chicago Press, 1987).

49. Jack Newfield, *Robert Kennedy: A Memoir* (New York: E. P. Dutton, 1969), 90–109.

50. Ibid., 95–96.

51. "50 Years of Restoration: Building Assets and Community Wealth," *Brooklyn Reader*, December 29, 2017, https://www.bkreader.com/2017/12/29/50-years-restoration-building-assets-community-wealth/.

52. Alexander von Hoffman, "The Past, Present, and Future of Community Development in the United States," Harvard Joint Center for Housing Studies, 2012, https://www. jchs.harvard.edu/sites/default/files/w12-6_von_hoffman.pdf.

53. Ibid., 26.

54. Mark Pinsky, "Taking Stock: CDFIs Look Ahead after 25 Years of Community Development Finance," Brookings Institution and Joint Center for Housing Studies, December 1, 2001, https://www.brookings.edu/articles/taking-stock-cdfis-look-ahead-after-25-years-of-community-development-finance/.

55. Louis Winnick, "The Triumph of Housing Allowance Programs: How a Fundamental Policy Conflict Was Resolved," *Cityscape: A Journal of Policy Development and Research* 1, no. 3 (1995): 95–121.

56. Molly W. Metzger and Danilo Pelletiere, "Patterns of Housing Voucher Use Revisited: Segregation and Section 8 Using Updated Data and More Precise Comparison Groups, 2013," Center for Social Development Research, 2013, https://www.stlouisfed.org/~/media/files/pdfs/community%20development/econ%20mobility/sessions/metzgerpaper508.pdf.

57. Alic MacGillis, "Jared Kushner's Other Real Estate Empire in Baltimore," *New York Times*, May 23, 2017, https://www.nytimes.com/2017/05//23/magazine/jared-kushners-other-real-estate-empire.html.

58. https://www.huduser.gov/portal/datasets/lihtc.html; David J. Erickson, *The Housing Policy Revolution: Networks and Neighborhoods* (Washington, DC: Urban Institute Press, 2009).

59. https://www.pbs.org/video/east-lake-meadows-full-film/; Shirley Franklin and David Edwards, *It Takes a Neighborhood: Purpose Built Communities and Neighborhood Transformation* (Purpose Built Communities, 2020), https://purposebuiltcommunities.org/takes-neighborhood-purpose-built-communities-neighborhood-transformation.

60. https://www.bloomberg.com/news/articles/2016-01-04/how-zoning-restrictions-make-segregation-worse. Increased density would also have environmental benefits.

61. Jesse Drucker and Eric Lipton, "How a Trump Tax Break to Help Poor Communities Became a Windfall for the Rich," *New York Times*, August 31, 2019, https://www.nytimes.com/2019/08/31/business/tax-opportunity-zones.html.

62. Matthew Desmond, *Evicted: Poverty and Profit in the American City* (New York: Crown, 2016), 308–12.

63. Bill Adair, "The Peace Dividend Began with a Bush," PolitiFact, January 24, 2008, https://www.politifact.com/truth-o-meter/statements/2008/jan/24/rudy-giuliani/the-peace-dividend-began-with-a-bush/.

64. https://theconversation.com/african-americans-economic-setbacks-from-the-great-recession-are-ongoing-and-could-be-repeated-109612.

65. William Jr. Darity, Darrick Hamilton, Mark Paul, Alan Aja, Anne Price, Antonio Moore, and Caterina Chiopris, *What We Get Wrong about Closing the Racial Wealth Gap*, Durham, NC: Samuel DuBois Cook Center on Social Equity Insight Center for Community Economic Development, 2018.

66. Justice Research and Statistics Association (JRSA), *Crime and Justice Atlas 2000* (Washington, DC, JRSA, 1998).

67. Alexia Cooper and Erica L. Smith, *Homicide Trends in the United States, 1980–2008: Annual Rates for 2009 and 2010* (Washington, DC: US Department of Justice Office of Justice Programs Bureau of Justice Statistics, 2011).

68. Walter Shapiro, "The Flawed Politics of a Law-and-Order Campaign," *New Republic*, May 31, 2020, https://newrepublic.com/article/157939/flawed-politucs-law-and-order-camapaign.

69. https://www.nixonfoundation.org/2017/08/nixons-record-civil-rights-2/.

70. Mark J. Perry, "The Shocking Story Behind Richard Nixon's 'War on Drugs' That Targeted Blacks and Anti-war Activists," American Enterprise Institute, June 14, 2018, https://www.aei.org/carpe-diem/the-shocking-and-sickening-story-behind-nixons-war-on-drugs-that-targeted-blacks-and-anti-war-activists/?fbclid=IwAR0kd4j7SVj9iY_Bmp1eoL_Kbo3Orx9zM7Hcbd2-6oboK1hDOZmhD5PIxe8.

71. Dan Baum, "Legalize It All: How to Win the War on Drugs," *Harper's Magazine*, April 2016, https://harpers.org/archive/2016/04/legalize-it-all/.

72. Steven Harmon Wilson, "Controlled Substances Act (1970)," Encyclopedia.com, 2020, https://www.encyclopedia.com/history/encyclopedias-almanacs-transcripts-and-maps/controlled-substances-act-1970.

73. https://web.archive.org/web/20070715212213/http://www.ussc.gov/r_congress/02crack/2002crackrpt.pdf; Daniel S. Lucks, *Reconsidering Reagan: Racism, Republicans, and the Road to Trump* (Boston: Beacon Press, 2020).

74. C. Peter Rydell and Susan S. Sohler Everingham, *Controlling Cocaine: Supply versus Demand Programs* (Santa Monica, CA: RAND Corporation, 1994), https://www.rand.org/pubs/monograph_reports/MR331.html.

75. Al From, *The New Democrats and the Return to Power* (New York: St. Martin's Press, 2013), 198.

76. Bill Clinton, *My Life* (New York: Alfred A. Knopf, 2004), 610–11.

77. The Senencing Project, "Federal Crack Cocaine Sentencing," Senencing Project, 2010, https://www.sentencingproject.org/publications/federal-crack-cocaine-sentencing/.

78. Samantha Michaels, "Trump Just Bragged about Criminal Justice Reform: Look Closer at How His Administration Is Undoing It," *Mother Jones*, February 4, 2020, https://www.motherjones.com/crime-justice/2020/02/trump-just-bragged-about-criminal-justice-reform-look-closer-at-how-his-administration-is-undoing-it/.

79. Nathan James, *The First Step Act of 2018: An Overview* (Washington, DC: Congressional Research Service, 2019), https://crsreports.congress.gov.

80. Michaels, "Trump Just Bragged."

81. Hailey Fuchs, "Law to Reduce Crack Cocaine Sentences Is No Help for Some Inmates," *New York Times*, August 2, 2020.

82. Baum, "Legalize It All."

83. Shaun L. Gabbidon, *Criminological Perspectives on Race and Crime*, 4th ed. (New York: Routledge, 2020).

84. Richard Wilkinson and Kate Pickett, *The Spirit Level: Why Greater Equality Makes Societies Stronger* (New York and London: Bloomsbury Press, 2010), 134–48.

85. Richard A. Cloward and Lloyd E. Ohlin, *Delinquency and Opportunity: A Theory of Delinquency Gangs* (Glencoe, IL: The Free Press, 1960).

86. George Guerrero, "Social Disorganization Theory," in *Encyclopedia of Race and Crime*, ed. Shaun L. Gabbidon and Helen T. Greene (Thousand Oaks: CA: SAGE, 2009).

87. Ashley Nellis, Judy Greene, and Marc Mauer, *Reducing Racial Disparity in the Criminal Justice System: A Manual for Practitioners and Policymakers* (Washington, DC: The Sentencing Project, 2016), https://www.sentencingproject.org/wp-cont ent/uploads/2016/01/Reducing-Racial-Disparity-in-the-Criminal-Justice-System- A-Manual-for-Practitioners-and-Policymakers.pdf.

88. Fredrik H. Leinfelt, "Racial Influences on the Likelihood of Police Searches and Search Hits: A Longitudinal Analysis from an American Midwestern City," *Police Journal* 79, no. 3 (2006): 238–57; Sarah A. Seo, "Racism on the Road," *New York Review of Books* 67, no. 12 (July 23, 2020): 45–46.

89. Beth Roy, *41 Shots . . . And Counting* (Syracuse, NY: Syracuse University Press, 2009).

90. Wendy Sawyer, *How Race Impacts Who Is Detained Pretrial* (Northampton, MA: Prison Policy Initiative, 2019).

91. Ellen A. Donnelly and John M. MacDonald, "The Downstream Effects of Bail and Pretrial Detention on Racial Disparities in Incarceration," *Journal of Criminal Law and Criminology* 108, no. 4 (2018): 775, https://scholarlycommons.law.northwest ern.edu/jclc/vol108/iss4/4.

92. M. Marit Rehavi and. Sonja B. Starr, "Racial Disparity in Federal Criminal Sentences," *Journal of Political Economy.* 122, no. 6 (2014): 1320–54.

93. The Sentencing Project, 2021. 1705 DeSales St. NW Washington, DC 20036. https:// www.sentencingproject.org

94. https://www.naacp.org/criminal-justice-issues/.

95. Samuel Walker, *In Defense of American Liberties: A History of the ACLU* (New York: Oxford University Press, 1990).

96. David Chalmers, *Backfire: How the Ku Klux Klan Helped the Civil Rights Movement* (Lanham, MD: Rowman & Littlefield, 2003).

97. Brian Lyman, "Southern Poverty Law Center Staff Vote to Unionize," *Montgomery Advertiser*, December 16, 2019, https://www.montgomeryadvertiser.com/story/ news/2019/12/16/southern-poverty-law-center-votes-unionize/2661726001/.

98. Alan Blinder, "Southern Poverty Law Center President Plans Exit amid Turmoil," *New York Times*, March 22, 2019.

99. https://people.com/archive/bryan-stevenson-vol-44-no-22/.

100. https://deathpenaltyinfo.org/state-and-federal-info/state-by-state/alabama.

101. Bryan A. Stevenson, *Just Mercy: A Story of Justice and Redemption* (New York: Spiegel & Grau, 2014).

102. https://atlantablackstar.com/2016/05/22/top-civil-rights-attorney-offers-compell ing-argument-on-why-the-justice-system-is-an-expensive-continuation-of-slav ery-and-jim-crow/.

103. Scott G. McNall, *Cultures of Defiance and Resistance: Social Movements in 21st- Century America* (New York and London: Routledge, 2018), 87–120.

104. Barbara Ransby, *Making All Black Lives Matter: Reimagining Freedom in the Twenty-First Century* (Oakland: University of California Press, 2018), 29–46.

105. Community Relations Service, *Annual Report 2018* (Washington, DC: Community Relations Service, Department of Justice, 2018).

106. Ron Mott, "Three Years after Michael Brown's Death, Has Ferguson Changed?," NBC News, August 9, 2017, https://www.nbcnews.com/storyline/michael-brown-shooting/three-years-after-michael-brown-s-death-has-ferguson-changed-n791081.

107. Jill Quadagno, *The Color of Welfare: How Racism Undermined the War on Poverty* (New York: Oxford University Press, 1994).

108. Community Relations Service, "Annual Report 2018," 6.

109. Jenna Wortham, "The Vision Forward," *New York Times Magazine*, August 30, 2020, 28–35, 45.

110. Jessica Byrd, "A Different Direction for Black Politics," *New York Times*, September 4, 2020, A27.

111. https://www.fox9.com/news/4-minneapolis-police-officers-fired-following-death-of-george-floyd-in-police-custody.

112. https://wbhm.org/npr_story_post/2020/vigilante-militia-confusion-and-politics-shape-how-shooting-suspect-is-labeled/.

113. I was a commissioner on the Human Rights Commission of Syracuse and Onondaga County (1989–1993). After public hearings and difficult negotiations, an oversight board was established, but it had little power and was generally ignored by the police department until recent years. Interestingly, we held public meetings in an effort to create a police review board and I noted that in African American neighborhoods, the residents' greatest complaint about the police was that they responded slowly if at all.

114. Ted Robert Gurr, *Why Men Rebel* (Princeton, NJ: Princeton University Press, 1970).

115. Astead W. Herndon and Dionne Searcey, "White Voters' Views on Race Show Change," *New York Times*, June 28, 2020, 24–25.

116. Elizabeth Arias and Jiaquan Xu, "United States Life Tables, 2017," in *National Vital Statistics Reports* (Washington, DC: National Vital Statistics System, 2019).

117. https://www.everycrsreport.com/reports/RL32792.html#_Ref194285325.

118. Sam Harper, Richard F. MacLehose, and Jay S. Kaufman, "Trends in the Black-White Life Expectancy Gap among US States, 1990–2009," *Health Affairs* 33, no. 8 (2014), https://www.healthaffairs.org/doi/full/10.1377/hlthaff.2013.1273; Arias and Xu, "United States Life Tables, 2017."

119. Tiffany Ford, Sarah Reber, and Richard V. Reeves, "Race Gaps in COVID-19 Deaths Are Even Bigger Than They Appear" (Washington, DC: Brookings, 2020), https://www.brookings.edu/blog/up-front/2020/06/16/race-gaps-in-covid-19-deaths-are-even-bigger-than-they-appear/.

120. https://www.theguardian.com/society/2018/mar/01/how-americas-identity-politics-went-from-inclusion-to-division.

121. https://www.poorpeoplescampaign.org/about/.

122. https://www.pewresearch.org/fact-tank/2015/09/30/how-u-s-immigration-laws-and-rules-have-changed-through-history/.

123. Isabel Wilkerson, *Caste: The Origins of Our Discontents* (New York: Random House, 2020).

124. Fabio Rojas, *From Black Power to Black Studies: How a Radical Social Movement Became an Academic Discipline* (Baltimore, MD: Johns Hopkins University Press, 2007).

125. https://en.wikipedia.org/wiki/Martin_Luther_King_Jr._Day.

126. https://nmaahc.si.edu/about/museum.

127. For an account of the revolt at the 1968 Olympics, see Harry Edwards, *The Revolt of the Black Athlete*, 50th Anniversary Edition (Urbana: University of Illinois Press, 2017).

128. Jack Hamilton, *Just Around Midnight: Rock and Roll and the Racial Imagination* (Cambridge, MA: Harvard University Press, 2016).

129. Diana C. Mutz, "Status Threat, Not Economic Hardship, Explains 2016 Presidential Vote," *PNAS* 115, no. 19 (2018): E4330–E4339, https://doi.org/10.1073/pnas.171 8155115.

130. https://www.huffpost.com/entry/remembering-dr-martin-lut_b_14153164.

131. https://www.npr.org/2020/09/05/910053496/trump-tells-agencies-to-end-traini ngs-on-White-privilege-and-critical-race-theora.

132. Robin DiAngelo, *White Fragility: Why It's So Hard for White People to Talk about Racism* (Boston: Beacon Press, 2018).

133. Randall Robinson, *The Debt: What America Owes to Blacks* (New York: Dutton, 2000); Nikole Hannah-Jones, "What Is Owed," *New York Times Magazine*, June 28, 2020, 30–35, 47–53; Jim Yardley, "Panel Recommends Reparations in Long-Ignored Tulsa Race Riot," *New York Times*, February 5, 2000, A1, 12.

134. https://www.congress.gov/bill/115th-congress/house-bill/40/text?format=txt.

Chapter 7

1. Patrick G. Coy and Lynne M. Woehrle, eds., *Social Conflicts and Collective Identities* (Lanham, MD: Rowman & Littlefield, 2000).

2. Karen DeCrow, *Sexist Justice: How Legal Sexism Affects You* (New York: Random House, 1974).

3. In the mid-1960s, my wife fought hard to get her Syracuse public library card in her name, Lois A. Kriesberg, not as Mrs. Louis Kriesberg. She eventually won.

4. Betty Friedan, *The Feminine Mystique* (New York: W. W. Norton, 1963).

5. http://now.org/about/history/founding-2/.

6. Nicholas Pedriana and Amanda Abraham, "Now You See Them, Now You Don't: The Legal Field and Newspaper Desegregation of Sex-Segregated Help Wanted Ads 1965–75," *Law & Social Inquiry* 31, no. 4 (2006): 905–38.

7. My wife, Lois Ablin Kriesberg, as a lawyer, was an active participant in the chapter activities. I attended meetings and supported its events and policies.

8. Zoe Cornwall, *Human Rights in Syracuse: Two Memorable Decades* (Syracuse, NY: Human Rights Commission of Syracuse and Onondaga County, 1989), 119–28.

9. Ibid., 122.

10. Robert Seidenberg, *Corporate Wives—Corporate Casualties?* (New York: Doubleday Anchor, 1973).

11. Cornwall, *Human Rights in Syracuse*, 32–133.

12. Carol Gilligan, "In a Different Voice: Women's Conceptions of Self and of Morality," *Harvard Educational Review* 47, no. 4 (1977): 481–51; Robin Morgan, ed. *Sisterhood Is Powerful: An Anthology of Writings from the Women's Liberation Movement* (New York: Vintage Books, 1970); Jo Freeman, ed. *Women: A Feminist Perspective* (Palo Alto, CA: Mayfield, 1975).

13. DeCrow, *Sexist Justice*.

14. Elise Boulding, *The Underside of History: A View of Women through Time* (Boulder, CO: Westview Press, 1976).

15. Patricia Hill Collins, *Black Feminist Thought: Knowledge, Consciousness, and the Politics of Empowerment* (New York: Routledge, 2000).

16. Carrie Lukas, ed. *The Politically Incorrect Guide to Women, Sex, and Feminism* (Washington, DC: Regency Publishing, 2006).

17. David Brady and Kent L. Tedin, "Ladies in Pink: Religion and Political Ideology in the Anti-ERA Movement," *Social Science Quarterly* 56, no. 4 (1976): 564–75.

18. Jessie Wegman, "Why Can't We Make Women's Equality the Law of the Land?," *New York Times*, January 28, 2022, https://www.nytimes.com/2022/01/28/opinion/equal-rights-amendment-ratification.html.

19. Jane J. Mansbridge, *Why We Lost the ERA* (Chicago: University of Chicago Press, 1986); Phyllis Schlafly, ed. *The Power of the Positive Woman* (New Rochelle, NY: Arlington House, 1977).

20. Donald T. Critchlow and Cynthia L. Stacheck, "The Equal Rights Amendment Reconsidered: Politics, Policy, and Social Mobilization in a Democracy," *Journal of Policy History* 20, no. 1 (2008): 157–76.

21. https://web.archive.org/web/20170410091210/http://equalrightsamendment.org/history.htm.

22. Kate Klonick, "Why America Needs More Housewives Like Phyllis Schlafly," *Atlantic*, September 7, 2016, https://www.theatlantic.com/politics/archive/2016/09/why-america-needs-more-housewives-like-phyllis-schlafly/498917/.

23. Cornwall, *Human Rights in Syracuse*, 135–36.

24. Kelsy Kretschmer, "Shifting Boundaries and Splintering Movements: Abortion Rights in the Feminist and New Right Movements," *Sociological Forum* 29, no. 4 (2014).

25. Rebecca E. Klatch, *Women of the New Right* (Philadelphia: Temple University Press, 1987).

26. https://www.washingtonpost.com/lifestyle/style/planned-parenthood-is-a-symbol-this-is-the-reality-of-one-ohio-clinic/2015/09/15/b97c465e-4774-11e5-846d-02792f854297_story.html. Accessed November 8, 2020.

27. https://www.plannedparenthood.org/about-us/facts-figures/annual-report.

28. Byron W. Daynes and Raymond Tatalovich, "Presidential Politics and Abortion, 1972–1988," *Presidential Studies Quarterly* 22, no. 3 (1992).

29. https://www.washingtonpost.com/blogs/ezra-klein/post/how-the-republican-party-became-pro-life/2012/03/09/gIQAZcD31R_blog.html; Daniel K. Williams, "The

GOP's Abortion Strategy: Why Pro-Choice Republicans Became Pro-Life in the 1970s," *Journal of Policy History* 23, no. 4 (2011): 513–39.

30. The Supreme Court ruled that a state could set restrictions on abortions, but did not disturb the *Roe v. Wade* decision. https://legaldictionary.net/webster-v-eproductive-health-services/.

    https://embryo.asu.edu/pages/webster-v-reproductive-health-services-1989#:~:text=In%20the%201989%20case%20Webster.

31. Before 2019, six states had passed legislation restricting abortions: Georgia, Kentucky, Louisiana, Missouri, Mississippi, and Ohio, stopping short of outright bans. https://www.nytimes.com/interactive/2019/us/abortion-laws-states.html.

32. https://prochoice.org/dramatic-increase-in-threats-violence-directly-correlates-with-release-of-misleading-videos/.

33. https://prochoice.org/naf-releases-2019-violence-disruption-statistics/.

34. Lara Whyte, "Has Trump's White House 'Resurrected' Army of God Anti-Abortion Extremists," *Open Democracy*, February 5, 2018, https://www.opendemocracy.net/en/5050/army-of-god-abortion-terrorists-emboldened-under-trump/.

35. Jennifer Jefferis, *The Army of God and Anti-Abortion Terror in the United States* (New York: Praeger, 2011), 40.

36. For text of FACE Act see: law.cornell.edu/uscode/text/18/248.

37. https://www.pewresearch.org/politics/2019/08/29/u-s-public-continues-to-favor-legal-abortion-oppose-overturning-roe-v-wade/.

38. Lawrence B. Finer, Lori F. Frohwirth, Lindsay A. Dauphinee, Susheela Singh, and Ann M. Moore. "Reasons U.S. Women Have Abortions: Quantitative and Qualitative Perspectives," *Perspectives on Sexual and Reproductive Health* 37, no. 3 (2007): 110–18. Also see: https://www.verywellhealth.com/reasons-for-abortion-906589.

39. https://www.theatlantic.com/health/archive/2019/05/why-more-women-dont-choose-adoption/589759/.

40. Robin West, Justin Murray, Meredith Esser, and Anna Grear, eds., *In Search of Common Ground on Abortion: From Culture War to Reproductive Justice* (New York: Routledge, 2016). Also see https://magazine.nd.edu/stories/arms-unfolded-the-search-for-common-ground-on-abortion/.

41. In the 2020 presidential election, of twelve important issues, abortion was the least likely to be chosen. htpps://www.pewresearch.org/politics/2020/08/13/important-issues-in-the-2020-election/.

42. https://www.americanbar.org/groups/crsj/publications/human_rights_magazine_home/human_rights_vol31_2004/summer2004/irr_hr_summer04_gaps/.

43. The Supreme Court's decision in *Franklin v. Gwinnett County Public Schools*, 503 U.S. 60 (1992), held that plaintiffs can obtain monetary damages from defendants for intentional discrimination.

44. https://www.justice.gov/ovw/blog/violence-against-women-act-ongoing-fixture-nation-s-response-domestic-violence-dating.

45. https://www.eeoc.gov/statutes/notice-concerning-lilly-ledbetter-fair-pay-act-2009.

46. https://www.pewresearch.org/fact-tank/2018/03/15/for-womens-history-month-a-look-at-gender-gains-and-gaps-in-the-u-s/.

47. https://www.Britannica.com/biography/Ruth-Bader-Ginsburg.

48. Ian D. Wyatt and Daniel E. Hecker, "Occupational Changes During the 20th Century," *Monthly Labor Review*, March, 2006, 35–57.

49. The success of the March of Dimes organization in mobilizing a movement and raising money is illustrative. When a vaccine was made to prevent infantile paralysis, it wisely commissioned research about what to do afterward. The research noted that the disease was not related to socioeconomic status and children of good families needed care. The organization leaders chose to focus on genetic defects as the never-ending human malady unrelated to socioeconomic rank. David L. Sills, *The Volunteers, Means and Ends in a National Organization* (Glencoe, IL: Free Press, 1957).

50. Sue Tolleson-Rinehart and Susan J. Carroll, "'Far from Ideal': The Gender Politics of Political Science," *American Political Science Review* 100, no. 4 (2006).

51. https://www.insidehighered.com/news/2019/07/30/political-science-association-pleases-and-surprises-members-its-flagship.

52. https://www.cambridge.org/core/journals/ps-political-science-and-politics/article/gender-diversity-and-methods-in-political-science-a-theory-of-selection-and-survival-biases/9AB3B75F3F6E47C3650ECFFB4872E69F.

53. https://data.lawschooltransparency.com/enrollment/all/.

54. Ruth Bader Ginsburg, "Women's Work: The Place of Women in Law Schools," *Journal of Legal Education* 32, no. 2 (1982): 272.

55. Cynthia Grant Bowman, "Women in the Legal Profession from the 1920s to the 1970s: What Can We Learn from Their Experience about Law and Social Change?," *Cornell Law Faculty Publications 12* (2009), http://scholarship.law.cornell.edu/facpub/12). Also see Erwin O. Smigel, *The Wall Street Lawyer: Professional Organization Man?* (New York and London: The Free Press of Glencoe and Collier-Macmillan, 1964).

56. Bowman, "Women in the Legal Profession," 15.

57. https://www.pewresearch.org/wp-content/uploads/sites/3/2011/12/women-in-the-military.pdf.

58. https://www.pbs.org/independentlens/films/9to5-the-story-of-a-movement/.

59. https://9to5.website/main.asp.

60. https://www.nber.org/digest/mar19/employment-growth-and-rising-womens-labor-force-participation.

61. https://hbr.org/2018/01/when-more-women-join-the-workforce-wages-rise-including-for-men.

62. There is some evidence that the amount of wife beating was less in the 1890s than in the 1970s, perhaps because community control was greater in the earlier years and the norms controlling men's treatment of women were stronger in the earlier period. David Peterson, "Wife Beating: An American Tradition," *Journal of Interdisciplinary History* 23, no. 1 (1992): 97–118.

63. Kathleen J. Tierney, "The Battered Women Movement and the Creation of the Wife Beating Problem," *Social Problems* 28, no. 3 (1982): 207–20.

64. Ibid.

65. Anthony Downs, "Up and Down with Ecology—the 'Issue Attention Cycle,'" *Public Interest* 28 (1972): 38–50; John D. McCarthy and Mayer Zald, *The Trend of Social Movements in America: Professionalization and Resource Mobilization* (Princeton, NJ: General Learning Press, 1973).

66. https://www.npr.org/2020/12/07/943938352/outside-in-everybody-knows-someb ody.

67. Ibid.

68. Ibid.

69. https://www.washingtonpost.com/politics/joe-biden-apologized-to-anita-hill-but-shes-not-satisfied/2019/04/25/0f2ff1d8-679a-11e9-82ba-fcfeff232e8f_story.html.

70. In the 117th Congress, women made up a much larger share of congressional Democrats (38%) than Republicans (14%). Across both chambers, there were 106 Democratic women and 38 Republican women in the new Congress. Women accounted for 40% of House Democrats and 32% of Senate Democrats, compared with 14% of House Republicans and 16% of Senate Republicans. https://www.pewr esearch.org/fact-tank/2021/01/15/a-record-number-of-women-are-serving-in-the-117th-congress/.

71. Brad Boserup, Mark McKenney, and Adel Elkbuli, "Alarming Trends in US Domestic Violence During the COVID-19 Pandemic," *American Journal of Emergency Medicine* 38, no. 12 (2020): 2753–55; Yasmin B. Kofman and Dana Rose Garfin, "Home Is Not Always a Haven: The Domestic Violence Crisis amid the COVID-19 Pandemic," *Psychological Trauma: Theory, Research, Practice, and Policy* 12, Suppl. 1 (2020): S199–S201.

72. Melen Ryzik and Katie Benner, "Biden's Aid Package Funnels Millions to Victims of Domestic Abuse," *New York Times*, March 19, 2021, A21.

73. https://www.dolanconsultinggroup.com/wp-content/uploads/2017/10/RB_Domes tic-Violence-Calls_Officer-Safety.pdf.

74. https://www.pewsocialtrends.org/2020/07/07/a-century-after-women-gained-the-right-to-vote-majority-of-americans-see-work-to-do-on-gender-equality/.

75. https://www.washingtonpost.com/news/the-intersect/wp/2017/10/19/the-woman-behind-me-too-knew-the-power-of-the-phrase-when-she-created-it-10-years-ago/. Accessed January 27, 2021.

76. https://www.rollcall.com/2018/12/21/donald-trump-signs-overhaul-of-anti-har assment-law-for-members-of-congress-staff/.

77. https://www.pewsocialtrends.org/2020/07/07/a-century-after-women-gained-the-right-to-vote-majority-of-americans-see-work-to-do-on-gender-equality/. Accessed January 29, 2021.

78. PBS's film *Cured* tells the story of how LGBTQ people convinced the APA that they were not mentally ill. https://www.lgbttqnation.com/2921/10/pbs-new-film.

79. Elizabeth A. Armstrong and Suzanna M. Crage, "Movements and Memory: The Making of the Stonewall Myth," *American Sociological Review* 71, no. 5 (2006): 724–51.

80. Erin Ruel and Richard T. Campbell, "Homophobia and HIV/AIDS: Attitude Change in the Face of an Epidemic," *Social Forces* 84, no. 4 (2006): 2167–78.

81. https://www.kff.org/wp-content/uploads/2013/07/8186-hiv-survey-report_fi nal.pdf.

82. https://www.pewforum.org/fact-sheet/changing-attitudes-on-gay-marriage/.

83. https://www.brennancenter.org/our-work/analysis-opinion/improbable-victory-marriage-equality/; Sasha Issenberg, *The Engagement: America's Quarter-Century Struggle over Same-Sex Marriage* (New York: Pantheon, 2021).

84. "How We Got Marriage Equality," *New York Times*, June 6, 2021, SR8.

85. Everett C. Hughes, "The Study of Ethnic Relations," *Dalhousie Review* 27, no. 4 (1948): 477–85.

86. Charles B. Keely, "The Development of U.S. Immigration Policy since 1965," *Journal of International Affairs* 33, no. 2 (1979): 249–63.

87. Louis Kriesberg, "Ethnicity, Nationalism, and Violent Conflict in the 1990's," *Peace Studies Bulletin* 2 (Winter 1992–93); "Identities in Intractable Conflicts," *Beyond Intractability (Intractable Conflicts Knowledge Base)*, 2003, https://www.crinfo.org/; W. A. Elliott, *Us and Them: A Study of Group Consciousness* (Aberdeen: Aberdeen University Press, 1986).

88. Robert A. Levine and Donald T. Campbell, *Ethnocentrism: Theories of Conflict, Ethnic Attitudes, and Group Behavior* (New York: John Wiley & Sons, 1972).

89. https://www.pewresearch.org/social-trends/2019/05/08/americans-see-advantages-and-challenges-in-countrys-growing-racial-and-ethnic-diversity/.

90. Ashley Jardina, *White Identity Politics* (Cambridge, UK: Cambridge University Press, 2019).

91. Jardina, *White Identity Politics*, Table 3.6.

Chapter 8

1. Max Weber, *The Theory of Social and Economic Organization*, trans. A. M. Henderson and Talcott Parsons (New York: Oxford University Press, 1947), 152.

2. Bertrand Russell, *Power: A New Social Analysis* (London: Allen and Unwin, 1938); Dennis Wrong, *Power: Its Forms, Bases, and Uses*, 2nd ed. (New Brunswick, NJ: Transaction, 1995).

3. Walter H. Conser Jr., Donald M. McCarthy, David J. Toscano, and Gene Sharp, eds., *Resistance, Politics, and the American Struggle for Independence, 1765–1775* (Boulder, CO: Lynne Rienner, 1986).

4. https://nvdatabase.swarthmore.edu/content/american-colonials-struggle-against-british-empire-1765-1775. Gene Sharp was the founder of an empirically based theory of the power of nonviolence.

5. Calvin C. Jillson, *American Government: Political Development and Institutional Change*, 5th ed. (New York: Taylor & Francis, 2009).

6. Garry Wills, *Lincoln at Gettysburg: The Words That Remade America* (New York: Touchstone Books, 1993).

7. Noah Feldman, *The Broken Constitution: Lincoln, Slavery, and the Refounding of America* (New York: Farrar, Straus & Giroux, 2021).

8. Robert D. Putnam, *The Comparative Study of Political Elites* (New Jersey: Prentice Hall, 1976).

9. C. Wright Mills, *The Power Elite* (New York: Oxford University Press, 1956), 9. Even earlier, Mills had published many works on social classes and related subjects. As a sociology instructor also at Columbia University, 1953–1956, I can report that he was highly attractive to many students. However, the senior distinguished sociology faculty were critical of him and dismissive of his research.

10. G. William Domhoff, *The Powers That Be: Processes of Ruling Class Domination in America* (New York: Vintage Books, 1978); Floyd Hunter, *Top Leadership U.S.A.* (Chapel Hill: University of North Carolina Press, 1959).

11. The text may be found in Appendix B of Seymour Melman, *Pentagon Capitalism: The Political Economy of War* (New York: McGraw Hill, 1970), 235–39.

12. Ibid. Also see, Sidney Lens, *The Military-Industrial Complex* (Philadelphia and Kansas City: Pilgrim Press & The National Catholic Reporter, 1970). For later report, see: https://www.npr.org/2011/01/17/132942244/ikes-warning-of-military-expansion-50-years-later.

13. Arnold M. Rose, *The Power Structure* (New York: Oxford University Press, 1967).

14. http://uweconsoc.com/trends-in-corporate-lobbying-its-incentives-and-socio-economic-effects/.

15. https://www.theatlantic.com/business/archive/2015/04/how-corporate-lobbyists-conquered-american-democracy/390822/.

16. Nicholas Confessore, Sarah Cohen, and Karen Yourish, "Small Pool of Rich Donors Dominates Election Giving," *New York Times*, August 1, 2015, https://www.nytimes.com/2015/08/02/us/small-pool-of-rich-donors-dominates-election-giving.html?referringSource=articleShare.

17. Bruce Cummings, *The Origins of the Korean War*: Vol. 2, *The Roaring of the Cataract, 1947–1950* (Princeton, NJ: Princeton University Press, 1990); Don Oberdorfer and Robert Carlin, *The Two Koreas*, 3rd ed..(New York: Basic Books, 2014).

18. https://www.britannica.com/event/Vietnam-War.

19. Vincent Boucher, Charles-Philippe David, and Karine Prémont, *National Security Entrepreneurs and the Making of American Foreign Policy* (Montreal: McGill-Queen's University Press 2021).

20. Rachel Maddow, *Drift: The Unmooring of American Military Power* (New York: Crown, 2012); Andrew J. Bacevich, *Washington Rules: America's Path to Permanent War* (New York: Metropolitan Books, 2010).

21. Seymour Martin Lipset, "The Radical Right: A Problem for U.S. Democracy," *British Journal of Sociology* 6, no. 2 (June 1955): 176–209.

22. Seymour Martin Lipset and Earl Raab, *The Politics of Unreason: Right Wing Extremism in America 1790–1970* (New York: Harper & Row, 1970).

23. http://www.electproject.org/home/voter-turnout/voter-turnout-data.

24. https://onlinepoliticalsciencedegree.eku.edu/insidelook/what-causes-low-voter-turnout-united-states.

25. https://www.politico.com/news/2021/01/24/republicans-voter-id-laws-461707.
26. https://www.pewresearch.org/fact-tank/2021/01/28/turnout-soared-in-2020-as-nea rly-two-thirds-of-eligible-u-s-voters-cast-ballots-for-president/.
27. Ezra Klein, *Why We're Polarized* (New York: Avid Reader Press, 2020), 2.
28. Julian Zelizer, *Burning Down the House: Newt Gingrich, the Fall of a Speaker, and the Rise of the New Republican Party* (New York: Penguin, 2020); Lilliana Mason, *Uncivil Agreement: How Politics Became Our Identity* (Chicago: University of Chicago Press, 2018); Steve Kornacki, *The Red and the Blue: The 1990s and the Birth of Political Tribalism* (New York: Harper Collins, 2018); Sam Rosenfeld, *The Polarizers: Postwar Architects of Our Partisan Era* (Chicago: University of Chicago Press, 2017).
29. https://www.nytimes.com/2015/07/12/us/the-best-way-to-vilify-clinton-gop-spe nds-heavily-to-test-it.html.
30. David Brock, *Blinded by the Right: The Conscience of an Ex-Conservative* (New York: Three Rivers Press, 2003).
31. https://www.theguardian.com/commentisfree/2017/nov/08/robert-mercer-offsh ore-dark-money-hillary-clinton-paradise-papers.
32. https://cnnpressroom.blogs.cnn.com/2021/05/13/cnn-special-report-a-radical-rebellion-the-transformation-of-the-gop-a-fareed-zakaria-special/.
33. Robert Draper, *To Start a War: How the Bush Administration Took America into Iraq* (New York: Penguin Press, 2021).
34. Louis Kriesberg, *Realizing Peace: A Constructive Conflict Approach* (New York: Oxford University Press, 2015), 186–232.
35. https://www.nytimes.com/2012/11/08/us/politics/little-to-show-for-cash-flood-by-big-donors.html.
36. https://www.nbcnews.com/id/wbna40007802; https://www.theatlantic.com/polit ics/archive/2017/01/no-one-knows-what-the-powerful-mercers-really-want/ 514529/.
37. Thomas E. Mann and Norman J. Ornstein, *It's Even Worse Than It Looks: How the American Constitutional System Collided with the New Politics of Extremism* (New York: Basic Books, 2012).
38. Heather McGhee, *The Sum of Us* (New York: One World, 2021), 10–15.
39. Louis Kriesberg, "Interactions among Populism, Peace, and Security in Contemporary America," *Sicherheit und Frieden (Security and Peace)* 37, no. 1 (2019): 1–7.
40. https://news.gallup.com/poll/197231/trump-clinton-finish-historically-poor-ima ges.aspx.
41. https://www.fec.gov/resources/cms-content/documents/federalelections2016.pdf.
42. https://edition.cnn.com/election/2016/results/exit-polls/national/president.
43. Thomas Frank, *Listen, Liberal: Or What Ever Happened to the Party of the People* (New York: Metropolitan Books, 2016).
44. Michael Lewis, *The Fifth Risk: Undoing Democracy* (New York: W. W. Norton, 2018).
45. Kriesberg, "Interactions among Populism, Peace, and Security."
46. https://news.gallup.com/poll/203198/presidential-approval-ratings-donald-trump.aspx.
47. Bob Woodward, *Rage* (New York: Simon & Schuster, 2020).

48. https://datausa.io/profile/cip/peace-studies-conflict-resolution#institutions.

49. https://www.scu.edu/ethics/all-about-ethics/top-5-dirty-tricks-in-politics/.

50. https://www.nytimes.com/2016/11/05/us/politics/dirty-tricks-vandalism-and-the-dark-side-of-politics.html.

51. https://www.history.com/this-day-in-history/truman-orders-loyalty-checks-of-fede ral-employees.

52. David. Oshinsky, *A Conspiracy So Immense: The World of Joe McCarthy* (Oxford: Oxford University Press, 2005), 109.

53. Sevmour Martin Lipset, "The Radical Right."

54. https://www.politico.com/story/2016/08/joe-mccarthys-dirty-tricks-upend-sen ate-race-aug-20-1951-2.

55. https://www.senate.gov/about/powers-procedures/investigations/mccarthy-heari ngs/have-you-no-sense-of-decency.htm.

56. Oshinsky, *A Conspiracy So Immense*, 503–4.

57. Everett C. Hughes, "The Professions in Society," in *The Sociological Eye*, ed. Everett C. Hughes (Chicago: Aldine, 1971), 364–73.

58. Colin Barker, "Some Reflections on Student Movements of the 1960s and Early 1970s," *Revista Critical de Ciencias Sociais* 81 (2008): 43–91, https://journals.openedit ion.org/rccs/646.

59. David Frum, *How We Got Here: The '70s.* (New York, NY: Basic Books, 2000); Todd Gittlin, *The Sixties: Years of Hope, Days of Rage* (New York: Bantam, 1993).

60. Robert Cohen, *Freedom's Orator: Mario Savio and the Radical Legacy of the 1960s* (New York: Oxford University Press, 2009).

61. Kirkpatrick Sale, *SDS: The Rise and Development of the Students for a Democratic Society.* (New York: Random House, 1973).

62. Mark Kurlansky, *1968: The Year That Rocked the World* (New York: Random House, 2904).

63. During the 1970 student protests at Syracuse University, some excited students thought they might be more powerful than the earlier protests in Paris.

64. On the basis of some experiences abroad and reports about other countries, it seems that in countries with languages that have formal and familiar forms of address, familiar forms became more widely used after 1968.

65. Rachel Carson, *Silent Spring* (Boston: Houghton Mifflin, 1962).

66. Ralph Nader, *Unsafe at Any Speed: The Designed-In Dangers of the American Automobile* (New York: Grossman, 1965).

67. Benjamin Epstein and Arnold Forster, *The Radical Right: Report on the John Birch Society and Its Allies* (New York: Vintage Books, 1966).

68. https://slate.com/news-and-politics/2018/02/right-wing-conspiracy-theories-from-the-1960s-to-today.html.

69. http://www.unkochmycampus.org/los-ch3-part-3-White-citizens-councils-and-jbs. William Simmons and Louis Hollis were leaders in both JBS and WCC.

70. https://www.huffpost.com/entry/john-birch-society_b_958207.

71. Andrew J. Bacevich, "Secrets That Were No Secret, Lessons That Weren't Learned," *New York Times*, June 13, 2021, SR2.

72. Michael R. Beschloss, *Reaching for Glory: Lyndon Johnson's Secret White House Tapes, 1964–1965* (New York: Simon & Schuster, 2001). In phone calls with senators and associates, Johnson expressed his dismay and lack of options. Nearly always, listeners expressed their sympathy for the problem, but offered no way out.

73. William L. Lunch and Peter W. Sperlich, "American Political Opinion and the War in Vietnam," *Western Political Quarterly* 32, no. 1 (1979): 21–44.

74. https://www.history.com/this-day-in-history/kennedy-proposes-plan-to-end-the-war.

75. The assassin was Sirhan Sirhan, a Palestinian-born, Christian, Jordanian citizen, living in the United States. He was angry about Kennedy's support of Israel.

76. https://www.washingyonpost.com/news/retropolis/wp/2018/08/24/a-party-that-lost-its-mind-in-1968-democrats-held-one-of-historys-most-disastrous-conventions/.

77. https://www.history.com/news/richard-nixon-kent-state-shootings-response.

78. Charles DeBenedetti and Charles Chatfield, *An American Ordeal: The Antiwar Movement of the Vietnam Era* (Syracuse, NY: Syracuse University Press, 1990).

79. https://eastbayyesterday.com/episodes/if-it-takes-a-bloodbath-lets-get-it-over-with/.

80. James Eric Eichsteadt, " 'Shut It Down': The May 1970 National Student Strike at the University of California at Berkeley, Syracuse University, and the University of Wisconsin-Madison" (PhD diss., Syracuse University, 2007), https://surface.syr.edu/hist_etd/7.

81. Carl Bernstein and Bob Woodward, *All the President's Men* (New York: Simon & Schuster., 1974).

82. https://www.washingtonpost.com/wp-srv/national/longterm/watergate/articles/072574-1.htm.

83. https://watergate.info/impeachment/articles-of-impeachment.

84. https://www.pewresearch.org/fact-tank/2019/09/25/how-the-watergate-crisis-eroded-public-support-for-richard-nixon/.

85. https://news.gallup.com/poll/3157/U.S.s-grew-accept-nixons-pardon.aspx.

86. Dunlap. Riley E. and Robert L. Wisniewski, "The Effect of Watergate on Political Party Identification: Results from a 1970–1974 Panel Study," *Sociological Focus* 11, no. 2 (1978): 69–80.

87. Troy A. Zimmer, "The Impact of Watergate on the Public's Trust in People and Confidence in the Mass Media," *Social Science Quarterly* 59, no. 4 (1979): 743–51.

88. https://www.nytimes.com/2022/02/06/opinion/covid-pandemic-policy-trust.html.

Chapter 9

1. Jeffrey Haydu, "Managing 'the Labor Problem' in the United States ca. 1897–1911," in *Intractable Conflicts and Their Transformation*, ed. Louis Kriesberg, T. A. Northrup, and S. J. Thorson (Syracuse, NJ: Syracuse University Press, 1989), 93–106.

2. https://detroithistorical.org/learn/encyclopedia-of-detroit/race-riot-1943.

3. Dominic J. Capeci Jr. and Martha Wilkerson, "The Detroit Rioters of 1943: A Reinterpretation," *Michigan Historical Review* 16, no. 1 (1990): 49–72.

4. Thomas Sugrue, *The Origins of the Urban Crisis* (Princeton, NJ: Princeton University Press, 2014), 7–74.

5. Martin Duberman, *Paul Robeson: A Biography* (New York: Random House, 1989), 364–74, 694–97.

6. William A. Gamson, *Power and Discontent* (Homewood, IL: Dorsey Press, 1968); *The Strategy of Social Protest*, 2nd ed. (Belmont, CA: Wadsworth, 1990); Louis Kriesberg and Bruce W. Dayton, *Constructive Conflicts: From Escalation to Resolution*, 5th ed. (Lanham, MD: Rowman & Littlefield, 2017), 117–46.

7. https://www.thoughtco.com/move-philadelphia-bombing-4175986.

8. Jayne Seminare Docherty, *Learning Lessons from Waco: When the Parties Bring Their Gods to the Negotiation Table* (Syracuse, NY: Syracuse University Press, 2001).

9. Hizkias Assefa and Paul Wahrhaftig, *Extremist Groups and Conflict Resolution: The MOVE Crisis in Philadelphia* (New York: Praeger, 1988).

10. My account is based largely on the PBS program, which produced a film called *Ruby Ridge* in 2021 (see https://www.pbs.org/wgbh/americanexperience/films/ruby-ridge/).

11. https://libertarianinstitute.org/articles/ruby-ridge-siege-forgotten-history-weaver-family-atf-standoff-militia/.

12. Docherty, *Learning Lessons from Waco*.

13. James R. Lewis, ed. *From the Ashes: Making Sense of Waco* (Lanham, MD: Rowman & Littlefield, 1994).

14. Bill Clinton, *My Life* (New York: Alfred A. Knopf, 2004), 497–98.

15. https://www.nbcdfw.com/news/local/25-years-later-the-waco-branch-davidian-raid/221352/; https://www.pbs.org/wgbh/pages/frontline/waco/timeline.html.

16. Clinton, *My Life*, 498–99.

17. https://www.pbs.org/wgbh/pages/frontline/waco/timeline.html.

18. Alexis de Tocqueville, *Democracy in America*, edited and translated by Harvey Claflin Mansfield, and Delba Winthrop (Chicago: University of Chicago Press, 2000).

19. https://cpb-us-e2.wpmucdn.com/sites.uci.edu/dist/5/2530/files/2017/01/Red_Scare_Grade11.pdf.

20. Wyn Craig Wade, *The Fiery Cross: The Ku Klux Klan in America* (New York: Oxford University Press, 1998), 393.

21. https://www.nytimes.com/2021/01/26/us/louis-beam-White-supremacy-internet.html.

22. https://www.adl.org/.

23. Janet Reitman, "State of Denial," *New York Times Magazine*, November 11, 2018, 38–49, 66–68.

24. https://www.washingtonpost.com/news/acts-of-faith/wp/2017/08/14/jews-will-not-replace-us-why-White-supremacists-go-after-jews/.

25. https://www.washingtonpost.com/local/one-dead-as-car-strikes-crowds-amid-protests-of-white-nationalist-gathering-in-charlottesville-two-police-die-in-helicopter-crash/2017/08/13/3590b3ce-8021-11e7-902a-2a9f2d808496_story.html.

26. Murray Edelman, *The Symbolic Uses of Politics* (Urbana: University of Illinois, 1985).

27. https://www.huffpost.com/entry/donald-trump-mark-milley-chiefs-of-staff-elect ion-military-interference_n_5f4ae0f9c5b697186e36fcd1.

28. https://www.nytimes.com/2021/01/06/us/politics/pence-rejects-trumps-pressure- to-block-certification-saying-he-loves-the-constitution.html.

29. https://www.politico.com/news/2020/12/21/trump-house-overturn-election- 449787.

30. Michael Wolff, *Landslide: The Final Days of the Trump White House* (New York: Henry Holt, 2021).

31. https://www.pbs.org/wgbh/frontline/article/several-well-known-hate-groups-ide ntified-at-capitol-riot/.

32. Daniel Bessner and Amber A'Lee Frost, "How the QAnon Cult Stormed the Capitol," *Jacobin*, January 19, 2021, https://jacobinmag.com/2021/01/q-anon-cult-capitol- hill-riot-trump/.

33. https://www.washingtonpost.com/opinions/2021/04/06/capitol-insurrection-arre sts-cpost-analysis/.

Chapter 10

1. William D. Hartung, "Profits of War: Corporate Beneficiaries of the Post-9/11 Pentagon Spending Surge," Brown University Center for International Policy, September 13, 2021, https://watson.brown.edu/costsofwar/files/cow/imce/papers/ 2021/Profits%20of%20War_Hartung_Costs%20of%20War_Sépt%2013%2C%202 021.pdf.

2. Louis Kriesberg, *Realizing Peace: A Constructive Conflict Approach* (New York: Oxford University Press, 2015); Andrew Bacevich, *Washington Rules: America's Path to Permanent War* (New York: Henry Holt, 2010).

3. https://www.politico.com/news/2021/12/27/biden-signs-defense-policy-bill- 526171.

4. Kriesberg, *Realizing Peace*, 2015.

5. Spencer Ackerman, *Reign of Terror: How the 9/11 Era Destabilized America and Produced Trump* (New York: Viking Press, 2021).

6. As a youth in the later 1930s and early 1940s, I did a great deal of hitch-hiking across the country, around Chicago and from coast to coast. Many of the people who gave me lifts or I met in towns on my travels spoke about how bankers controlled the economy and their lives. They were expressing commonsense knowledge, a kind of unschooled Marxism.

7. Nicholas Wapshott, *Samuelson Friedman: The Battle over the Free Market* (New York: W. W. Norton, 2021); Paul Krugman, *Arguing with Zombies: Economics, Politics, and the Fight for a Better Future* (New York: W. W. Norton, 2020).

8. I recall when the older Chicago mayor, Richard J. Daley, was criticized for getting a job for a son. He responded, "Now, what kind of a father would I be, if I didn't help my own son get a job?" That ended the criticism.

9. Jesse Drucker and Danny Hakim, "A Revolving Door Keeps Tax Policy on Clients' Side," *New York Times* September 20, 2021, A1, A15.

10. https://www.jec.senate.gov/public/index.cfm/democrats/2016/6/the-economy-under-democratic-vs-republican-presidents.

11. Dionne, E. J., Jr., Norman J. Orenstein, and Thomas E. Mann, *One Nation after Trump* (New York: St. Martin's Press, 2017).

12. Ben Casselman and Jeanna Smailer, "U.S. Poverty Rate Falls to a Record Low as Aid Helps Offset Job Losses," *New York Times*, September 15, 2021, A1, A13.

13. https://www.hklaw.com/-/media/files/insights/publications/2021/03/americanresc ueplankeyprovisions.pdf?la=en.

14. https://www.vox.com/policy-and-politics/2021/3/6/22315536/stimulus-package-passes-checks-unemployment.

15. https://www.nytimes.com/2021/03/06/business/economy/biden-economy.html.

16. https://www.povertycenter.columbia.edu/news-internal/monthly-pove rty-july-2021.

17. https://www.washingtonpost.com/politics/biden-poised-to-sign-12-trillion-infrast ructure-bill-fulfilling-campaign-promise-and-notching-achievement-that-eluded-trump/2021/11/15/1b69f9a6-4638-11ec-b8d9-232f4afe4d9b_story.html.

18. https://nymag.com/intelligencer/2021/07/5-reasons-biden-got-his-bipartisan-inf rastructure-deal.html.

19. https://www.washingtonpost.com/politics/biden-poised-to-sign-12-trillion-infrast ructure-bill-fulfilling-campaign-promise-and-notching-achievement-that-eluded-trump/2021/11/15/1b69f9a6-4638-11ec-b8d9-232f4afe4d9b_story.html.

20. Christopher Pulliam, "Tax Wealth, Reward Work," Brookings Up Front, (Brookings Institution, July 14, 2021), https://www.brookings.edu/blog/up-front/2021/07/14/ tax-wealth-reward-work/?utm_campaign=Economic%20Studies&utm_medium= email&utm_content=142104627&utm_source=hs_email). Greg Leiserson, "Taxing Wealth," Washington Center for Equitable Growth, 2020, https://equitablegrowth. org/taxing-wealth/.

21. Elizabeth Warren, "Ultra-Millionaire Tax," https://elizabethwarren.com/plans/ultra-millionaire-tax.

22. Robert W. McGee, ed. *The Ethics of Tax Evasion* (Switzerland: Springer, 2011).

23. https://newsroom.ibm.com/2021-08-19-American-Council-on-Education-Valida tes-IBM-Apprenticeship-Program,-Recommends-College-Credit-to-Participants.

24. Suzanne Kahn and Steph Sterling, "Supply-Side Childcare Investments: Policies to Develop an Equitable and Stable Childcare Industry," Roosevelt Institute, 2021, https://rooseveltinstitute.org/wp-content/uploads/2021/07/RI_Childcare_IssueB rief_202108.pdf.

25. Occupational Safety and Health Administration, "Meatpacking: Overview," https:// www.osha.gov/meatpacking.

26. https://www.intheblack.com/articles/2020/10/01/protecting-workers-gig-economy.

27. David Autor, "The Labor Shortage Has Empowered Workers," *New York Times*, September 5, 2021, SR2.

28. Robert Reich, "Is America Experiencing an Unofficial General Strike?," *Guardian*, October 13, 2021, https://www.theguardian.com/prohttps://www.theguardian.com/profile/robert-reichfile/robert-reich.

29. www.dol.gov/general/labortaskforce.

30. https://uniontrack.com/blog/anti-union-legislation.

31. Subodh Mishra, "U.S. Board Diversity Trends in 2019," Harvard Law School Forum on Corporations, 2019.

32. Rebecca Page, *Co-determination in Germany* (Dusseldorf, Germany: Hans-Bockler-Stiftung, 2018).

33. nkofinfo.com/success-story-of-germany-as-a-result-of-co-determination/:~:text = Gradual evaluation of Co-determination.

34. https://www.usworker.coop/what-is-a-worker-cooperative/.

35. https://eos.com/blog/agricultural-cooperatives/. As a college student at the University of Chicago, I lived in a co-op house and found it to be an important educational experience. They exist on many campuses.

36. https://www.bing.com/search?q=rochdale+society+principles&FORM=AFSCVO&PC=AFSC.

37. https://roseferro.com/2021/10/a-rescue-plan-built-on-community-wealth-not-corporate-giveaways/.

38. Progressive Democrats had organized as such beginning in 2004. See Progressive Democrats of America, https://www.bing.com/search?q=+Progressive+Democrats+of+America&FORM=AFSCVO&PC=AFSC.

39. Patrick Adams, "Abortion Pills Kept at Home?" *New York Times*, October 14, 2021, A20.

40. https://thehill.com/homenews/senate/593697-senate-passes-bill-ending-forced-arbitration-in-sexual-misconduct-cases, #3464. Gretchen Carlson's role was widely reported, including on the PBS *NewsHour* program; see "What Ending Forced Arbitration for Sexual Assault Claims Could Mean for Survivors," PBS *NewsHour*, February 10, 2022, https://www.pbs.org/newshour/show/what-ending-forced-arbitration-for-sexual-assault-claims-could-mean-for-survivors.

41. https://www.whitehouse.gov/briefing-room/statements-releases/2022/03/16/fact-sheet-reauthorization-of-the-violence-against-women-act-vawa/.

42. Adam Harris, "The GOP's 'Critical Race Theory' Obsession," *Atlantic*, May 7, 2021, https://www.bing.com/search?q=Adam+Harris++the+Atlantic++GOP%27s+Critical+Race+theory&FORM=AFSCVO&PC=AFSC.

43. https://www.brighthubpm.com/methods-strategies/119172-tips-for-handling-disorderly-conduct-in-work-groups/.

    https://institute.crisisprevention.com/De-Escalation-Tips.html?code=BSIT01DT&src=PPC&utm_source=bing&utm_medium=cpc&utm_campaign=dt_resource202011&utm_content=tofu_gen&msclkid=f76106b7e9dc17545cdf67da08836679.

44. https://www.intelligentliving.co/trees-greenery-reduce-crime-improve-health/.
    https://brockleytree.com/can-large-trees-reduce-crime-in-cities/.

45. https://www.theatlantic.com/ideas/archive/2019/12/historians-clash-1619-project/604093/.

46. https://thehill.com/homenews/administration/534823-trumps-1776-project-relea sed-on-mlk-day-receives-heavy-backlash.

47. Neil MacFarquhar, "Far-Right Organizations Intended to Foment Violence, Victims Say," *New York Times*, October 25, 2021, A13.

48. https://www.theguardian.com/music/2020/mar/18/daryl-davis-black-musician-who-converts-ku-klux-klan-members.

49. https://www.reuters.com/investigates/special-report/usa-trump-georgia-threats/.

50. https://www.reuters.com/investigates/special-report/usa-election-threats-law-enfo rcement/.

51. https://www.washingtonpost.com/s/opinions/2021/09/23/robert-kagan-constitutio nal-crisis/?.

52. Johnathan Hopkin, *Anti-System Politics: The Crisis of Market Liberalism in Rich Democracies* (New York: Oxford University Press, 2021). Also see https://www. newyorker.com/news/daily-comment/why-democracy-is-on-the-decline-in-the-united-states.

53. https://www.nytimes.com/interactive/2022/04/03/magazine/thomas-piktty-interv iew.htm/.

54. https://www.politico.com/story/2018/04/11/ryan-in-interview-im-done-seeking-elected-office-515678.

55. Max Boot, *The Corrosion of Conservatism: Why I Left the Right* (New York: W. W. Norton, 2018); Stuart Stevens, *It Was All a Lie: How the Republican Party Became Donald Trump* (New York: Knopf, 2020).

56. Andrew L. Pieper and Jeff R. DeWitt, eds., *The Republican Resistance: #Nevertrump Conservatives and the Future of the GOP* (Lanham, MD: Lexington Books, 2021. https://en.wikipedia.org/wiki/Never_Trump_movement.

57. https://www.deseret.com/2021/8/25/22580333/the-rise-and-fall-of-the-trumpiest-never-trumper-lincoln-project-steve-schmidt-donald-trump.

58. https://news.gallup.com/poll/203198/presidential-approval-ratings-donald-trump.aspx.

59. https://www.bing.com/search?q=call+for+american+renewal&FORM=AFS CVO&PC=AFSC https://www.cnet.com/how-to/new-party-for-anti-trump-repu blicans-heres-what-is-happening/.

60. Miles Taylor and Christine Todd Whitman, "We Are Republicans. There's Only One Way to Save Our Party from Pro-Trump Extremists," *New York Times*, October 11, 2021, https://www.nytimes.com/2021/10/11/opinion/2022-house-senate-trump.html.

61. https://news.gallup.com/poll/329639/support-third-political-party-high-point.aspx.

62. https://unicornriot.ninja/2021/school-board-disruptions-escalate-funded-by-conse rvative-dark-money/.

63. https://www.nationalpopularvote.com/written-explanation#:~:text=Learn%20M ore%20%20%20Electoral%20votes%20%20,%20%20%20%2048%20more%20r ows%20.

64. https://www.msn.com/en-us/news/politics/the-fight-for-dc-statehood-gets-its-best-chance-yet/ar-BB1fVK30#:~:text=The%20fight%20for%20statehood%20

has%20been%20ongoing%20since,the%20House%20of%20Representatives%20pr
oportionate%20to%20its%20population.%E2%80%9D.

65. https://www.nytimes.com/2021/04/22/us/politics/dc-statehood-vote.html.

66. https://repository.usfca.edu/cgi/viewcontent.cgi?article=1205&context=usfla
wreview.
https://repository.uchastings.edu/cgi/viewcontent.cgi?article=3495&context=
hastings_law_journal.

67. Charlie Savage, "Supreme Court Panel Shows Interest in Calls for Judges' Term
Limits," *New York Times*, November 19, 2021, A20.

68. Dino P. Christenson and Douglas L. Kriner, *The Myth of the Imperial Presidency: How
Public Opinion Checks the Unilateral Executive* (Chicago: University of Chicago Press,
2020).https://www.politico.com/magazine/story/2016/04/barack-obama-gop-most-
powerful-213814/.

69. Nick Corasaniti, Reid J. Epstein, Taylor Johnston, Rebecca Lieberman, and Eden
Weingart, "How Maps Reshape American Politics," *New York Times*, November 12,
2021,   https://www.nytimes.com/interactive/2021/11/07/us/politics/redistricting-
maps-explained.html.

70. https://navigatingthroughturbulence.com/.

71. Lee Rainie, Scott Keeter, and Andrew Perrin, "Trust and Distrust in America," Pew
Research Center, July 22, 2019, https://www.pewresearch.org/politics/2019/07/22/
trust-and-distrust-in-america/.

72. https://www.pewresearch.org/politics/2021/05/17/public-trust-in-government-
1958-2021/.

73. Tom W. Smith and Jaesok Son, "Trends in Public Attitudes about Confidence in
Institutions," National Opinion Research Center, University of Chicago, 2013.

74. https://www.newsmax.com/us/gallup-poll-americans-politicians/2021/10/07/id/
1039510/.

75. https://www.pewresearch.org/politics/2021/05/17/public-trust-in-government-
1958-2021/.

76. https://news.gallup.com/poll/352316/americans-confidence-major-institutions-
dips.aspx.

77. https://www.mediamatters.org/fox-news/30-reasons-why-fox-news-not-legit.

78. https://www.pewresearch.org/journalism/fact-sheet/network-news/.

79. https://www.pewresearch.org/journalism/fact-sheet/network-news/.

80. https://www.bing.com/search?q=PEW+PUBLIC+BROADCASTING+FACT+
SHEET&FORM=AFSCVO&PC=AFSC.

81. MoveOn.org has a petition supporting reestablishment of a fairness doctrine.

82. https://www.congress.gov/bill/116th-congress/house-bill/4401?s=1&r=1saddam82.

83. For example, in 2013, the *Washington Post* was sold to Nash Holdings, a holding company established by Jeff Bezos, for $250 million.

84. https://niemanreports.org/articles/4-ways-to-fund-and-save-journalism/.

85. https://www.psychologytoday.com/us/blog/cultural-psychiatry/202011/how-social-
media-algorithms-inherently-create-polarization.

86. Michael D'Antonio, *The Hunting of Hillary: The Forty-Year Campaign to Destroy Hillary Clinton* (New York: Thomas Dunne Books, 2020).

87. https://www.newyorker.com/magazine/2021/08/09/the-big-money-behind-the-big-lie?utm_source=onsite-share&utm_medium=email&utm_campaign=onsite-share&utm_brand=the-new-yorker.

88. https://www.vox.com/2021/1/6/22218058/republicans-objections-election-results.

89. https://www.insider.com/capitol-rioters-who-pleaded-guilty-updated-list-2021-5#:~:text=Jon%20Schaffer%2C%20the%20guitarist%20and%20founder%20of%20the,their%20attack%20on%20the%20Capitol%20weeks%20in%20advance.

90. https://www.nytimes.com/2021/11/23/us/politics/jan-6-proud-boys-oathkeepers.html.

91. Rebecca Davis O'Brien, "Man Who Threatened Democrats Is Sentenced to 19 Months in Prison," *New York Times*, November 23, 2021, A17.

92. https://www.buzzfeednews.com/article/kadiagoba/gosar-aoc-censure-house-anime-twitter-biden.

93. https://slate.com/news-and-politics/2021/11/trump-giuliani-ellis-sued-defamation-pennsylvania.html.

94. https://www.nytimes.com/2021/11/15/us/politics/alex-jones-sandy-hook.html) and https://www.theguardian.com/us-news/2021/nov/15/alex-jones-liable-damages-sandy-hook-school-shooting-connecticut.

95. For example, immediately after the 2020 election, a Trump lawyer spliced together video footage that purported to show two Georgia election workers adding 18,000 fraudulent ballots. This was investigated and found by election officials in Georgia to be untrue. Nevertheless, a right-wing website, the Gateway Pundit, continued the false accusations. Trump, Rudolph Giuliani, and others also continued the allegations. The two election workers were severely harassed by phone calls and death threats; they filed a defamation suit against the owners of the website, the Hoft brothers, in December 2021; see https://www.nytimes.com/2021/12/02/us/politics/gateway-pundit-defamation-lawsuit.html.

96. Neil MacFarquhar, "Far-Right Organizations Intended to Foment Violence, Victims Say," *New York Times*, October, 25, 2021, A13.

97. Michael J. Sandel, *The Tyranny of Merit: What's Become of the Common Good?* (Stuttgart, Germany: Farrar, Straus and Giroux, 2020).

98. Roger Fisher and William Ury, *Getting to Yes* (New York: Penguin, 1981)

99. Louis Kriesberg, Terrell A. Northrup, and Stuart J. Thorson, eds., *Intractable Conflicts and Their Transformation* (Syracuse, NY: Syracuse University Press, 1989); Chester A. Crocker, Fen Osler Hampson, and Pamela R. Aall, *Taming Intractable Conflicts: Mediation in the Hardest Cases* (Washington, DC: United States Institute of Peace Press, 2004); Miriam F. Elman et al., eds., *Overcoming Intractable Conflicts: New Approaches to Constructive Transformations* (London: Rowman & Littlefield, 2019).

References

"50 Years of Restoration: Building Assets and Community Wealth." *Brooklyn Reader*, December 29, 2017. https://www.bkreader.com/2017/12/29/50-years-restoration-building-assets-community-wealth/.

Ackerman, Spencer. *Reign of Terror: How the 9/11 Era Destabilized U.S. and Produced Trump*. New York: Viking Press, 2021.

Adair, Bill. "The Peace Dividend Began with a Bush." *Politifact*, January 24, 2008. https://www.politifact.com/truth-o-meter/statements/2008/jan/24/rudy-giuliani/the-peace-dividend-began-with-a-bush/.

Adams, Patrick. "Abortion Pills Kept at Home?" *New York Times*, October 14, 2021, A20.

Alinsky, Saul. *Reveille for Radicals*. New York: Vintage, 1946.

Alinsky, Saul D. *Rules for Radicals*. New York: Random House, 1971.

Anderson, Benedict. *Imagined Communities: Reflections on the Origin and Spread of Nationalism*. Revised ed. London: Verso, 1991.

Arias, Elizabeth, and Jiaquan Xu. "United States Life Tables, 2017." *National Vital Statistics Reports*. Washington, DC: National Vital Statistics System, 2019.

Armstrong, Elizabeth A., and Suzanna M. Crage. "Movements and Memory: The Making of the Stonewall Myth." *American Sociological Review* 71, no. 5 (2006): 724–51.

Arnesen, Eric. "A. Philip Randolph: Labor and the New Black Politics." In *The Human Tradition in the Civil Rights Movement*, edited by Susan M. Glisson, 80–82. Lanham, MD: Rowman & Littlefield, 2006.

Arsenault, Raymond. *Freedom Riders: 1961 and the Struggle for Racial Justice*. New York: Oxford University Press, 2006.

Assefa, Hizkias, and Paul Wahrhaftig. *Extremist Groups and Conflict Resolution: The MOVE Crisis in Philadelphia*. New York: Praeger, 1988.

The Associated Press-NORC Center for Public Affairs Research and the General Social Survey (GSS). "Issue Brief: Changing Attitudes about Racial Inequality." Chicago: AP-NORC Center, 2019. https://apnorc.org/projects/changing-attitudes-about-racial-ine quality/.

Autor, David. "The Labor Shortage Has Empowered Workers." *New York Times*, September 5, 2021, SR2.

Bacevich, Andrew J. "Secrets That Were No Secret, Lessons That Weren't Learned." *New York Times*, June 13, 2021, SR2.

Bacevich, Andrew J. *Washington Rules: America's Path to Permanent War*. New York: Metropolitan Books, 2010.

Baker, Bruce, Danielle Farrie, Theresa Luhm, and David G. Sciarra. "Is School Funding Fair? A National Report Card." In *National Report Card*. Newark, NJ: Education Law Center, and Rutgers University Graduate School of Education, 2016. https://edlawcen ter.org/assets/files/pdfs/publications/National_Report_Card_2016.pdf.

Barker, Colin. "Some Reflections on Student Movements of the 1960s and Early 1970s." *Revista Critical de Ciencias Sociais* 81 (2008): 43–91. https://journals.openedition.org/rccs/646).

Barnard, John. *American Vanguard: The United Auto Workers during the Reuther Years, 1935–1970*. Detroit: Wayne State University Press, 2004.

Baum, Dan. "Legalize It All: How to Win the War on Drugs." *Harper's Magazine*, April 2016. https://harpers.org/archive/2016/04/legalize-it-all/.

Bernstein, Carl, and Bob Woodward. *All the President's Men*. New York: Simon & Schuster, 1974.

Beschloss, Michael R. *Reaching for Glory: Lyndon Johnson's Secret White House Tapes, 1964–1965*. New York: Simon & Schuster, 2001.

Bessner, Daniel, and Amber A'Lee Frost. "How the QAnon Cult Stormed the Capitol." *Jacobin*, January 19, 2021. https://jacobinmag.com/2021/01/q-anon-cult-capitol-hill-riot-trump/.

Bivens, Josh, and Heidi Shierholz. *What Labor Market Changes Have Generated Inequality and Wage Suppression?* Washington, DC: Economic Policy Institute, 2018. https://www.epi.org/publication/what-labor-market-changes-have-generated-inequality-and-wage-suppression-employer-power-is-significant-but-largely-constant-whereas-workers-power-has-been-eroded-by-policy-actions/.

Blinder, Alan. "Southern Poverty Law Center President Plans Exit amid Turmoil." *New York Times*, March 22, 2019.

Boehm, Eric. "After the Supreme Court Said Unions Can't Force Non-members to Pay Dues, Almost All of Them Stopped." *Reason: Free Minds and Free Markets*, April 9, 2019. https://reason.com/2019/04/09/janus-211000-workers-fled-seiu-afscme.

Bogenschneider, Karen, and Carol Johnson. "Family Involvement in Education." Madison, WI: Policy Institute for Family Impact Seminars, 2015. https://www.purdue.edu/hhs/hdfs/fii/wp-content/uploads/2015/07/s_wifis20c02.pdf.

Boot, Max. *The Corrosion of Conservatism: Why I Left the Right*. New York: W. W. Norton, 2018.

Boserup, Brad, Mark McKenney, and Adel Elkbuli. "Alarming Trends in US Domestic Violence during the COVID-19 Pandemic." *American Journal of Emergency Medicine* 38, no. 12 (2020): 2753–55.

Bottari, Mary. "The Two Faces of Janus." *In These Times*, March 2018, 18–27.

Boucher, Vincent, Charles-Philippe David, and Karine Prémont. *National Security Entrepreneurs and the Making of American Foreign Policy*. Montreal: McGill-Queen's University Press 2021.

Boulding, Elise. *The Underside of History: A View of Women through Time*. Boulder, CO: Westview Press, 1976.

Bowman, Cynthia Grant. "Women in the Legal Profession from the 1920s to the 1970s: What Can We Learn from Their Experience about Law and Social Change?" *Cornell Law Faculty Publications 12* (2009). http://scholarship.law.cornell.edu/facpub/12.

Boyle, Kevin. *The UAW and the Heyday of American Liberalism, 1945–1968*. Ithaca, NY: Cornell University Press, 1995.

Brady, David, and Kent L. Tedin. "Ladies in Pink: Religion and Political Ideology in the Anti-ERA Movement." *Social Science Quarterly* 56, no. 4 (1976): 564–75.

Brady, Marion. *What's Worth Learning?* Charlotte, NC: Information Age, 2010.

Brock, David. *Blinded by the Right: The Conscience of an Ex-Conservative*. New York: Three Rivers Press, 2003.

Bundy, William. *A Tangled Web: The Making of Foreign Policy in the Nixon Presidency*. New York: Hill and Wang, 1998.

Busch, Andrew E. *Truman's Triumphs: The 1948 Election and the Making of Postwar America*. Lawrence: University Press of Kansas, 2012.

Byrd, Jessica. "A Different Direction for Black Politics." *New York Times*, September 4, 2020, A27.

Callahan, David. "Systemic Failure: Four Reasons Philanthropy Keeps Losing the Battle against Inequality." *Inside Philanthropy*, August 22, 2018. https://www.insidephilanthr opy.com/home/2018/1/10/systemic-failure-four-reasons-philanthropy-keeps-losing-the-battle-against-inequality.

Canning, Elizabeth A., Katherine Muenks, Dorainne J. Green, and Mary C. Murphy. "Faculty Who Believe Ability Is Fixed Have Larger Racial Achievement Gaps and Inspire Less Student Motivation in Their Classes." *Science Advances* 5, no. 2 (2019).https://advances.sciencemag.org/content/5/2/eaau4734.

Capeci, Dominic J., Jr., and Martha Wilkerson. "The Detroit Rioters of 1943: A Reinterpretation." *Michigan Historical Review* 16, no. 1 (1990): 49–72.

Carey, Alex. "The Hawthorne Studies: A Radical Criticism." *American Sociological Review* 32, no. 3 (1967): 403–16.

Carey, Kevin. "A Detailed Look at the Downsides of California's Ban on Affirmative Action." *New York Times*, August 21, 2020. https://www.nytimes.com/2020/08/21/ups hot/00up-affirmative-action-california-study.html.

Caro, Robert. *The Power Broker*. New York: Vintage, 1974.

Carson, Rachel. *Silent Spring*. Boston: Houghton Mifflin, 1962.

Cashin, Sheryll. *The Failures of Integration: How Race and Class Are Undermining the American Dream*. New York: Public Affairs, 2004.

Casselman, Ben, and Jeanna Smailer. "U.S. Poverty Rate Falls to a Record Low as Aid Helps Offset Job Losses." *New York Times*, September 15, 2021, A1, A13.

Cassidy, John. "An Inconvenient Truth: It Was George W. Bush Who Bailed out the Automakers." *New Yorker*, March 15, 2012. https://www.newyorker.com/news/john-cassidy/an-inconvenient-truth-it-was-george-w-bush-who-bailed-out-the-aut omakers.

Chalmers, David. *Backfire: How the Ku Klux Klan Helped the Civil Rights Movement*. Lanham, MD: Rowman & Littlefield, 2003.

Christenson, Dino P., and Douglas L. Kriner. *The Myth of the Imperial Presidency: How Public Opinion Checks the Unilateral Executive*. Chicago: University of Chicago Press, 2020.

Clinton, Bill. *My Life*. New York: Alfred A. Knopf, 2004.

Cloward, Richard A., and Lloyd E. Ohlin. *Delinquency and Opportunity: A Theory of Delinquency Gangs*. Glencoe, IL: Free Press, 1960.

Cohen, Robert. *Freedom's Orator: Mario Savio and the Radical Legacy of the 1960s*. New York: Oxford University Press, 2009.

Coleman, Peter T. *The Way Out: How to Overcome Toxic Polarization*. New York: Columbia University Press, 2021.

Collins, Randall. *Conflict Sociology*. New York: Academic Press, 1975.

Comaroff, John L. "Humanity, Ethnicity, Nationality: Conceptual and Comparative Perspectives on the U.S.S.R." *Theory and Society* 20 (1991): 661–87.

Community Relations Service. "Annual Report 2018." Washington, DC: Community Relations Service, Department of Justice, 2018.

Confessore, Nicholas, Sarah Cohen, and Karen Yourish. "Small Pool of Rich Donors Dominates Election Giving." *New York Times*, August 1, 2015. https://www.nytimes.

com/2015/08/02/us/small-pool-of-rich-donors-dominates-election-giving.html?refe
rringSource=articleShare.

Conser, Walter H., Jr., Donald M. McCarthy, David J. Toscano, and Gene Sharp, eds. *Resistance, Politics, and the American Struggle for Independence, 1765–1775.* Boulder, CO: Lynne Rienner, 1986.

Cooper, Alexia, and Erica L. Smith. *Homicide Trends in the United States, 1980–2008: Annual Rates for 2009 and 2010.* Washington, DC: US Department of Justice Office of Justice Programs, Bureau of Justice Statistics, 2011.

Corasaniti, Nick, Reid J. Epstein, Taylor Johnston, Rebecca Lieberman, and Eden Weingart. "How Maps Reshape American Politics." *New York Times,* November 12, 2021, A18–A19. https://www.nytimes.com/interactive/2021/11/07/us/politics/redist ricting-maps-explained.html.

Cornwall, Zoe. *Human Rights in Syracuse: Two Memorable Decades.* Syracuse, NY: Human Rights Commission of Syracuse and Onondaga County, 1989.

Cowan, Tadlock. "Military Base Closures: Socioeconomic Impacts." Washington. DC: Congressional Research Service, 2012. https://fas.org/sgp/crs/natsec/RS22147.pdf.

Coy, Patrick G, and Lynne M. Woehrle, eds. *Social Conflicts and Collective Identities.* Lanham, MD: Rowman & Littlefield, 2000.

Crenshaw, Kimberle. "Demarginalizing the Intersection of Race and Sex: A Black Feminist Critique of Antidiscrimination Doctrine, Feminist Theory and Antiracist Politics." *University of Chicago Legal Forum* 1 (1989): 139–67.

Critchlow, Donald T., and Cynthia L. Stacheck. "The Equal Rights Amendment Reconsidered: Politics, Policy, and Social Mobilization in a Democracy." *Journal of Policy History* 20, no. 1 (2008): 157–76.

Crocker, Chester A., Fen Osler Hampson, and Pamela R. Aall. *Taming Intractable Conflicts: Mediation in the Hardest Cases.* Washington, DC: United States Institute of Peace Press, 2004.

Cruse, Harold. *Plural but Equal: A Critical Study of Blacks and Minorities and America's Plural Society.* New York: William Morrow, 1987.

Cumings, Bruce. *The Origins of the Korean War.* Vol. 2, *The Roaring of the Cataract, 1947–1950.* Princeton: Princeton University Press, 1990.

Cutcher-Gershenfeld, Joel, Dan Brook, and Martin Mulloy. "The Decline and Resurgence of the U.S. Auto Industry." Washington, DC: Economic Policy Institute, May 6, 2015. https://Www.Epi.Org/Publication/the-Decline-and-Resurgence-of-the-U-S-Auto-Industry/.

D'Antonio, Michael. *The Hunting of Hillary: The Forty-Year Campaign to Destroy Hillary Clinton.* New York: Thomas Dunne Books, 2020.

Dahrendorf, Ralf. *Class and Class Conflict in Industrial Society.* Stanford, CA: Stanford University Press, 1959.

Darity, William, Jr., Darrick Hamilton, Mark Paul, Alan Aja, Anne Price, Antonio Moore, and Caterina Chiopris. *What We Get Wrong about Closing the Racial Wealth Gap.* Durham, NC: Samuel DuBois Cook Center on Social Equity Insight Center for Community Economic Development, 2018.

Daynes, Byron W., and Raymond Tatalovich. "Presidential Politics and Abortion, 1972–1988." *Presidential Studies Quarterly* 22, no. 3 (1992): 545–61.

Dayton, Bruce W., and Louis Kriesberg. *Constructive Conflicts: From Emergence to Transformation.* 6th ed. Lanham, MD: Rowman & Littlefield, 2022.

De Palma, Claudia. "A Decade after the 2008 Foreclosure Crisis, Northwest Philadelphia Is Fighting Back." *Montgomery News*, July 24, 2019. https://voicesforciviljustice.org/organization/26729/montgomery-news/?source_geo_paged=2.

DeBenedetti, Charles, and Charles Chatfield. *An American Ordeal: The Antiwar Movement of the Vietnam Era*. Syracuse, NY: Syracuse University Press, 1990.

DeCrow, Karen. *Sexist Justice: How Legal Sexism Affects You*. New York: Random House, 1974.

Deming, D. "Early Childhood Intervention and Life-Cycle Skill Development: Evidence from Head Start." *American Economic Journal: Applied Economics* 1, no. 3 (2009): 111–34.

DeSilver, Drew. "Global Inequality: How the U.S. Compares." Washington, DC: Pew Research Center, December 19, 2013. http://www.pewresearch.org/fact-tank/2013/12/19/global-inequality-how-the-u-s-compares.

Desmond, Matthew. *Evicted: Poverty and Profit in the American City*. New York: Crown Publishers, 2016.

Deutscher, Irwin, and Elizabeth J. Thompson, eds. *Among the People: Encounters with the Poor*. New York and London: Basic Books, 1968.

DiAngelo, Robin *White Fragility: Why It's So Hard for White People to Talk about Racism* Boston: Beacon Press, 2018.

Dillon, Erin, and Andy Rotherham. "States' Evidence: What It Means to Make 'Adequate Yearly Progress' under NCLB." *Education Sector*, December 7, 2007. http://www.educationsector.org/research/research_show.htm?doc_id=511096.

Dionne, E. J., Jr., Norman J. Orenstein, and Thomas E. Mann. *One Nation after Trump*. New York: St. Martin's Press, 2017.

Docherty, Jayne Seminare. *Learning Lessons from Waco: When the Parties Bring Their Gods to the Negotiation Table*. Syracuse, NY: Syracuse University Press, 2001.

Domhoff, G. William. *The Powers That Be: Processes of Ruling Class Domination in America*. New York: Vintage Books, 1978.

Donnelly, Ellen A., and John M. MacDonald. "The Downstream Effects of Bail and Pretrial Detention on Racial Disparities in Incarceration." *Journal of Criminal Law and Criminology* 108, no. 4 (2018). https://scholarlycommons.law.northwestern.edu/jclc/vol108/iss4/4).

Downs, Anthony. "Up and Down with Ecology—the 'Issue Attention Cycle.'" *Public Interest* 28 (1972): 38–50.

Drake, Laura F., and William A. Donohue. "Communicative Framing Theory in Conflict Resolution." *Communication Research* 23, no. 3 (1996): 297–322.

Draper, Robert. *Dead Certain: The Presidency of George W. Bush*. New York: Free Press, 2007.

Draper, Robert. *To Start a War: How the Bush Administration Took America into Iraq*. New York: Penguin Press, 2021.

Driver, Justin. *The Schoolhouse Gate: Public Education, the Supreme Court, and the Battle for the American Mind*. New York: Pantheon, 2018.

Drucker, Jesse, and Danny Hakim. "A Revolving Door Keeps Tax Policy on Clients' Side." *New York Times*, September 20, 2021, A1, A15.

Drucker, Jesse, and Eric Lipton. "How a Trump Tax Break to Help Poor Communities Became a Windfall for the Rich." *New York Times*, August 31, 2019. https://www.nytimes.com/2019/08/31/business/tax-opportunity-zones.html.

Du Bois, W.E.B. *The Souls of Black Folk*. Chicago: A.C. McClurg, 1903.

Duberman, Martin. *Paul Robeson: A Biography*. New York: Random House, 1989.

Dunlap. Riley E., and Robert L. Wisniewski. "The Effect of Watergate on Political Party Identification: Results from a 1970–1974 Panel Study." *Sociological Focus* 11, no. 2 (1978): 69–80.

Edelman, Murray. *The Symbolic Uses of Politics.* Urbana: University of Illinois, 1985.

Edwards, Harry. *The Revolt of the Black Athlete.* 50th Anniversary Edition. Urbana: University of Illinois Press, 2017.

Eichsteadt, James Eric. "'Shut It Down': The May 1970 National Student Strike at the University of California at Berkeley, Syracuse University, and the University of Wisconsin-Madison." PhD diss., History, Syracuse University, 2007. https://surface.syr.edu/hst_etd/7.

Elliott, W. A. *Us and Them: A Study of Group Consciousness.* Aberdeen: Aberdeen University Press, 1986.

Elman, Miriam F., Catherine Gerard, Galia Golan, and Louis Kriesberg, eds. *Overcoming Intractable Conflicts: New Approaches to Constructive Transformations.* London: Rowman & Littlefield, 2019.

Elngo, Sneha, Jorge Luis Garcia, Heckman James J, and Andres Hojman. "Early Childhood Education." NBER Working Paper 21766. Cambridge, MA: National Bureau of Economic Research, 2015. https://www.nber.org/papers/w21766.pdf).

Epstein, Benjamin, and Arnold Forster. *The Radical Right: Report on the John Birch Society and Its Allies.* New York: Vintage Books, 1966.

Erickson, David J. *The Housing Policy Revolution: Networks and Neighborhoods.* Washington, DC: Urban Institute Press, 2009.

Falk, Gene, and Karen Spar. *Poverty: Major Themes in Past Debates and Current Proposals.* Washington, DC: Congressional Research Service, 2014.

Farber, Henry S., Daniel Herbst, Ilyana Kuziemko, and Suresh Naidu. *Unions and Inequality over the Twentieth Century: New Evidence from Survey Data.* Cambridge, MA: National Bureau of Economic Research, 2018.

Federal Reserve Bank of St. Louis. "Foreign Automobile Sales in the United States." St. Louis, MO: Federal Reserve Bank of St. Louis, 1970. https://fraser,stlouisfed.org/title/960/item/37804/loc/174839.

Feenberg, Daniel R., and James M. Poterba. "Income Inequality and the Incomes of Very High-Income Taxpayers: Evidence from Tax Returns." Tax Policy and the Economy 7 (1993). https://www.journals.uchicago.edu/doi/pdfplus/10.1086/tpe.7.20060632.

Feldman, Noah. *The Broken Constitution: Lincoln, Slavery, and the Refounding of America.* New York: Farrar, Straus & Giroux, 2021.

Finer, Lawrence B., Lori F. Frohwirth, Lindsay A. Dauphinee, Susheela Singh, and Ann M. Moore. "Reasons U.S. Women Have Abortions: Quantitative and Qualitative Perspectives." *Perspectives on Sexual and Reproductive Health* 37, no. 3 (2007): 110–18.

Fischer, Martina, Joachim Giessmann, and Beatrix Schmelzle, eds. *Berghof Handbook for Conflict Transformation.* Farmington Hills, MI: Barbara Budrich Publishers, 2011.

Fisher, Roger, William Ury, and Bruce Patton. *Getting to Yes: Negotiating Agreement without Giving In.* 2nd ed. New York: Penguin, 1991.

Follett, Mary P. *Freedom and Co-ordination: Lectures in Business Organization.* New York: Management Publications Trust Limited, 1949.

Foner, Eric. *Reconstruction, America's Unfinished Revolution, 1863-1877.* New York: HarperCollins, 1988.

Ford, Tiffany, Sarah Reber, and Richard V. Reeves. "Race Gaps in COVID-19 Deaths Are Even Bigger Than They Appear." Washington, DC: Brookings, 2020. https://www.

brookings.edu/blog/up-front/2020/06/16/race-gaps-in-covid-19-deaths-are-even-big
ger-than-they-appear/.

Frank, Thomas. *Listen, Liberal: Or What Ever Happened to the Party of the People*. New York: Metropolitan Books, 2016.

Franken, Al. *Rush Limbaugh Is a Big Fat Idiot and Other Observations*. New York: Delacorte Press, 1996.

Frankenberg, Erica, Jongyeon Ee, Jennifer B. Ayscue, and Gary Orfield. *Harming Our Common Future: America's Segregated Schools 65 Years after Brown*. Los Angeles: Civil Rights Project, University of California, 2019. www.civilrightsproject.ucla.edu.

Franklin, Shirley, and David Edwards. *It Takes a Neighborhood: Purpose Built Communities and Neighborhood Transformation*. Purpose Built Communities, 2020. https://purp osebuiltcommunities.org/takes-neighborhood-purpose-built-communities-neigh borhood-transformation.

Freeman, Jo., ed. *Women: A Feminist Perspective*, Palo Alto, CA: Mayfield, 1975.

Friedan, Betty. *The Feminine Mystique*. New York: W. W. Norton, 1963.

From, Al. *The New Democrats and the Return to Power*. New York: St. Martin's, 2013.

Frum, David. *How We Got Here: The '70s*. New York: Basic Books, 2000.

Fuchs, Hailey. "Law to Reduce Crack Cocaine Sentences Is No Help for Some Inmates." *New York Times*, August 2, 2020, 18.

Furman, Jason. "How Obama Has Narrowed the Income Inequality Gap." *Washington Post*, September 23, 2016. http://wapo.st/2dg6njf?tid=ss_mail&utm_term=.9655f 73451c5.

Gabbidon, Shaun L. *Criminological Perspectives on Race and Crime*. 4th ed. New York: Routledge, 2020.

Gamson, William A. *Power and Discontent*. Homewood, IL: Dorsey Press, 1968.

Gamson, William A. *The Strategy of Social Protest*. 2nd ed. Belmont, CA: Wadsworth, 1990.

Ganley, Joseph V. "City Whipping Racial Housing Woes." *Herald American*, September 22, 1963, 1, 13.

Gelb, Leslie H. "Foreign Affairs; What Peace Dividend?" *New York Times*, February 21, 1992, A31.

Gerard, Catherine, and Louis Kriesberg, eds. *Conflict and Collaboration: For Better or Worse*. Abingdon, UK: Routledge, 2018.

Gilligan, Carol. "In a Different Voice: Women's Conceptions of Self and of Morality." *Harvard Educational Review* 47, no. 4 (1977): 481–51.

Gillon, Steven M. *Separate and Unequal: The Kerner Commission and the Unraveling of American Liberalism*. New York: Basic Books, 2018.

Ginsburg, Ruth Bader. "Women's Work: The Place of Women in Law Schools." *Journal of Legal Education* 32, no. 2 (1982): 272–75.

Gittlin, Todd. *The Sixties: Years of Hope, Days of Rage*. New York: Bantam, 1993.

Glass, Andrew. "Truman Desegregates Armed Forces on Feb. 2, 1948." *Politico*, February 2, 2008, https://www.politico.com/story/2008/02/truman-desegregates-armed-forces- on-feb-2-1948-008258.

Goldin, Claudia, and Katz Lawrence. *The Race between Education and Technology*. Cambridge, MA: Harvard University Press, 2008.

Gonda, Jeffrey D. *Unjust Deeds: The Restrictive Covenant Cases and the Making of the Civil Rights Movement*. Chapel Hill: University of North Carolina Press, 2015.

Green, Emma. "Are Jews White?" *Atlantic*, December 5, 2016. https://www.theatlantic. com/politics/archive/2016/12/are-jews-White/509453/.

Greenblatt, Alan. "Alec Enjoys a New Wave of Influence and Criticism." *Governing*, November 29, 2011. https://www.governing.com/topics/politics/ALEC-enjoys-new-wave-influence-criticism.html.

Guerrero, George. "Social Disorganization Theory." In *Encyclopedia of Race and Crime*, edited by Shaun L. Gabbidon and Helen T. Greene, 762–63. Thousand Oaks, CA: SAGE, 2009.

Gurr, Ted Robert. *Why Men Rebel.* Princeton, NJ: Princeton University Press, 1970.

Hamilton, Jack. *Just Around Midnight: Rock and Roll and the Racial Imagination.* Cambridge, MA: Harvard University Press, 2016.

Hannah-Jones, Nikole. "It Was Never about Busing." *New York Times*, July 14, 2019, 6–7.

Hannah-Jones, Nikole. "What Is Owed." *New York Times Magazine*, June 28, 2020, 30–35, 47–53.

Hansen, Michael, Elizabeth Mann Levesque, Jon Valant, and Diana Quintero. *2018 Brown Center Report on American Education: Trends in NAEP Math, Reading, and Civics Scores.* Washington, DC: Brookings Institution, 2018. https://www.brookings.edu/research/2018-brown-center-report-on-american-education-trends-in-naep-math-reading-and-civics-scores/.

Harper, Sam, Richard F. MacLehose, and Jay S. Kaufman. "Trends in the Black-White Life Expectancy Gap among US States, 1990–2009." *Health Affairs* 33, no. 8 (2014). https://www.healthaffairs.org/doi/full/10.1377/hlthaff.2013.1273.

Harrington, Michael. *The Other America: Poverty in the United States.* New York: Macmillan, 1962.

Harris, Adam "The GOP's 'Critical Race Theory' Obsession." *Atlantic*, May 7, 2021. https://www.bing.com/search?q=Adam+Harris++the+Atlantic++GOP%27s+Critical+Race+theory&FORM=AFSCVO&PC=AFSC.

Hartung, William D. *Profits of War: Corporate Beneficiaries of the Post-9/11 Pentagon Spending Surge.* Providence, RI: Brown University Center for International Policy, September 13, 2021. https://watson.brown.edu/costsofwar/files/cow/imce/papers/2021/Profits%20of%20War_Hartung_Costs%20of%20War_Sept%2013%2C%202021.pdf.

Haydu, Jeffrey. "Managing 'the Labor Problem' in the United States ca. 1897–1911." In *Intractable Conflicts and Their Transformation*, edited by Louis Kriesberg, T. A. Northrup and S. J. Thorson, 93–106. Syracuse, NY: Syracuse University Press, 1989. http://www.historycommons.org/context.jsp?item=a050270rhodeseradicate#a050270rhodeseradicate.

Herbold, Hilary. "Never a Level Playing Field: Blacks and the GI Bill." *Journal of Blacks in Higher Education* 6 (Winter 1994): 104–8.

Herndon, Astead W., and Dionne Searcey. "White Voters' Views on Race Show Change." *New York Times*, June 28, 2020, 24–25.

Hertel-Fernandex, Alex. *State Capture: How Conservative Activists, Big Businesses, and Wealthy Donors Reshaped the American States—and the Nation.* New York: Oxford University Press, 2019.

Hertel-Fernandez, Alexander, Caroline Tervo, and Theda Skocpol. "How the Koch Brothers Built the Most Powerful Rightwing Group You've Never Heard Of." *Guardian*, September 26, 2018. https://www.theguardian.com/us-news/2018/sep/26/koch-brothers-americans-for-prosperity-rightwing-political-group.

Hill Collins, Patricia. *Black Feminist Thought: Knowledge, Consciousness, and the Politics of Empowerment.* New York: Routledge, 2000.

Hochschild, Arlie Russell. *Strangers in Their Own Land: Anger and Mourning on the American Right*. New York: New Press, 2016.

Hofacker, Paul W. "The Elevation of the Elite: Historical Trends and Complicity of the Masses." *Public Organization Review* 5 (2005): 3–33. https://doi.org/10.1007/s11 115-004-6132-6.

Hopkin, Johnathan. *Anti-System Politics: The Crisis of Market Liberalism in Rich Democracies*. New York: Oxford University Press, 2021.

https://en.wikipedia.org/wiki/Martin_Luther_King_Jr._Day.

https://people.com/archive/bryan-stevenson-vol-44-no-22/.

https://theconversation.com/african-americans-economic-setbacks-from-the-great-re-cession-are-ongoing-and-could-be-repeated-109612.

https://web.archive.org/web/20070715212213/http://www.ussc.gov/r_congress/02crack/2002crackrpt.pdf.

https://www.bloomberg.com/news/articles/2016-01-04/how-zoning-restrictions-make-segregation-worse.

https://www.documentcloud.org/documents/1283358-school-desegregation-in-boston-a-staff-report.html. "School Desegregation in Boston: A Staff Report Prepared for the Hearing the U S Commission on Civil Rights June 1975." 1975.

https://www.nixonfoundation.org/2017/08/nixons-record-civil-rights-2/.

https://www.poorpeoplescampaign.org/about/.

https://www.pewresearch.org/politics/2020/08/13/important-issues-in-the-2020-election/.

https://embryo.asu.edu/pages/webster-v-reproductive-health-services-1989#:~:text=In%20the%201989%20case%20Webster,   provide%20abortion%20counseling%20or%20services.

https://hbr.org/2018/01/when-more-women-join-the-workforce-wages-rise-includ ing-for-men.

https://prochoice.org/dramatic-increase-in-threats-violence-directly-correlates-with-release-of-misleading-videos/.

https://prochoice.org/naf-releases-2019-violence-disruption-statistics/.

https://web.archive.org/web/20170410091210/http://equalrightsamendment.org/hist ory.htm.

https://www.americanbar.org/groups/crsj/publications/human_rights_magazine_h ome/human_rights_vol31_2004/summer2004/irr_hr_summer04_gaps/.

https://www.dolanconsultinggroup.com/wp-content/uploads/2017/10/RB_Domestic-Violence-Calls_Officer-Safety.pdf.

https://www.fbi.gov/investigate/civil-rights.

https://www.kff.org/wp-content/uploads/2013/07/8186-hiv-survey-report_final.pdf.

https://www.lgbttqnation.com/2921/10/pbs-new-film.

https://www.nber.org/digest/mar19/employment-growth-and-rising-womens-labor-force-participation.

https://www.npr.org/2020/12/07/943938352/outside-in-everybody-knows-somebody.

https://www.nytimes.com/interactive/2019/us/abortion-laws-states.html.

https://www.pewforum.org/fact-sheet/changing-attitudes-on-gay-marriage/.

https://www.pewresearch.org/fact-tank/2018/03/15/for-womens-history-month-a-look-at-gender-gains-and-gaps-in-the-u-s/.

https://www.pewresearch.org/politics/2019/08/29/u-s-public-continues-to-favor-legal-abortion-oppose-overturning-roe-v-wade/.

https://www.pewresearch.org/social-trends/2019/05/08/americans-see-advantages-and-challenges-in-countrys-growing-racial-and-ethnic-diversity/.

https://www.pewresearch.org/wp-content/uploads/sites/3/2011/12/women-in-the-military.pdf.

https://www.pewsocialtrends.org/2020/07/07/a-century-after-women-gained-the-right-to-vote-majority-of-americans-see-work-to-do-on-gender-equality/.

https://www.plannedparenthood.org/about-us/facts-figures/annual-report.

https://www.rollcall.com/2018/12/21/donald-trump-signs-overhaul-of-anti-harassment-law-for-members-of-congress-staff/.

https://www.washingtonpost.com/blogs/ezra-klein/post/how-the-republican-party-became-pro-life/2012/03/09/gIQAZcD31R_blog.html.

https://www.washingtonpost.com/lifestyle/style/planned-parenthood-is-a-symbol-this-is-the-reality-of-one-ohio-clinic/2015/09/15/b97c465e-4774-11e5-846d-02792f854297_story.html.

https://www.washingtonpost.com/news/the-intersect/wp/2017/10/19/the-woman-behind-me-too-knew-the-power-of-the-phrase-when-she-created-it-10-years-ago/.

https://www.washingtonpost.com/politics/joe-biden-apologized-to-anita-hill-but-shes-not-satisfied/2019/04/25/0f2ff1d8-679a-11e9-82ba-fcfeff232e8f_story.html.

http://uweconsoc.com/trends-in-corporate-lobbying-its-incentives-and-socio-economic-effects/.

http://www.electproject.org/home/voter-turnout/voter-turnout-data.

https://cnnpressroom.blogs.cnn.com/2021/05/13/cnn-special-report-a-radical-rebellion-the-transformation-of-the-gop-a-fareed-zakaria-special/.

CNN Exit Polls 2016

https://edition.cnn.com/election/2016/results/exit-polls/national/president.

https://news.gallup.com/poll/203198/presidential-approval-ratings-donald-trump.aspx.

https://nvdatabase.swarthmore.edu/content/american-colonials-struggle-against-british-empire-1765-1775.

https://onlinepoliticalsciencedegree.eku.edu/insidelook/what-causes-low-voter-turnout-united-states.

https://www.britannica.com/event/Vietnam-War.

https://www.npr.org/2011/01/17/132942244/ikes-warning-of-military-expansion-50-years-later.

https://www.nytimes.com/2015/07/12/us/the-best-way-to-vilify-clinton-gop-spends-heavily-to-test-it.html.

https://www.nytimes.com/2022/02/06/opinion/covid-pandemic-policy-trust.html. (2022).

https://www.pewresearch.org/fact-tank/2021/01/28/turnout-soared-in-2020-as-nearly-two-thirds-of-eligible-u-s-voters-cast-ballots-for-president/.

https://www.politico.com/news/2021/01/24/republicans-voter-id-laws-461707.

https://www.politico.com/story/2016/08/joe-mccarthys-dirty-tricks-upend-senate-race-aug-20-1951-2.

https://www.theatlantic.com/business/archive/2015/04/how-corporate-lobbyists-conquered-american-democracy/390822/.

https://www.theatlantic.com/politics/archive/2017/01/no-one-knows-what-the-powerful-mercers-really-want/514529/.

https://www.washingyonpost.com/news/retropolis/wp/2018/08/24/a-party-that-lost-its-mind-in-1968-democrats-held-one-of-historys-most-disastrous-conventi ons/.Hughes, Everett C. "The Study of Ethnic Relations." *Dalhousie Review* 27, no. 4 (1948): 477–85.

https://www.pbs.org/wgbh/pages/frontline/waco/timeline.html.

https://www.politico.com/news/2020/12/21/trump-house-overturn-election-449787.

https://www.washingtonpost.com/local/one-dead-as-car-strikes-crowds-amid-protests-of-white-nationalist-gathering-in-charlottesville-two-police-die-in-helicopter-crash/2017/08/13/3590b3ce-8021-11e7-902a-2a9f2d808496_story.html.

https://www.deseret.com/2021/8/25/22580333/the-rise-and-fall-of-the-trumpiest-never-trumper-lincoln-project-steve-schmidt-donald-trump.   https://en.dgb.de/fie lds-of-work/german-codetermination.

https://eos.com/blog/agricultural-cooperatives/.

https://nymag.com/intelligencer/2021/07/5-reasons-biden-got-his-bipartisan-infrast ructure-deal.html.

https://thehill.com/homenews/administration/534823-trumps-1776-project-released-on-mlk-day-receives-heavy-backlash.

https://www.bing.com/search?q=rochdale+society+principles&FORM=AFS CVO&PC=AFSC.

https://www.buzzfeednews.com/article/kadiagoba/gosar-aoc-censure-house-anime-twit ter-biden.

https://www.intheblack.com/articles/2020/10/01/protecting-workers-gig-economy.

https://www.newyorker.com/news/daily-comment/why-democracy-is-on-the-decline-in-the-united-states.

https://www.pewresearch.org/journalism/fact-sheet/network-news/.

https://www.politico.com/story/2018/04/11/ryan-in-interview-im-done-seeking-elec ted-office-515678.

https://www.theatlantic.com/ideas/archive/2019/12/historians-clash-1619-project/604093/.

https://www.usworker.coop/what-is-a-worker-cooperative/.

https://www.washingtonpost.com/s/opinions/2021/09/23/robert-kagan-constitutional-crisis/?pwapi_token=eyJ0eXAiOiJKV1QiLCJhbGciOiJIUzI1NiJ9.eyJzdWJpZCI6IjI 0Mjk2MTMiLCJyZWFfb24iOiJnaWZ0IiwibmJmIjoxNjMyNDQ1NTg3LCJpc3MiO iJzdWJz.

Huffman, John Pearley. "5 Most Notorious Recalls of All Time." *Popular Mechanics*, February 12, 2010. https://www.popularmechanics.com/cars/g261/4345725/.

Hughes, Everett C. "The Professions in Society." In *The Sociological Eye*, edited by Everett C. Hughes, 364–73. Chicago: Aldine, 1971.

Hunter, Floyd *Top Leadership U.S.A.* Chapel Hill: University of North Carolina Press, 1959.

Issenberg, Sasha. *The Engagement: America's Quarter-Century Struggle over Same-Sex Marriage*. New York: Pantheon, 2021.

Issenberg, Sasha. "How We Got Marriage Equality." *New York Times*, June 6, 2021, SR8.

Jackson, Brooks. "Obama's Final Numbers." FactCheck.org, September 29, 2017. www. factcheck.org/2017/09/obamas-final-numbers/.

Jacobson, Matthew Frye. *Whiteness of a Different Color: European Immigrants and the Alchemy of Race*. Cambridge, MA: Harvard University Press, 1999.

James, Nathan. "The First Step Act of 2018: An Overview." Washington, DC: Congressional Research Service, 2019. https://crsreports.congress.gov.

Jardina, Ashley. *White Identity Politics*. Cambridge, UK: Cambridge University Press, 2019.

Jefferis, Jennifer. *The Army of God and Anti-abortion Terror in the United States*. New York: Praeger, 2011.

Jencks, Christopher, and Meredith Phillips. "The Black-White Test Score Gap: Why It Persists and What Can Be Done." Washington, DC: Brookings Institution, March 1, 1998. https://www.brookings.edu/articles/the-Black-white-test-score-gap-why-it-persists-and-what-can-be-done/.

Jillson, Calvin C. *American Government: Political Development and Institutional Change*. 5th ed. New York: Taylor & Francis, 2009.

Johnson, Rucker. *Children of the Dream: Why School Integration Works*. New York: Basic Books and the Russell Sage Foundation, 2019.

Jorgensen, Margaret A., and Jenny Hoffmann. "History of the No Child Left Behind Act of 2001 (NCLB)." In *Assessment Report*. San Antonia, TX: Pearson Education, 2003.

Justice Research and Statistics Association. "Crime and Justice Atlas 2000." Washington, DC: JRSA, 1998.

Kahn, Suzanne, and Steph Sterling. "Supply-Side Childcare Investments: Policies to Develop an Equitable and Stable Childcare Industry." Issue Brief. New York: Roosevelt Institute, 2021. https://rooseveltinstitute.org/wp-content/uploads/2021/07/RI_Childcare_IssueBrief_202108.pdf.

Kamper, Dave. "Facing Wisconsin-Style Attack, Iowans Stick to the Union." *Labornotes*, November 16, 2017. https://labornotes.org/2017/11/facing-wisconsin-style-attack-iowans-stick-union.

Katznelson, Ira. *When Affirmative Action Was White: An Untold History of Racial Inequality in Twentieth-Century America*. New York: W. W. Norton, 2006.

Keely, Charles B. "The Development of U.S. Immigration Policy since 1965." *Journal of International Affairs* 33, no. 2 (1979): 249–63.

Kiner, Douglas L., and Francis X. Shen. *The Casualty Gap: The Causes and Consequences of American Wartime Inequalities*. New York: Oxford University Press, 2010.

King, Martin Luther, Jr. "Martin Luther King Jr. on the Vietnam War." *The Atlantic* (March 2018). https://www.theatlantic.com/magazine/archive/2018/02/martin-luther-king-jr-vietnam/552521/.

Kirp, David. *The College Dropout Scandal*. New York: Oxford University Press, 2020.

Klatch, Rebecca E. *Women of the New Right*. Philadelphia: Temple University Press, 1987.

Klein, Ezra. *Why We're Polarized*. New York: Avid Reader Press, 2020.

Klonick, Kate. "Why America Needs More Housewives Like Phyllis Schlafly." *Atlantic*, September 7, 2016. https://www.theatlantic.com/politics/archive/2016/09/why-america-needs-more-housewives-like-phyllis-schlafly/498917/.

Kluger, Richard. *Simple Justice: The History of "Brown V. Board of Education" and Black America's Struggle for Equality*. New York: Knopf, 1975.

Kofman, Yasmin B., and Dana Rose Garfin. "Home Is Not Always a Haven: The Domestic Violence Crisis amid the COVID-19 Pandemic." *Psychological Trauma: Theory, Research, Practice, and Policy* 12, Suppl. 1 (2020): S199–S201.

Koch, James V. *The Impoverishment of the American College Student*. Washington, DC: Brookings Institution, 2020.

Kornacki, Steve. *The Red and the Blue: The 1990s and the Birth of Political Tribalism*. New York: Harper Collins, 2018.

Kovick, David. "The Hewlett Foundation's Conflict Resolution Program: Twenty Years of Field-Building, 1984–2004." Menlo Park, CA: Hewlett Foundation, 2005.

Kozol, Jonathan. *The Shame of the Nation: The Restoration of Apartheid Schooling in America*. New York: Three Rivers Press, 2006.

Kramer, Ralph M. *Participation of the Poor: Comparative Community Case Studies in the War on Poverty*. Englewood Cliffs, NJ: Prentice-Hall, 1969.

Kretschmer, Kelsy "Shifting Boundaries and Splintering Movements: Abortion Rights in the Feminist and New Right Movements." *Sociological Forum* 29, no. 4 (2014): 893–915.

Kriesberg, Louis. "The Evolution of Conflict Resolution." In *SAGE Handbook of Conflict Resolution*, edited by Jacob Bercovitch, Victor Kremenyuk, and I. William Zartman, 15–32. London: SAGE, 2009.

Kriesberg, Louis. "Ethnicity, Nationalism, and Violent Conflict in the 1990's." *Peace Studies Bulletin* 2 (Winter 1992–93): 24–28.

Kriesberg, Louis. "Identities in Intractable Conflicts." Beyond Intractability (Intractable Conflicts Knowledge Base), 2003. https://www.crinfo.org/.

Kriesberg, Louis. "Interactions among Populism, Peace, and Security in Contemporary America." *Sicherheit und Frieden (Security and Peace)* 37, no. 1 (2019): 1–7.

Kriesberg, Louis. *International Conflict Resolution: The U.S.-USSR and Middle East Cases*. New Haven, CT: Yale University Press, 1992.

Kriesberg, Louis. *Louis Kriesberg: Pioneer in Peace and Constructive Conflict Resolution Studies*. Edited by Hans Guenter Brauch. Cham, Switzerland: Springer International, 2016.

Kriesberg, Louis. *Mothers in Poverty: A Study of Fatherless Families*. Chicago: Aldine, 1970.

Kriesberg, Louis. "Noncoercive Inducements in U.S.-Soviet Conflicts: Ending the Occupation of Austria and Nuclear Weapons Tests." *Journal of Political and Military Sociology* 9, no. 1 (1981): 1–16.

Kriesberg, Louis. "Peace Movements and Government Peace Efforts." In *Research in Social Movements Conflicts and Change*, edited by Louis Kriesberg, Bronislaw Misztal, and Janusz Mucha, 57–75. Greenwich, CT: JAI Press, 1988.

Kriesberg, Louis. "Policy Continuity and Change." *Social Problems* 32 (December 1984): 89–102.

Kriesberg, Louis. "The Relationship between Socio-economic Rank and Behavior." *Social Problems* 10, no. 4 (1963): 334–53.

Kriesberg, Louis. *Realizing Peace: A Constructive Conflict Approach*. New York: Oxford University Press, 2015.

Kriesberg, Louis. *Social Inequality*. Englewood Cliffs, NJ: Prentice-Hall, 1979.

Kriesberg, Louis, Bronislaw Miszta, and Janusz Mucha, eds. *Social Movements as a Factor of Change in the Contemporary World*. Edited by Louis Kriesberg. Research in Social Movements, Conflicts and Change, Vol. 10. Greenwich, CT: JAI Press, 1988.

Kriesberg, Louis, and Bruce W. Dayton. *Constructive Conflicts: From Escalation to Resolution*. 5th ed. Lanham, MD: Rowman & Littlefield, 2017.

Kriesberg, Louis, and Seymour S. Bellin. "On the Relationship between Attitudes, Circumstances, and Behavior: The Case of Applying for Public Housing." *Sociology and Social Research* 51, no. 4 (July 1967): 453–67.

Kriesberg, Louis, Terrell A. Northrup, and Stuart J. Thorson, eds. *Intractable Conflicts and Their Transformation*. Syracuse, NY: Syracuse University Press, 1989.

Krugman, Paul. *Arguing with Zombies: Economics, Politics, and the Fight for a Better Future*. New York: Norton & Company, 2020.

Kull, Steven. *Misperceptions, the Media and the Iraq War*. College Park, MD: Center for International Security Studies at Maryland, 2003.

Kurlansky, Mark. *1968: The Year That Rocked the World*. New York: Random House, 2904.

Kurtz, Howard. "Doing Something Right; Fox News Sees Ratings Soar, Critics Sore." *Washington Post*, February 5, 2001.

Lane, Robert E. "The Politics of Consensus in an Age of Affluence." *American Political Science Review* 59, no. 4 (December 1965): 874–95.

Lederach, John Paul. *Preparing for Peace: Conflict Transformation across Cultures*. Syracuse, NY: Syracuse University Press, 1995.

Leinfelt, Fredrik H. "Racial Influences on the Likelihood of Police Searches and Search Hits: A Longitudinal Analysis from an American Midwestern City." *Police Journal* 79, no. 3 (2006): 238–57.

Leiserson, Greg. "Taxing Wealth." Washington, DC: Washington Center for Equitable Growth, 2020. https://equitablegrowth.org/taxing-wealth/.

Lessig, Lawrence. *Republic, Lost: How Money Corrupts Congress—And a Plan to Stop It*. 2nd ed. New York: Hachette Book Group, 2016.

Levine, Robert A., and Donald T. Campbell. *Ethnocentrism: Theories of Conflict, Ethnic Attitudes, and Group Behavior*. New York: John Wiley & Sons, 1972.

Lewicki, Roy J., David M. Saunders, and John W. Minton. *Essentials of Negotiation*. New York: McGraw-Hill, 2000.

Lewis, James R., ed. *From the Ashes: Making Sense of Waco*. Lanham, MD: Rowman & Littlefield, 1994.

Lewis-Kraus, Gideon. "The Change Artists." *New York Times Magazine*, July 24, 2016, 30–35, 47.

Lewis, Michael. *The Big Short: Inside the Doomsday Machine*. London: Allen Lane, 2010.

Lewis, Michael. *The Fifth Risk: Undoing Democracy*. New York: W. W. Norton, 2018.

Lipset, Seymour Martin, and Earl Raab. *The Politics of Unreason: Right Wing Extremism in America 1790–1970*. New York: Harper & Row, 1970.

Long, Heather, and Andrew Van Dam. "Financial Chasm Separates Blacks from Whites." *New York Times*, June 7, 2020, B2.

Lucks, Daniel S. *Reconsidering Reagan: Racism, Republicans, and the Road to Trump*. Boston: Beacon Press, 2020.

Lukas, Carrie, ed. *The Politically Incorrect Guide to Women, Sex, and Feminism*. Washington, DC: Regency Publishing, 2006.

Lunch, William L., and Peter W. Sperlich. "American Political Opinion and the War in Vietnam." *Western Political Quarterly* 32, no. 1 (1979): 21–44.

Lyman, Brian. "Southern Poverty Law Center Staff Vote to Unionize." *Montgomery Advertiser*, December 16, 2019. https://www.montgomeryadvertiser.com/story/news/2019/12/16/southern-poverty-law-center-votes-unionize/2661726001/.

MacFarquhar, Neil. "Far-Right Organizations Intended to Foment Violence, Victims Say." *New York Times*, October 25, 2021, A13.

MacFarquhar, Larissa "What Money Can Buy: Darren Walker and the Ford Foundation Set Out to Conquer Inequality." *New Yorker*, December 27, 2015. https://www.newyorker.com/magazine/2016/01/04/what-money-can-buy-profiles-larissa-macfarquhar.

MacGillis, Alic. "Jared Kushner's Other Real Estate Empire in Baltimore." *New York Times*, May 23, 2017. https://www.nytimes.com/2017/05/23/magazine/jared-kushners-other-real-estate-empire.html.

Maddow, Rachel. *Drift: The Unmooring of American Military Power*. New York: Crown, 2012.

Madland, David, and Alex Rowell. *Attacks on Public-Sector Unions Harm States: How Act 10 Has Affected Education in Wisconsin*. Washington, DC: Center for American Progress Action Fund, 2017. https://www.americanprogressaction.org/issues/econ omy/reports/2017/11/15/169146/attacks-public-sector-unions-harm-states-act-10-affected-education-wisconsin/.

Mahler, Jonathan, and Jim Rutenberg. "Planet Fox." *New York Times Magazine*, April 7, 2019, 32–55, 62–73.

Mann, Charles C. *1491: New Revelations of the Americas before Columbus*. New York: Random House, 2011.

Mann, Thomas E., and Norman J. Ornstein. *It's Even Worse Than It Looks: How the American Constitutional System Collided with the New Politics of Extremism*. New York: Basic Books, 2012.

Mansbridge, Jane J. *Why We Lost the ERA*. Chicago: University of Chicago Press, 1986.

Martin, Carmel, Ulrich Boser, Meg Benner, and Perpetual Baffour. "A Quality Approach to School Funding; Lessons Learned from School Finance Litigation." Washington, DC: Center for American Progress, 2018. https://www.americanprogress.org/article/quality-approach-school-funding/.

Mason, Lilliana. *Uncivil Agreement: How Politics Became Our Identity*. Chicago: University of Chicago Press, 2018.

Massey, Douglas S., and Andrew B. Gross. "Explaining Trends in Racial Segregation, 1970–1980." *Urban Affairs Quarterly* 27, no. 1 (1991): 13–35.

Mayer, Gerald. *Union Membership Trends in the United States*. Washington, DC: Congressional Research Service, 2004.

Mayer, Jane. *Dark Money: The Hidden History of the Billionaires*. New York: Anchor Books, 2016.

McAdams, John. "Eisenhower's Farewell Address to the Nation."

McCarthy, John D., and Mayer Zald. *The Trend of Social Movements in America: Professionalization and Resource Mobilization*. Princeton, NJ: General Learning Press, 1973.

McGhee, Heather. *The Sum of Us*. New York: One World, 2021.

McGill, Andrew. "The Missing Black Students at Elite American Universities." *Atlantic*, November 24, 2015. https://www.theatlantic.com/notes/2015/11/the-missing-Black-students-at-elite-american-universities-contd/417585/.

McKee, Guian A. "Lyndon B. Johnson and the War on Poverty, Vol. 1." Presidential Recordings, Digital Edition. Charlottesville: Miller Center, University of Virginia, 2014. http://prde.upress.virginia.edu/content/WarOnPoverty.

McNall, Scott G. *Cultures of Defiance and Resistance: Social Movements in 21st-Century America*. New York and London: Routledge, 2018.

Meier, August, and Elliott M. Rudwick. *CORE: A Study in the Civil Rights Movement, 1942–1968*. Urbana: University of Illinois Press, 1975.

Merton, Robert K. *Social Theory and Social Structure*. New York: Free Press, 1949. Rev. and enl. ed. 1957; 3rd ed., enl., 1968.

Metzger, Molly W., and Danilo Pelletiere. "Patterns of Housing Voucher Use Revisited: Segregation and Section 8 Using Updated Data and More Precise Comparison Groups, 2013." St. Louis, MO: Center for Social Development Research, 2013. https://www.stlouisfed.org/~/media/files/pdfs/community%20development/econ%20mobility/sessions/metzgerpaper508.pdf.

Michaels, Samantha. "Trump Just Bragged about Criminal Justice Reform: Look Closer at How His Administration Is Undoing It." *Mother Jones*, February 4, 2020. https://www.motherjones.com/crime-justice/2020/02/trump-just-bragged-about-criminal-just ice-reform-look-closer-at-how-his-administration-is-undoing-it/.

Mills, C. Wright. *The Power Elite*. New York: Oxford University Press, 1956.

Mishel, Lawrence, Elise Gould, and Josh Bivens. "Wage Stagnation in Nine Charts." Washington, DC: Economic Policy Institute, 2015. https://www.epi.org/publication/charting-wage-stagnation/.

Mishel, Lawrence, John Schmitt, and Heidi Shierholz. "Wage Inequality: A Story of Policy Choices." *New Labor Forum* 23 (2014): 26–31.

Mishra, Subodh. "U.S. Board Diversity Trends in 2019." Cambridge, MA: Harvard Law School Forum on Corporations, 2019. https://corpgov.law.harvard.edu/2019/06/18/u-s-board-diversity-trends-in-2019/.

Montopoli, Brian. "Poll: 43 Percent Agree with Views of 'Occupy Wall Street.'" CBS News, October 26, 2011.

Morgan, Robin, ed. Sisterhood Is Powerful: *An Anthology of Writings from the Women's Liberation Movement*. New York: Vintage Books, 1970.

Morris, Aldon D. "Birmingham Confrontation Reconsidered: An Analysis of the Dynamics and Tactics of Mobilization." *American Sociological Review* 58, no. 5 (1993): 621–36.

Moynihan, Daniel P. *The Politics of a Guaranteed Income: The Nixon Administration and the Family Assistance Plan*. New York: Vintage Books, 1973.

Mott, Ron. "Three Years after Michael Brown's Death, Has Ferguson Changed?" NBC News, August 9, 2017. https://www.nbcnews.com/storyline/michael-brown-shooting/three-years-after-michael-brown-s-death-has-ferguson-changed-n791081.

Mulholland, Quinn "The Case against Standardized Testing." *Harvard Political Review*, May 14, 2015. http://harvardpolitics.com/united-states/case-standardized-testing/.

Mutz, Diana C. "Status Threat, Not Economic Hardship, Explains 2016 Presidential Vote." *PNAS* 115, no. 19 (2018): E4330–39. https://doi.org/10.1073/pnas.1718155115.

Myrdal, Gunnar, Richard Sterner, and Arnold Rose. *An American Dilemma: The Negro Problem and American Democracy*. New York and London: Harper & Brothers, 1944.

Nader, Ralph. *Unsafe at Any Speed: The Designed-In Dangers of the American Automobile*. New York: Grossman, 1965.

National Advisory Commission on Civil Disorders (Kerner Commission). *Report of the National Commission on Civil Disorders*. New York: Bantam, 1968.

Nellis, Ashley. *The Color of Justice: Racial and Ethnic Disparity in State Prisons*. Washington, DC: The Sentencing Project, 2016. https://www.sentencingproject.org/publications/color-of-justice-racial-and-ethnic-disparity-in-state-prisons/.

Nellis, Ashley, Judy Greene, and Marc Mauer. *Reducing Racial Disparity in the Criminal Justice System: A Manual for Practitioners and Policymakers*. Washington, DC: The Sentencing Project, 2016. https://www.sentencingproject.org/wp-content/uploads/2016/01/Reducing-Racial-Disparity-in-the-Criminal-Justice-System-A-Manual-for-Practitioners-and-Policymakers.pdf.

Nelson, F. Howard, and Michael Rosen. "Are Teachers' Unions Hurting American Education?" Milwaukee, WI: The Institute for Wisconsin's Future, 1996.

Newfield, Jack. *Robert Kennedy: A Memoir*. New York: E. P. Dutton, 1969.

Noble, Kenneth B. "Reagan Vetoes Measure to Affirm Fairness Policy for Broadcasters." *New York Times*, June 21, 1987. https://nyti.ms/29zHkoV, https://nyti.ms/29zHkoV.

O'Brien, Rebecca Davis. "Man Who Threatened Democrats Is Sentenced to 19 Months in Prison." *New York Times*, November 23, 2021, A17.

O'Connor, John. "US Social Welfare Policy: The Reagan Record and Legacy." *Journal of Social Policy* 27, no. 1 (1998): 37–61.

Oberdorfer, Don, and Robert Carlin. *The Two Koreas*. 3rd ed. New York: Basic Books, 2014.

Orton, Kathy. "Homeowners Get More Time to Take Advantage of HAMP, HARP." *Washington Post*, May 8, 2015. http://wapo.st/1IWnFIO?tid=ss_mail&utm_term=.c2333298288b.

Oshinsky, David. *A Conspiracy So Immense: The World of Joe McCarthy*. Oxford: Oxford University Press, 2005.

Palmer, John N., and Isabel V. Sawhill. "Perspectives on the Reagan Experiment." In *The Reagan Experiment*, edited by John N. Palmer and Isabel V. Sawhill, 1–28. Washington, DC: Urban Institute Press, 1982.

Parenti, Michael. *Dirty Truths*. San Francisco: City Lights Books, 1996.

Parsons, Talcott. *The Social System*. Glencoe, IL: The Free Press, 1951.

Patterson, James T. *Restless Giant: The United States from Watergate to "Bush v. Gore."* New York: Oxford University Press, 2005.

Pedriana, Nicholas, and Amanda Abraham. "Now You See Them, Now You Don't: The Legal Field and Newspaper Desegregation of Sex-Segregated Help Wanted Ads 1965–75." *Law & Social Inquiry* 31, no. 4 (2006): 905–38.

Peristein, Rick. *Reganland: America's Right Turn 1976–1980*. New York: Simon & Schuster, 2020.

Perry, Mark J. "The Shocking Story behind Richard Nixon's 'War on Drugs' That Targeted Blacks and Anti-War Activists." Washington, DC: American Enterprise Institute, June 14, 2018. https://www.aei.org/carpe-diem/the-shocking-and-sickening-story-behind-nixons-war-on-drugs-that-targeted-Blacks-and-anti-war-activists/?fbclid=IwAR0kd4j7SVj9iY_Bmp1eoL_Kbo3Orx9zM7Hcbd2-6oboK1hDOZmhD5PIxe8.

Peterson, David. "Wife Beating: An American Tradition." *Journal of Interdisciplinary History* 23, no. 1 (1992): 97–118.

Pettigrew, Thomas F. "Intergroup Contact Theory." *Annual Review of Psychology* 49 (1998): 65–85.

Pew Research Center. "Demographic Trends and Economic Well-Being." Washington, DC: Pew Research Center, 2016. https://www.pewsocialtrends.org/2016/06/27/1-demographic-trends-and-economic-well-being/.

Pieper, Andrew L., and Jeff R. DeWitt, eds. *The Republican Resistance: #Nevertrump Conservatives and the Future of the GOP*. Lanham, MD: Lexington Books, 2021.

Piketty, Thomas. *Capital in the Twenty-First Century*. Translated by Arthur Goldhammer. Cambridge, MA: Belknap Press, 2014.

Pilkington, Ed, and Suzanne Goldenberg. "State Conservative Groups Plan US-Wide Assault on Education, Health and Tax." *Guardian*, December 5, 2013. https://www.theguardian.com/world/2013/dec/05/state-conservative-groups-assault-education-health-tax.

Pinsky, Mark. "Taking Stock: CDFIs Look Ahead after 25 Years of Community Development Finance." Washington, DC: Brookings Institution and Joint Center for Housing Studies, 2001.

Piven, Frances Fox, and Richard A. Cloward. *Poor People's Movements*. New York: Vintage Books, 1979.

Piven, Frances Fox, and Richard A. Cloward. *Regulating the Poor: The Functions of Public Welfare*. Updated edition. New York: Vintage Books, 1993.

Plotnick, Robert D. "Changes in Poverty, Income Inequality, and the Standard of Living in the United States during the Reagan Years." *International Journal of Health Services* 23 (April 1993): 347–358.

Polkinghorn, Brian D., Haleigh La Chance, and Robert La Chance. "Constructing a Baseline Understanding of Developmental Trends in Graduate Conflict Resolution Programs in the United States." In *Pushing the Boundaries: New Frontiers in Conflict Resolution and Collaboration*, edited by Rachel Fleishman, Catherine Gerard, and Rosemary O'Leary, 233–65. Bingley, UK: Emerald Group, 2008.

Powell, John A. *Racing to Justice: Transforming Our Conceptions of Self and Other to Build an Inclusive Society*. Bloomington: Indiana University Press, 2012.

Prewitt, Kenneth. *What Is Your Race? The Census and Our Flawed Efforts to Clarify Americans*. Princeton, NJ: Princeton University Press, 2013.

Pulliam, Christopher. "Tax Wealth, Reward Work." Brookings Up Front. Washington, DC: Brookings Institution, July 14, 2021. https://www.brookings.edu/blog/up-front/2021/07/14/tax-wealth-reward-work/?utm_campaign=Economic%20Studies&utm_medium=email&utm_content=142104627&utm_source=hs_email.

Puma, Mike, Stephen Bell, Ronna Cook, Camilla Heid, Pam Broene, Frank Jenkins, Mashburn, Andrew, and Downer Jason. *Third Grade Follow-Up to the Head Start Impact Study Final Report, Executive Summary*. Washington, DC: Administration for Children and Families, US Department of Health and Human Services 2012.

Quadagno, Jill. *The Color of Welfare: How Racism Undermined the War on Poverty*. New York: Oxford University Press, 1994.

Quadagno, Jill. "Race, Class, and Gender in the U.S. Welfare State: Nixon's Failed Family Assistance Plan." *American Sociological Review* 55, no. 1 (1990): 11–28.

Raiffa, Howard. *The Art and Science of Negotiation*. Cambridge, MA: Harvard University Press, 1982.

Rainie, Lee, Scott Keeter, and Andrew Perrin. "Trust and Distrust in America." Washington, DC: Pew Research Center, July 22, 2019. https://www.pewresearch.org/politics/2019/07/22/trust-and-distrust-in-america/.

Ransby, Barbara. *Making All Black Lives Matter: Reimagining Freedom in the Twenty-First Century*. Oakland, CA: University of California Press, 2018.

Reardon, Sean F., and Ann Owens. "60 Years after *Brown*: Trends and Consequences of School Segregation." *Annual Review of Sociology* 40 (2014): 199–218.

Rehavi, M. Marit, and Sonja B. Starr. "Racial Disparity in Federal Criminal Sentences." *Journal of Political Economy* 122, no. 6 (2014): 1320–54.

Reich, Robert. "Is U.S. Experiencing an Unofficial General Strike?" *Guardian*, October 13, 2021. https://www.theguardian.com/prohttps://www.theguardian.com/profile/robert-reichfile/robert-reich.

Reich, Robert B. "Foreword." In *The Spirit Level: Why Greater Equality Makes Societies Stronger*. Edited by Richard Wilkinson and Karte Pickett, IX–XII. New York and London: Bloomsbury Press, 2010.

Reitman, Janet. "State of Denial." *New York Times Magazine*, November 11, 2018, 38–49, 66–68.

Reitzes, Donald C., and Dietrich C. Reitzes. *The Alinsky Legacy: Alive and Kicking*. Greenwich, CT: JAI Press, 1987.

Robinson, Jo Ann, and David J. Garrow. *The Montgomery Bus Boycott and the Women Who Started It*. Knoxville: University of Tennessee Press, 1986.

Robinson, Randall. *The Debt: What America Owes to Blacks*. New York: Dutton, 2000.

Roethliberger, P. J., William J. Dickson, and Harold A. Wright. *Management and the Worker*. Cambridge, MA: Harvard University Press, 1939.

Rojas, Fabio. *From Black Power to Black Studies: How a Radical Social Movement Became an Academic Discipline*. Baltimore, MD: Johns Hopkins University Press, 2007.

Rose, Arnold M. *The Power Structure*. New York: Oxford University Press, 1967.

Rosenfeld, Sam. *The Polarizers: Postwar Architects of Our Partisan Era*. Chicago: University of Chicago Press, 2017.

Rothstein, Richard. *The Color of Law: A Forgotten History of How Our Government Segregated America*. New York and London: Liveright, 2017.

Roy, Beth. *41 Shots . . . And Counting*. Syracuse, NY: Syracuse University Press, 2009.

Ruel, Erin, and Richard T. Campbell. "Homophobia and HIV/AIDS: Attitude Change in the Face of an Epidemic." *Social Forces* 84, no. 4 (2006): 2167–78.

Runciman, W. G. *Relative Deprivation and Social Justice*. Berkeley: University of California Press, 1966.

Rupesinghe, Kumar, ed. *Conflict Transformation*. New York: St. Martin's, 1995.

Russell, Bertrand. *Power: A New Social Analysis*. London: Allen and Unwin, 1938.

Rydell, C. Peter, and Susan S. Sohler Everingham. "Controlling Cocaine: Supply versus Demand Programs." Santa Monica, CA: RAND Corporation, 1994 https://www.rand.org/pubs/monograph_reports/MR331.html. Also available in print form.

Ryzik, Melen, and Katie Benner. "Biden's Aid Package Funnels Millions to Victims of Domestic Abuse." *New York Times*, March 19, 2021, A21.

Sachs, Jeffrey D. *The End of Poverty: Economic Possibilities for Our Time*. New York: Penguin, 2005.

Sader, Jonathan, and Shannon Rieger. "Patterns and Trends of Residential Integration in the United States since 2000." Research Brief. Cambridge, MA: Harvard Joint Center for Housing Studies, 2017. https://www.jchs.harvard.edu/sites/default/files/09192017_spader-rieger_integration_brief.pdf.

Saez, Emmanuel. "Striking It Richer: The Evolution of Top Incomes in the United States." Berkeley: University of California, 2013.

Sale, Kirkpatrick. *SDS: The Rise and Development of the Students for a Democratic Society*. New York: Random House, 1973.

Samuels, Robert. "Walker's Anti-Union Law Has Labor Reeling in Wisconsin." *The Washington Post*, February 22, 2015. https://www.google.com/search?q=washington+Post+Samuels+Walker%27s+ani-union&rlz=1C1GCEV_enUS899US899&oq=washington+Post+Samuels+Walker%27s+ani-union&aqs=chrome..69i57.31179j0j7&sourceid=chrome&ie=UTF-8.

Sandel, Michael J. *The Tyranny of Merit: What's Become of the Common Good?* Stuttgart, Germany: Farrar, Straus and Giroux, 2020.

Savage, Charlie. "Supreme Court Panel Shows Interest in Calls for Judges' Term Limits." *New York Times*, November 19, 2021, A20.

Sawyer, Wendy "How Race Impacts Who Is Detained Pretrial." Northampton, MA: Prison Policy Initiative, 2019.

Schlafly, Phyllis, ed. *The Power of the Positive Woman*. New Rochelle, NY: Arlington House, 1977.

Schlesinger, Arthur M., Jr. *A Thousand Days*. Boston: Houghton Mifflin, 1965.

Scott, Robert E. "The U.S. Trade Deficit: Are We Trading Away Our Future?" Washington, DC: Economic Policy Institute, 2002. https://www.epi.org/publication/webfeatures_viewpoints_tradetestimony/.

Seidenberg, Robert. *Corporate Wives—Corporate Casualties?* New York: Doubleday Anchor, 1973.

Semuels, Alana. "The End of Welfare as We Know It." *Atlantic*, April 1, 2016. https://www.theatlantic.com/business/archive/2016/04/the-end-of-welfare-as-we-know-it/476322/.

The Sentencing Project. "Fact Sheet." Washington, DC: The Sentencing Project, 2019.

The Sentencing Project. "Federal Crack Cocaine Sentencing." Washington, DC: The Sentencing Project, 2010. file://hd.ad.syr.edu/01/9e375f/Documents/Downloads/Federal-Crack-Cocaine-Sentencing%20(1).pdf.

Seo, Sarah A. "Racism on the Road." *New York Review of Books* 67, no. 12 (July 23, 2020): 45–46.

Shapiro, Walter. "The Flawed Politics of a Law-and-Order Campaign." *New Republic*, May 31, 2020. https://newrepublic.com/article/157939/flawed-politucs-law-and-order-camapaign.

Sills, David L. *The Volunteers, Means and Ends in a National Organization.* Glencoe, IL: Free Press, 1957.

Sitkoff, Harvard. "Harry Truman and the Election of 1948: The Coming of Age of Civil Rights in American Politics." *Journal of Southern History* 37, no. 4 (1971): 597–616.

Skocpol, Theda, Ariane Liazos, and Marshall Ganz. *What a Mighty Power We Can Be: African American Fraternal Groups and the Struggle for Racial Equality.* Princeton, NJ: Princeton University Press, 2006.

Skocpol, Theda, and Vanessa Williamson. *The Tea Party and the Remaking of Republican Conservatism.* New York: Oxford University Press, 2016.

Smigel, Erwin O. *The Wall Street Lawyer: Professional Organization Man?* New York and London: The Free Press of Glencoe and Collier-Macmillan, 1964.

Smith, Allen W. "Ronald Reagan and the Greet Social Security Heist." *FedSmith.com*, October 11, 2013. https://www.fedsmith.com/2013/10/11/ronald-reagan-and-the-great-social-security-heist.

Smith, Christian. *Resisting Reagan: The U.S. Central America Peace Movement.* Chicago: University of Chicago Press, 1996.

Smith, Tom W., and Jaesok Son. "Trends in Public Attitudes about Confidence in Institutions." Chicago: National Opinion Research Center, University of Chicago, 2013.

Stahl, Jason. *Right Moves: The Conservative Think Tank in American Political Culture since 1945.* Chapel Hill: University of North Carolina Press, 2016.

Startz, Dick. "When It Comes to Students, HBCUs Do More with Less." Washington, DC: Brookings Institution, January 18, 2021. https://www.brookings.edu/blog/brown-center-chalkboard/2021/01/18/when-it-comes-to-student-success-hbcus-do-more-with-less/#:~:text=HBCUs%20receive%20much%20less%20revenue,to%20only%20%244%2C900%20at%20HBCUs.

Stevens, Stuart. *It Was All a Lie: How the Republican Party Became Donald Trump.* New York: Knopf, 2020.

Stevenson, Bryan A. *Just Mercy: A Story of Justice and Redemption.* New York: Spiegel & Grau, 2014.

Stone, Chad, Danilo Trisi, Arloc Sherman, and Roderick Taylor. "A Guide to Statistics on Historical Trends in Income Inequality." Washington, DC: Center on Budget and

Policy Priorities, May 15, 2018. https://www.cbpp.org/research/poverty-and-inequal ity/a-guide-to-statistics-on-historical-trends-in-income-inequality.

Strauss, Valerie. "34 Problems with Standardized Tests." *Washington Post*, April 19, 2017. https://www.washingtonpost.com/news/answer-sheet/wp/2017/04/19/34-problems-with-standardized-tests/.

Sugrue, Thomas. *The Origins of the Urban Crisis*. Princeton, NJ: Princeton University Press, 2014.

Taylor, Miles, and Christine Todd Whitman. "We Are Republicans. There's Only One Way to Save Our Party from Pro-Trump Extremists." *New York Times*, October 11, 2021. https://www.nytimes.com/2021/10/11/opinion/2022-house-senate-trump.html.

Tierney, Kathleen J. "The Battered Women Movement and the Creation of the Wife Beating Problem." *Social Problems* 28, no. 3 (February 1982): 207–20.

Tobin, James "Voodoo Curse: Exorcising the Legacy of Reaganomics." *Harvard International Review* 14, no. 4 (1992): 10–13.

Tocqueville, Alexis de. *Democracy in America*. Edited and translated by Harvey Claflin Mansfield and Delba Winthrop. Chicago: University of Chicago Press, 2000.

Tolleson-Rinehart, Sue, and Susan J. Carroll. " 'Far from Ideal': The Gender Politics of Political Science." *American Political Science Review* 100, no. 4 (2006): 507–13.

Troy, Gil. *The Age of Clinton: America in the 1990s*. New York: Thomas Dunne Books, 2015.

Truman, Harry S. *Memoirs by Harry S. Truman*, Vol. 2, *Years of Trial and Hope*. Garden City, NY: Doubleday, 1956.

Tuchman, Barbara W. *The March of Folly: From Troy to Vietnam*. New York: Ballantine, 1984.

US Commission on Civil Rights. "Process of Change: The Story of School Desegregation in Syracuse, New York." Washington, DC: US Government Printing Office, 1968.

von Hoffman, Alexander. "The Past, Present, and Future of Community Development in the United States." Cambridge, MA: Harvard Joint Center for Housing Studies, 2012. https://www.jchs.harvard.edu/sites/default/files/w12-6_von_hoffman.pdf.

Wade, Wyn Craig. *The Fiery Cross: The Ku Klux Klan in America*. New York: Oxford University Press, 1998.

Walker, Samuel. *In Defense of American Liberties: A History of the ACLU*. New York: Oxford University Press, 1990.

Wanderer, Jules J. "An Index of Riot Severity and Some Correlates." *American Journal of Sociology* 74 (March 1969): 500–505.

Wanis-St. John, Anthony. *Back Channel Negotiations: Secrecy in the Middle East Peace Process*. Syracuse, NY: Syracuse University Press, 2011.

Wapshott, Nicholas. *Samuelson Friedman: The Battle over the Free Market*. New York: W. W. Norton, 2021.

Weber, Max. *The Theory of Social and Economic Organization*. Translated by A. M. Henderson and Talcott Parsons. New York: Oxford University Press, 1947.

Wegman, Jessie. "Why Can't We Make Women's Equality the Law of the Land?" *New York Times*, January 28, 2022. https://www.nytimes.com/2022/01/28/opinion/equal-rights-amendment-ratification.html.

West, Robin, Justin Murray, Meredith Esser, and Anna Grear, eds. In Search of Common Ground on Abortion: *From Culture War to Reproductive Justice*. New York: Routledge, 2016.

White, Micah. *The End of Protest: A Playbook for Revolution*. Toronto: Knopf Canada, 2016.

Whyte, Lara. "Has Trump's White House 'Resurrected' Army of God Anti-Abortion Extremists." *Open Democracy*, February 5, 2018. https://www.opendemocracy.net/en/5050/army-of-god-abortion-terrorists-emboldened-under-trump/.

Wilkerson, Isabel. *Caste: The Origins of Our Discontents*. New York: Random House, 2020.

Wilkinson, Richard, and Kate Pickett. *The Spirit Level: Why Greater Equality Makes Societies Stronger*. New York and London: Bloomsbury Press, 2010.

Williams, Daniel K. "The GOP's Abortion Strategy: Why Pro-Choice Republicans Became Pro-Life in the 1970s." *Journal of Policy History* 23, no. 4 (2011): 513–39.

Wills, Garry. *Lincoln at Gettysburg: The Words That Remade America*. New York: Touchstone Books, 1993.

Wilson, Steven Harmon "Controlled Substances Act (1970)." Encyclopedia.com, 2020. https://www.encyclopedia.com/history/encyclopedias-almanacs-transcripts-and-maps/controlled-substances-act-1970).

Wilson, William Julius. *The Truly Disadvantaged: The Inner City, the Underclass, and Public Policy*. Chicago: University of Chicago Press, 1987.

Windmuller, John P. *American Labor and the International Labor Movement, 1940–1953*. Ithaca, NY: Institute of International Industrial and Labor Relations, Cornell University, 1954.

Winnick, Louis "The Triumph of Housing Allowance Programs: How a Fundamental Policy Conflict Was Resolved." *Cityscape: A Journal of Policy Development and Research* 1, no. 3 (1995): 95–121.

Wolff, Michael. *Landslide: The Final Days of the Trump White House*. New York: Henry Holt & Company, 2021.

Woodward, Bob. *Rage*. New York: Simon & Schuster, 2020.

Woodward, C. Vann. *The Strange Career of Jim Crow*. 2nd ed. New York: Oxford University Press, 1966.

Wortham, Jenna. "The Vision Forward." *New York Times Magazine*, August 30 2020, 28–35, 45.

Wrong, Dennis *Power: Its Forms, Bases, and Uses*. 2nd ed. New Brunswick, NJ: Transaction, 1995.

Wyatt, Ian D., and Daniel E. Hecker. "Occupational Changes during the 20th Century." *Monthly Labor Review*, March 2006, 35–57.

Wydra, Abigail. "Teachers' Unions Improve Student Achievement: Insights from California Charter Schools." *Chicago Policy Review*, January 20, 2018. https://chicagopolicyreview.org/2018/01/20/teachers-unions-improve-student-achievement-insights-from-california-charter-schools/.

Yardley, Jim. "Panel Recommends Reparations in Long-Ignored Tulsa Race Riot." *New York Times*, February 5, 2000, A1, 12.

Zartman, I. William, and Guy Faure. *Escalation and Negotiation in International Conflicts*. Cambridge: Cambridge University Press, 2005.

Zelizer, Julian. *Burning Down the House: Newt Gingrich, the Fall of a Speaker, and the Rise of the New Republican Party*. New York: Penguin, 2020.

Zimmer, Troy A. "The Impact of Watergate on the Public's Trust in People and Confidence in the Mass Media." *Social Science Quarterly* 59, no. 4 (1979): 743–51.

Index